AF251447

ADVANCES IN SECOND MESSENGER
AND PHOSPHOPROTEIN RESEARCH

Volume 32

Adenylyl Cyclases

Advances in Second Messenger and
Phosphoprotein Research

Series Editors

Paul Greengard, *New York, New York*
Angus C. Nairn, *New York, New York*
Shirish Shenolikar, *Durham, North Carolina*

International Advisory Board

Michael J. Berridge, *Cambridge, England (United Kingdom)*
Ernesto Carafoli, *Zurich, Switzerland*
E. Costa, *Washington, D.C.*
Pedro Cuatrecases, *Ann Arbor, Michigan*
Raymond L. Erikson, *Cambridge, Massachusetts*
Alfred G. Gilman, *Dallas, Texas*
Joel G. Hardman, *Nashville, Tennessee*
Tony Hunter, *San Diego, California*
Claude B. Klee, *Bethesda, Maryland*
Edwin G. Krebs, *Seattle, Washington*
Yasutomi Nishizuka, *Kobe, Japan*
Ira H. Pastan, *Bethesda, Maryland*
G. Alan Robinson, *Houston, Texas*
Martin Rodbell, *Research Triangle Park, North Carolina*
Michael J. Welsh, *Iowa City, Iowa*
Keith R. Yamamoto, *San Francisco, California*

ADVANCES IN SECOND MESSENGER AND PHOSPHOPROTEIN RESEARCH

Volume 32

Adenylyl Cyclases

Editor

Dermot M. F. Cooper, Ph.D.

Department of Pharmacology
University of Colorado Health Sciences Center
Denver, Colorado

Acquisitions Editor: Mark Placito
Manufacturing Manager: Dennis Teston
Production Manager: Jodi Borgenicht
Production Editor: Kimberly Monroe
Cover Designer: Patricia Gast
Indexer: Cynthia Bertelsen
Compositor: Eastern Composition
Printer: Maple Press

Printed in the United States of America

9 8 7 6 5 4 3 2 1

International Standard Book Number 0-397-51850-1

Care has been taken to confirm the accuracy of the information presented and to describe generally accepted practices. However, the authors, editor, and publisher are not responsible for errors or omissions or for any consequences from application of the information in this book and make no warranty, express or implied, with respect to the contents of the publication.

The authors, editor, and publisher have exerted every effort to ensure that drug selection and dosage set forth in this text are in accordance with current recommendations and practice at the time of publication. However, in view of ongoing research, changes in government regulations, and the constant flow of information relating to drug therapy and drug reactions, the reader is urged to check the package insert for each drug for any change in indications and dosage and for added warnings and precautions. This is particularly important when the recommended agent is a new or infrequently employed drug.

Some drugs and medical devices presented in this publication have Food and Drug Administration (FDA) clearance for limited use in restricted research settings. It is the responsibility of the health care provider to ascertain the FDA status of each drug or device planned for use in their clinical practice.

Contents

Contributing Authors vii

Preface ix

1. Mammalian Adenylyl Cyclases 1
   Martine J. Smit and Ravi Iyengar

2. Ca^{2+}-Sensitive Adenylyl Cyclases 23
   *Dermot M. F. Cooper, Jeffrey W. Karpen, Kent A. Fagan,
   and Nicole E. Mons*

3. Molecular Diversity of the Adenylyl Cyclases 53
   John Krupinski and James J. Cali

4. Type-Specific Regulation of Mammalian Adenylyl Cyclases
   by G Protein Pathways 81
   Ronald Taussig and Gregor Zimmermann

5. Regulation of cAMP Signaling by Phosphorylation 99
   Yoshihiro Ishikawa

6. Genetic Characterization of Adenylyl Cyclase Function 121
   Martin J. Cann and Lonny R. Levin

7. Class III Adenylyl Cyclases: Regulation and Underlying
   Mechanisms 137
   Wei-Jen Tang, Shuizhong Yan, and Chester L. Drum

8. Calcium Control of Adenylyl Cyclase: The Calcineurin
   Connection 153
   *Ferenc A. Antoni, Susan M. Smith, James Simpson,
   Roberta Rosie, George Fink, and Janice M. Paterson*

9. Adenylyl Cyclases and Alcohol 173
   Boris Tabakoff and Paula L. Hoffman

10. Simultaneous Fluorescence Ratio Imaging of Cyclic AMP and
    Calcium Kinetics in Single Living Cells 195
    Maria A. DeBernardi and Gary Brooker

Subject Index 215

Contributing Authors

Ferenc A. Antoni, M.D., Ph.D. *Senior Scientist, MRC Brain Metabolism Unit, Department of Pharmacology, University of Edinburgh, 1 George Square, Edinburgh EH8 9JZ, Scotland, United Kingdom*

Gary Brooker, Ph.D. *Department of Cell Biology, Georgetown University School of Medicine, 3900 Reservoir Road, NW, Washington, DC 20007; and Atto Instruments, Inc., 1450 Research Boulevard, Rockville, Maryland 20850*

James J. Cali, Ph.D. *Instructor, Weis Center for Research, Pennsylvania State University College of Medicine, 100 N. Academy Avenue, Danville, Pennsylvania 17822*

Martin J. Cann, Ph.D. *Postdoctoral Fellow, Department of Pharmacology, Cornell University Medical College, 1300 York Avenue, New York, New York 10021*

Dermot M. F. Cooper, Ph.D. *Professor, Department of Pharmacology, University of Colorado Health Sciences Center, 4200 East Ninth Avenue, Denver, Colorado 80262*

Maria A. DeBernardi, Ph.D. *Research Assistant Professor of Cell Biology, Department of Cell Biology, Georgetown University School of Medicine, 3900 Reservoir Road, NW, Washington, DC 20007; and Atto Instruments, Inc., 1450 Research Boulevard, Rockville, Maryland 20850*

Chester L. Drum, B.A. *Graduate Student, Department of Pharmacological and Physiological Sciences, University of Chicago, 5543 S. Kimbark, Apartment 1, Chicago, Illinois 60637*

Kent A. Fagan, B.S. *Graduate Student, Neuroscience Program, University of Colorado Health Sciences Center, 4200 East Ninth Avenue, Denver, Colorado 80262*

George Fink, M.D., D.Phil, F.R.S.E. *Honorary Professor, MRC Brain Metabolism Unit, Department of Pharmacology, University of Edinburgh, 1 George Square, Edinburgh EH8 9JZ, Scotland, United Kingdom*

Paula L. Hoffman, B.A., M.S., Ph.D. *Professor of Pharmacology, Department of Pharmacology, University of Colorado Health Sciences Center, 4200 East Ninth Avenue, C-236, Denver, Colorado 80262*

Yoshihiro Ishikawa, M.D., Ph.D. *Associate Professor of Medicine, Cardiovascular and Pulmonary Research Institute, Allegheny University of the Health Sciences, 320 North East Avenue, Pittsburgh, Pennsylvania 15212*

Ravi Iyengar, Ph.D. *Professor of Pharmacology, Department of Pharmacology, The Mount Sinai Medical Center, One Gustave L. Levy Place, Box 1215, New York, New York 10029*

viii   *CONTRIBUTING AUTHORS*

Jeffrey W. Karpen, Ph.D.   *Associate Professor, Department of Physiology, Campus Box C240, University of Colorado Health Sciences Center, 4200 East Ninth Avenue, Denver, Colorado 80262*

John Krupinski, Ph.D.   *Senior Research Investigator, Bristol-Myers Squibb, P. O. Box 400, K13-09, Princeton, New Jersey 08543*

Lonny R. Levin, Ph.D.   *Assistant Professor, Department of Pharmacology, Cornell University Medical College, 1300 York Avenue, New York, New York 10021*

Nicole E. Mons, Ph.D.   *Laboratory of Functional Neurobiology, University of Bordeaux I, URA-CNRS 339, Avenue des Facultés, Talence 33405, France*

Janice M. Paterson, B.Sc. (Hons), Ph.D.   *MRC Brain Metabolism Unit, Department of Pharmacology, University of Edinburgh, 1 George Square, Edinburgh, EH8 9JZ, Scotland, United Kingdom*

Roberta Rosie, M.Phil, C.Biol, M.I.Biol   *MRC Brain Metabolism Unit, Department of Pharmacology, University of Edinburgh, 1 George Square, Edinburgh, EH8 9JZ, Scotland, United Kingdom*

James Simpson, B.Sc., M.Phil   *Higher Scientific Officer, MRC Brain Metabolism Unit, Department of Pharmacology, University of Edinburgh, 1 George Square, Edinburgh, EH8 9JZ, Scotland, United Kingdom*

Martine J. Smit, Ph.D.   *Department of Pharmacology, The Mount Sinai Medical Center, P. O. Box 1215, One Gustave L. Levy Place, New York, New York 10029*

Susan M. Smith, B.Sc.   *MRC Brain Metabolism Unit, Department of Pharmacology, University of Edinburgh, 1 George Square, Edinburgh, EH8 9JZ, Scotland, United Kingdom*

Boris Tabakoff, Ph.D.   *Professor and Chairman, Department of Pharmacology, University of Colorado Health Sciences Center, 4200 East Ninth Avenue, C-236, Denver, Colorado 80262*

Wei-Jen Tang, Ph.D.   *Department of Pharmacological and Physiological Sciences, University of Chicago, 947 East 58th Street, MC 0926, Chicago, Illinois 60637*

Ronald Taussig, Ph.D.   *Assistant Professor, Department of Biological Chemistry, The University of Michigan Medical School, 1301 Catherine Road, Ann Arbor, Michigan 48109-0636*

Shuizhong Yan, Ph.D., M.D.   *Research Associate, Department of Pharmacological and Physiological Sciences, University of Chicago, 947 East 58th Street, MC 0926, Chicago, Illinois 60637*

Gregor Zimmermann, B.S.   *Graduate Student, Department of Biological Sciences, The University of Michigan Medical School, 1301 Catherine Road, Ann Arbor, Michigan 48109-0636*

Preface

Cyclic adenosine monophosphate, the archetypical second messenger, has formed the basis for much of our thinking on hormone and neurotransmitter action for 40 years now. The study of "adenyl cyclase" or "adenylate cyclase" or, more recently, "adenylyl cyclase"[1] has led to the discovery of guanosine triphosphate–regulatory proteins and has greatly facilitated our understanding of the actions of the so-called serpentine receptors. The concepts and technical strategies developed around G protein regulation of adenylyl cyclase laid the groundwork for uncovering the regulation of phosphoinositidase C and many ion channels. However, it is only over the last six or seven years that detailed information has developed on the enzymes responsible for the synthesis of this second messenger. Since the cloning of the first adenylyl cyclase in 1989, our understanding of these enzymes has exploded. It is now clear that there are multiple forms of this elementary signaling enzyme, which are, in turn, multiply regulated and discretely placed in specific cell types and even within discrete cellular domains. This explosion of information is difficult for experts to keep up with, but it must be even more bewildering for peripheral observers.

The purpose of *Adenylyl Cyclases* is to summarize most of the current developments in this rapidly moving area. Chapters deal with structural features of the adenylyl cyclases, their regulation by G protein [α and $\beta\gamma$] subunits, protein kinases, and calcium ions. The potential physiologic roles that they play is explored and a new technique for their study at the single cell level is presented. The latter approach may go some way toward breaking the technical impasse that separates thinking about cAMP signaling from $[Ca^{2+}]_i$-signaling. It is hoped that the book will stimulate and facilitate research on this critical group of signaling molecules in the coming years. The authors have aimed to provide a synthesis of their chosen topic as it fits into the overall subject of adenylyl cyclases. The volume is aimed both at researchers active in the area and investigators who wish to become current in this field.

Twelve years ago, with Ken Seamon, I edited Volume 19 in this series, entitled *Dual Regulation of Adenyl Cyclase*. That volume captured the excitement that prevailed at that time with the idea that in many sources adenylyl cyclase could be either stimulated or inhibited by different classes of receptors by two G proteins, "G_s/N_s" and "G_i/N_i." There were controversies on the precise mechanism of inhibition of adenylyl cyclase by G_i, which now, with the benefit of hindsight, vary de-

---

[1]The name, *adenylyl cyclase*, comes from the fact that it is the adenylyl (A-r-P) rather than the adenylate (A-r-PO) moiety that is cyclized; a full pyrophosphate rather than a pyrophosphoryl group is released from adenosine triphosphate.

pending on the species of adenylyl cyclase. Indeed, many early findings can now be reinterpreted and reconciled in terms of the species of adenylyl cyclase present in particular sources and the other regulatory mechanisms that may modulate these effects.

This volume documents the multiple regulation of nine adenylyl cyclases. Many of these cyclases have putative regulatory domains and binding domains that promise yet-to-be-discovered interactions with other signaling systems. It seems safe to predict that 12 years from now the area will look even more complex, but hopefully this volume will still accurately reflect the sense of excitement that researchers in the area currently feel at being on a new threshold of understanding the steps taken by nature to exploit this ancient signaling device.

Dermot M. F. Cooper, Ph.D.

ADVANCES IN SECOND MESSENGER
AND PHOSPHOPROTEIN RESEARCH

Volume 32

Adenylyl Cyclases

*Advances in Second Messenger and
Phosphoprotein Research,* Vol. 32,
edited by Dermot M. F. Cooper
Lippincott–Raven Publishers, Philadelphia © 1998

1

Mammalian Adenylyl Cyclases

Martine J. Smit and Ravi Iyengar

Department of Pharmacology, The Mount Sinai Medical Center, New York, New York 10029

Biologic systems possess highly efficient signal transduction systems to regulate and coordinate a variety of functions within a single cell and among different cells. Hormones, neurotransmitters, and autocrine and paracrine factors serve as extracellular messengers in biologic communication networks. The extracellular information is transferred into the cell by transmembrane signaling systems. One ubiquitous transmembrane signal transduction system uses heterotrimeric G proteins as the signal (1,2). These systems are composed of multiple components. A minimal composition of the system includes a receptor, a G protein, and an effector. Signal transfer is initiated by interaction of the extracellular agonist with the receptor. An agonist-occupied receptor activates the heterotrimeric G proteins composed of α, β, and γ subunits (2,3). Upon activation, the guanosine triphosphate (GTP)–bound G_α subunit dissociates from the $G_{\beta\gamma}$ complex. Both G_α and $G_{\beta\gamma}$ regulate effectors such as ion channels or enzymes, which are responsible for the production or degradation of second messengers within the cell. Thereafter, the intracellular second messengers propagate the signals, resulting in a cellular response.

The first identified effector for a G protein signaling system is the enzyme adenylyl cyclase (AC), which produces the second messenger cyclic adenosine monophosphate (cAMP). ACs are stimulated by members of G_s class of α subunits, resulting in elevated levels of intracellular cAMP. The elevation of cAMP results in stimulation of protein kinase A, which is capable of phosphorylating numerous metabolic enzymes, ion channels, and transcription factors, resulting in a change in their functional status. Most effects of cAMP are mediated through the protein kinase A, although in sensory organs cAMP directly regulates ion channels.

Molecular cloning has enhanced our understanding of the complexity of the AC system. Nine mammalian ACs have been identified by means of molecular cloning (4,5). They are all stimulated by $G_{s\alpha}$. All ACs are also stimulated by the diterpene forskolin, albeit to varying degrees. In addition to their core signaling capability in response to signals from $G_{s\alpha}$ the different ACs are capable of receiving signals from a variety of sources, including other G protein subunits ($G_{i\alpha}$ and $G_{\beta\gamma}$), protein kinases (protein kinase A [PKA], protein kinase C [PKC], and calmodulin [CaM] kinase), Ca^{2+}/CaM, and Ca^{2+} by itself. A unique feature of mammalian ACs is that they are differentially regulated: some are stimulated, others are inhibited, and still others are not affected by these different signaling entities. The different ACs can be assigned to distinct groups based on their regulatory properties. One group of ACs is activated by Ca^{2+}/CaM (AC1, AC3, and AC8), a second group is conditionally activated by $G_{\beta\gamma}$ subunits (AC2, AC4, and AC7), and a third is activated by PKC (AC2, AC5, and AC7). Similarly, different groups of ACs are inhibited by differing signals. Most ACs seem susceptible to inhibition by $G_{i\alpha}$. These include AC1, AC2, AC3, AC5, and AC6. More selective groups are inhibited by PKA (AC5 and AC6), CaM kinase (AC1 and AC3), and Ca^{2+} (AC5 and AC6). These properties are summarized in Table 1.

There are substantial differences between the tissue distributions of the different ACs (4). All isoforms are expressed in the brain, but the various ACs appear to be differentially expressed in defined areas within the brain (6). In peripheral tissues, the patterns of AC expression are also quite diverse. Some ACs are widely expressed throughout the periphery and brain (AC5, AC6, and AC9), whereas the expression of others is primarily restricted to designated areas (AC1, AC2, and AC8 in brain; AC3 in olfactory neuroepithelium).

In addition, some form of ACs appears to arise from alternative splicing. Comparison of AC5 and AC6 cloned in our laboratory (7) with sequences reported by Ishikawa and co-workers (8) and Yoshimura and Cooper (9) suggested that they were N-terminal splice variants of AC5 and AC6. Recently, also for AC8 three splice variants—termed AC8-A, B, and C—have been identified (10). The splice sites of these AC8 variants are found at different positions lacking various ranges of amino acids. It is important to note that the splice variants also display different functional properties, although all are stimulated by Ca^{2+}/CaM.

Taken together, these observations indicate that all nine ACs are regulated differently. This varied pattern of regulation allows the cell to change its cAMP levels in response to a variety of signals. It is therefore not surprising that the AC isoforms have distinct physiologic functions. In this review, we will focus on the structural features of mammalian ACs and the differential regulation of the AC isoforms.

TABLE 1. *Regulation of the various adenylyl cyclase isoforms*

Regulatory entity	Regulation adenylyl cyclase isoforms	
	Stimulatory	Inhibitory
$G_{s\alpha}$	1–9	
$G_{i\alpha}$		1–3, 5,6
$G_{\beta\gamma}$	2,4,7	1,8
Forskolin	1–9	
Ca^{2+}/calmodulin	1,3,8	
Ca^{2+}		5,6
PKC	2,5,7	
PKA		5,6
CaM kinase		1,3

Thereafter, we will discuss the functional consequences of this differential regulation with regard to evoked cellular responses.

STRUCTURAL FEATURES

Hydropathy analysis of the amino acid sequences indicates that all mammalian ACs share a common structure similar to that of transporters and some ion channels (11). The predicted secondary structure includes 12 transmembrane spans in two sets of six, a large central cytoplasmic loop between the sixth and seventh transmembrane regions, and a long cytoplasmic tail. These two intracellular domains are similar and are highly conserved between the different ACs. Initial evidence that the cytoplasmic domains are involved in catalysis came from studies on the *Drosophila* homolog of AC1. The gene for AC1-like AC maps to the rutabaga genetic locus (12), which is known to cause a defect in associative learning in flies and is correlated with a loss of CaM-stimulated AC activity in the heads of affected flies (13). The loss of AC activity results from a point mutation in the C-terminal cytoplasmic domain of the *Drosophila* type I AC. This mutation (Gly→Arg) changes a residue absolutely conserved among known G protein–regulated ACs. Further evidence that the core functions of AC are contained within the cytoplasmic domains has come from construction of a soluble AC using the the central cytoplasmic domain of AC1 and the cytoplasmic tail of AC2 (14) that is stimulated by $G_{s\alpha}$ and forskolin. Extension of these experiments independently in the laboratories of Tang and Gilman has shown that $G_{s\alpha}$ and forskolin act to increase the affinity of the two domains of AC for each (15,16). Although such studies that use a reductionist approach are useful in identifying key functional regions within AC, the quantitative and mechanistic conclusions arising from them need to be confirmed by studies on the intact enzymes. It is noteworthy that no reports document a functional regulated enzyme composed of both domains of AC1 or AC2. Recent studies with AC5 have shown that a construct containing the central cytoplasmic domain and C-terminal tail exhibits $G_{s\alpha}$- and forskolin-stimulated activities (17).

The crystal structure of the truncated cytoplasmic tail of AC2 complexed with forskolin has been solved (18). Both in solution (15,16,19) and in crystal, the cytoplasmic tail exists as a dimer. The dimer has a wreathlike structure with a cleft in the center. Each dimeric structure contains two molecules of forskolin. Several residues that line the center of the ventral cavity have been shown by site-directed mutagenesis and chemical fragmentation to bind either adenosine triphosphate (ATP) or ATP analogs (20,21). Hence, the ventral cavity may contain the ATP binding site. Surprisingly, the forskolin binding site is close to the proposed ATP binding site. Forskolin binds to the two ends of the ventral cavity. As expected, the forskolin binding pockets are highly hydrophobic. This dimer of the cytoplasmic tail has been proposed as a model for the interaction between the central cytoplasmic loop and the C-terminal tail in the formation of an active enzyme. Further studies are required to determine the validity of this proposal. The cartoon in Fig. 1 summarizes our current

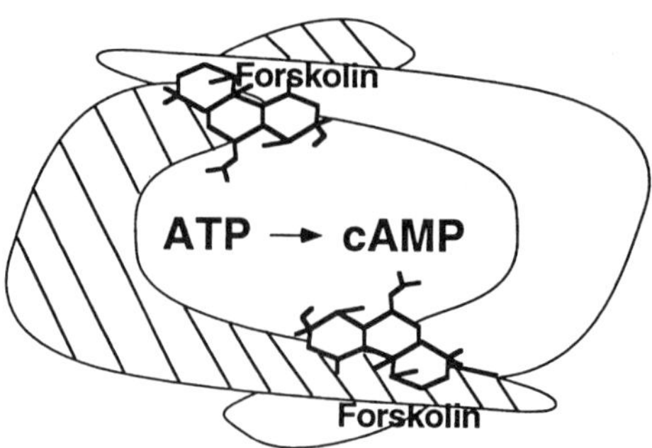

FIG. 1. Schematic representation of the interactions between the cytoplasmic domains of AC2 to form the catalytic site. The cartoon is based on the recently solved crystal structure of the C-terminal tail of AC2. The loop (hatched) and the C-terminal tail would form each half of a wreath-like structure such that the ATP binding site would be in the central cavity. Forskolin is proposed to bind to the interface between the two domains and thus would stabilize the interactions between the central cytoplasmic loop and C-terminal tail to form the catalytic site.

knowledge of structural features of the C-terminal tail of AC2. A surprising and perhaps disappointing aspect of this crystal structure is that even in the presence of forskolin, a clear outline of ATP binding pocket is not obvious. Perhaps this will emerge as structures of the cytoplasmic loop-C-terminal tail complex are solved.

REGULATION OF ADENYLYL CYCLASE ISOFORMS

Basal Activities

The different ACs have distinct basal activities in the presence of Mg^{2+}. AC1 and AC2 show much higher basal activities when compared with AC3 and AC6 (20,22,23). In an explicit comparison, AC2, for example, was found to display a 25-fold greater activity than AC6 (23). This difference in basal activity between AC2 and AC6 cannot be explained solely by differences in their catalytic capacity because in the presence of Mn^{2+} and forskolin the difference in activity is only twofold. The differences between the basal activities of AC2 and AC6 are most pronounced at low physiological relevant free Mg^{2+} ion concentrations. AC1 was found to have high basal activities in the presence of Mg^{2+} (20). Similarly, AC8 has a relatively high basal activity, whereas the newly cloned AC9 has basal activities comparable to AC6 (5).

In many peripheral tissues, relatively low levels of basal intracellular cAMP are observed. As AC6 and AC9 are widely expressed (7,24), their low basal activity may account for the low observed levels of cAMP in most peripheral tissues. On the other hand, AC1, AC2, and AC8—which are enzymes that have high basal activity—are most abundantly expressed in the brain (6), where relatively high basal cAMP levels are detected. The presence of AC1 and AC2 may be reflective of a cell's need to have elevated cAMP levels in response to multiple stimuli, whereas the presence of AC6 and AC9 would allow the cells to maintain low levels of intracellular cAMP.

The structural basis for differences in basal activity is not known. It may be partly ascribed to differences in the catalytic site of the different ACs. Regulation of catalytic activity by divalent cations is also possible. Earlier studies in our laboratory had shown that in S49 cyc⁻ cells divalent cations allosterically regulate AC activity by interactions at the level of AC (25). In addition, point mutations in AC1 resulted in a selective decrease in Mg^{2+}-dependent basal activity, whereas no change in Mn^{2+} plus forskolin activity was observed (20). Preliminary predictions from the crystal structure suggest that there may be a divalent cation binding site (18). The proposed location of this binding site at the interface between the two monomers suggests a mechanism by which divalent cations can allosterically regulate catalytic activity. The divalent cations may promote dimerization and thus stabilize the catalytic site. The structural features that result in the distinct basal activities of the different ACs need to be explored further.

Regulation by G Protein Subunits

$G_{s\alpha}$ and Forskolin

All the cloned mammalian ACs are regulated by $G_{s\alpha}$. Regulation by forskolin is somewhat variable, with the recently cloned AC9 least susceptible to regulation (5). Kinetic analysis of the $G_{s\alpha}$ stimulation of the different ACs indicate that there may be mechanistic differences in the regulation of the different ACs by $G_{s\alpha}$. Stimulation of AC1 appears to involve a single site, but stimulation of AC2 and AC6 appear to involve multiple sites (26). For all three ACs, forskolin not only increases the affinity for $G_{s\alpha}$ but also changes the profile of the curve such that stimulation by $G_{s\alpha}$ appears to involve a single high-affinity site. Thus, it is possible that the $G_{s\alpha}$ and forskolin sites are proximal to each other. A forskolin binding site has been identified in the crystal structure of the cytoplasmic tail of AC2 (18). The site is at the interface of the two monomers and involves residues from both monomers (Figs. 1 and 2). Forskolin-interacting residues are tightly clustered within the 895–1020 region and form a hydrophobic pocket with a crevice into which the forskolin appears to fit. It remains to be determined if this location for the forskolin binding site is preserved in a complex between the central cytoplasmic loop and the C-terminal tail and among the various ACs. At this time, a reasonable working hypothesis is that forskolin binding promotes and stabilizes interactions between the central cytoplasmic loop and C-terminal tail.

Some of the regions of ACs involved in interactions with $G_{s\alpha}$ are also now being identified. Recent studies in our laboratory have identified the region 660–682 of AC6 as being involved in $G_{s\alpha}$ interactions (27) (see Fig. 2). However, this region appears to be involved only in stimulation of ACs by high concentrations of $G_{s\alpha}$ and may thus represent a "low-affinity" binding site for $G_{s\alpha}$. The region contains a motif conserved among the various ACs. The location of the "high-affinity" site has yet to be fully established. Point mutations in the cytoplasmic tail in the 950–1050 region of AC1 result in abolishment of $G_{s\alpha}$ binding (20). The involvement of this region is in agreement with the recently identified forskolin binding site (18). Since $G_{s\alpha}$ and

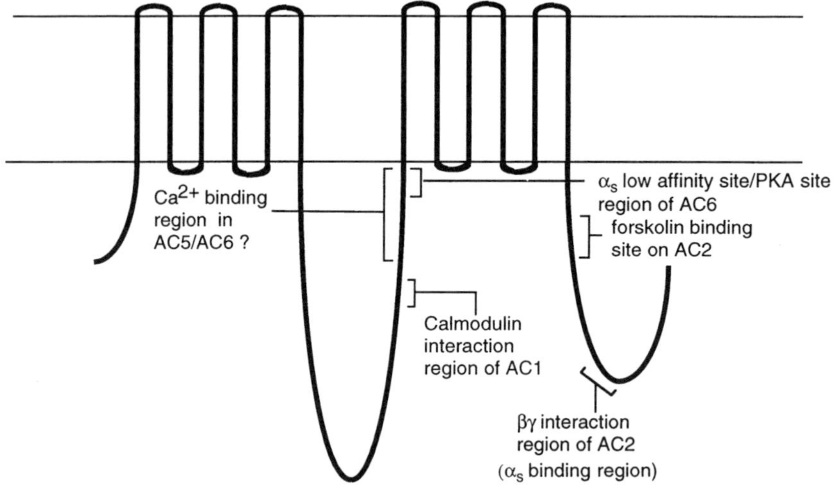

FIG. 2. Regions of ACs involved in interactions with different signaling molecules. The regions are shown within the predicted secondary structure of ACs. The domains identified in the different ACs as being involved in interactions with Ca^{2+}/CaM, Ca^{2+}, G$_{s\alpha}$, and G$_{\beta\gamma}$ are indicated. Also shown is the location of the PKA site in AC6. Mutational analysis of chimeric AC1/AC2 soluble AC has identified a region close to the G$_{\beta\gamma}$ binding region as being involved in α_s interaction (26a). The importance of these residues in native ACs remains to be established.

forskolin show strong positive cooperative heterotrophic interactions, it would be reasonable to assume that some of the residues involved in G$_{s\alpha}$ interactions are close to the forskolin binding site. Such a location would also be in agreement with our recent observation that occupancy of the "high-affinity" site by G$_{s\alpha}$ is sufficient for full stimulation of AC2 by G$_{\beta\gamma}$ (26).

$$G_{i\alpha}$$

Inhibition of cAMP production by receptors that couple to pertussis toxin substrates has been known for the past 15 years. However, extensive inhibition of AC is only seen in some cells and tissues, whereas in others much more modest inhibition is observed. This confusing picture became clearer with the cloning of the various ACs. It was found that the different ACs were susceptible to differing extent of inhibition by activated G$_{i\alpha}$. AC6 was most extensively inhibited in intact (28) as well as *in vitro* assays (29). AC5, which is very similar to AC6, is also extensively inhibited by G$_{i\alpha}$. In contrast to the different ACs tested, AC2 was the least inhibitable in the intact cell (28). In the *in vitro* assays, no inhibition of AC2 was observed (30). A unique feature of G$_{i\alpha}$ inhibition of AC2 in the intact cell is that activation of AC2 by PKC results in abolishing its ability to be regulated by G$_{i\alpha}$ (28). AC3 and AC1 can also be inhibited by G$_{i\alpha}$. Inhibition of AC shows an interesting dichotomy. The Ca^{2+}/CaM-stimulated activity of AC1 can be extensively inhibited, but the activation by G$_{s\alpha}$ is far less susceptible to inhibition (29). All three G$_i$s are capable of inhibiting ACs both in the intact cell (31) and *in vitro* (30). Additionally, Bourne and co-workers (31) have shown that the $G_{\alpha z}$, a G_i-related protein that is not pertussis toxin sensitive, can inhibit ACs. This has been confirmed in *in vitro* assays as well (32).

The regions of ACs involved in $G_{i\alpha}$ interactions are currently unknown. On the basis of kinetic analysis of the S49 cell cyc⁻, Hildebrandt and co-workers (33) had proposed that there may be distinct sites of interactions for G_s and G_i. More recently, Gilman and co-workers have come to a similar conclusion based on the observation that $G_{i\alpha}$ stimulates AC2 in the presence of forskolin (30). It is possible that one of the $G_{s\alpha}$ sites on AC6 also interacts with $G_{i\alpha}$.

$$G_{\beta\gamma}$$

Regulation of ACs by the $G_{\beta\gamma}$ subunits is among the most complex phenomena studied in this system. AC2 and AC4 are conditionally stimulated by $G_{\beta\gamma}$ subunits. For $G_{\beta\gamma}$ stimulation the enzymes need to be first stimulated by activated $G_{s\alpha}$ (30,34,35). Since activated $G_{s\alpha}$ has low affinity for $G_{\beta\gamma}$ subunits, it appears likely that $G_{s\alpha}$ and $G_{\beta\gamma}$ bind to different sites on AC2. Purified AC2 is regulated by $G_{s\alpha}$ and $G_{\beta\gamma}$ subunits (36), and hence regulation involves direct interaction between the G protein subunits and AC2. The notion that $G_{s\alpha}$ and $G_{\beta\gamma}$ bind to different sites on AC2 gained further support with the identification of the region 956–982 of AC2 as being involved in $G_{\beta\gamma}$ signaling (37) (see Fig. 2). A peptide encoding this region blocked stimulation by $G_{\beta\gamma}$ but did not affect stimulation by $G_{s\alpha}$. The peptide also blocked $G_{\beta\gamma}$ regulation of several other effectors, including phospholipase C-β, potassium channels, AC1, and the β-adrenergic receptor kinase. Direct interactions between the region 956–982 of AC2 and $G_{\beta\gamma}$ subunits have been shown by use of the yeast two-hybrid system. These studies have shown that the 956–982 region of AC2 interacts directly with G_β but not with G_γ subunits (38). Cross-linking between the peptide from AC2 and $G_{\beta\gamma}$ subunits has also shown that the peptide interacts with G_β when it is part of the free $G_{\beta\gamma}$ complex but not when it is part of the heterotrimeric complex with G_α. Molecular modeling studies based on the cross-linking analysis have resulted in identification of regions in G_β involved in signal transfer to effectors (39). Within the 956–982 region we had identified a QXXER motif that appeared to be important for $G_{\beta\gamma}$ regulation of AC2. Recent studies on the Ca^{2+} channels have shown that a similar region is involved in $G_{\beta\gamma}$ inhibition of the neuronal Ca^{2+} channels (40,41).

In addition to stimulating AC2 and AC4, $G_{\beta\gamma}$ subunits inhibit AC1 and AC8. $G_{\beta\gamma}$ inhibition of CaM-stimulated AC activity had been described previously but had been ascribed to $G_{\beta\gamma}$ sequestration of CaM (42). Analysis with the cloned AC1 showed that $G_{\beta\gamma}$ subunits interacted directly with AC1 as well and inhibited both $G_{s\alpha}$ and CaM-stimulated AC (36). However, the region in AC2 identified as being involved in $G_{\beta\gamma}$ signaling is not conserved in AC1 or AC8 and a motif similar to the QXXER motif is not obvious. Hence different ACs may contain several distinct regions with which $G_{\beta\gamma}$ subunits can interact. It is currently not known where on AC1 the $G_{\beta\gamma}$ complex binds. The other ACs—AC3, AC5, AC6, and AC9—are not directly regulated by $G_{\beta\gamma}$ subunits. AC7, which is closely related to AC2 and AC4, is likely to be stimulated by $G_{\beta\gamma}$ subunits, though this has not been reported as yet.

Regulation by Ca²⁺/Calmodulin

Changes in intracellular Ca^{2+} concentrations have been shown to modulate AC activity of some AC isoforms positively (AC1, AC3, and AC8), others negatively (AC5 and AC6), and some not at all (AC2, AC4, AC7, and AC9). AC1 and AC8 are markedly stimulated by nanomolar concentrations of Ca^{2+}/CaM (43,44). AC1 and AC8 predominantly respond to capacitative Ca^{2+} entry and not to the release of Ca^{2+} from intracellular stores or to a nonspecific increase in Ca^{2+} elicited by an ionophore (45). These results indicate a high degree of spatial co-localization of ACs and Ca^{2+} entry channels. Moreover, AC1 is synergistically activated by Ca^{2+} and G_s-coupled receptors (46). Interestingly, AC8 is not synergistically activated by neurotransmitters and Ca^{2+}, indicating an important regulatory distinction between these two Ca^{2+}-stimulated ACs.

There is now considerable evidence that the carboxy terminal region of the central cytoplasmic loop of Ca^{2+}/CaM-sensitive AC1 is involved in the regulation by Ca^{2+} and CaM. A peptide corresponding to residues 495–522 of the C_{1b} region of AC1 was shown to bind to CaM and to inhibit CaM-stimulated enzyme activity (47) (see Fig. 2). Point mutagenesis within this putative CaM-binding domain diminishes the ability of Ca^{2+}/CaM to stimulate enzyme activity, indicating that AC1 can be directly regulated by Ca^{2+} and CaM *in vivo* (48). However, other regions may be involved as well. Tang and co-workers identified two amino acids, K350 and K923, that when modified to Ala residues resulted in lack of Ca^{2+}/CaM stimulation (20).

Recently, three alternatively spliced AC8 isoforms have been identified, referred to as AC8-A, AC8-B, and AC8-C (10). Type AC8-A corresponds to the earlier reported AC8. AC8-B lacks amino acids 802–831 of AC8-A, and AC8-C lacks amino acids 666–682 of AC8-A corresponding to the C_{1b} region. There is no apparent sequence homology between the CaM-binding domain of AC1 and the corresponding sequence in AC8-A (10). If AC8-A is aligned with AC1, the CaM-binding domain terminates seven amino acids before the start of the 66-amino-acid region missing in AC8-C. Interestingly, AC8-C has a threefold higher affinity for Ca^{2+}/CaM (10). These data suggest that the 66-amino-acid region in AC8-A and AC8-B affects the conformational change or plays an inhibitory role in the affinity toward Ca^{2+}/CaM.

All ACs are inhibited by high concentrations of Ca^{2+} as a result of competition for Mg^{2+}, which is required for catalysis or formation of complexes between ATP and Ca^{2+} (49). AC3, however, is insensitive toward Ca^{2+} and might therefore allow cAMP synthesis to persist even in the presence of very high concentrations of Ca^{2+}. AC5 and AC6 are inhibited by micromolar amounts of Ca^{2+} elevated through capacitive Ca^{2+} entry (9,50). Inhibition of AC5 and AC6 activities by Ca^{2+} does not require exogenously added CaM (9,17,51). Truncation analysis during the construction of soluble forms of AC5 indicates that a 113-amino-acid stretch from 571–683 is involved in Ca^{2+} inhibition of AC5 (17) (see Fig. 2). Soluble AC5 lacking this region is not susceptible to inhibition by Ca^{2+}, and when the entire central cytoplasmic loop is expressed by itself it binds Ca^{2+} in overlay assays, but such binding is not observed in the truncated forms. Finer analysis is required to determine the residues involved in Ca^{2+} binding.

AC2, AC4, AC7, and AC9 appear to be insensitive to changes in intracellular Ca^{2+} concentrations (52,53).

Regulation by Protein Kinases

Protein Kinase C

Stimulation of ACs by PKC has been reported in many but not all cellular systems. Initially, reports suggested that such regulation of ACs by PKC may occur at the level of the G protein (54,55), whereas others suggested the ACs as the target for regulation (56,57). Using expression of AC in mammalian cell lines, we and others showed that AC2 but not other isoforms of AC are extensively stimulated by activation of PKC (58–60). In particular, the basal activity of AC2 was shown to be increased upon PKC activation. At low Mg^{2+} concentrations, which mimic intracellular conditions, a five- to sixfold increase in AC2 activity was observed upon PKC activation. Forskolin- and $G_{s\alpha}$-stimulated activity of AC2 was also increased, but to a lesser extent. Further studies in our laboratory showed that the increase in AC2 activity correlated with an increase in phosphorylation of AC2 in Sf9 cells, when the native PKCs in Sf9 cells were activated by phorbol esters (61). Recently, Zimmermann and Taussig showed that *in vitro* treatment of AC2 with PKCα leads to increases in basal-, forskolin-, and βγ-stimulated activity and direct phosphorylation of AC2 (62). Surprisingly, in these *in vitro* experiments, PKCα decreases βγ potentiation of $G_{s\alpha}$-stimulated activity. In contrast, in transient transfection experiments activation of PKC augmented $G_{\beta\gamma}$ stimulation of AC2 (63). AC4, displaying almost identical regulatory properties as AC2 with respect to their responsiveness toward G protein subunits, is not stimulated by PKC (59). This has recently been confirmed by Zimmerman and co-workers (62). PKCα activation has little or no effect on the basal-, forskolin-, or βγ-stimulated activities but leads to reduced responsiveness of AC4 to $G_{s\alpha}$ and blocks the ability of βγ to superstimulate AC4 in the presence of $G_{s\alpha}$. AC7, an isoform that is structurally related to AC2, is also increased through activation of PKC (52,53). The activity of AC2 is more gradually increased upon phorbol ester exposure, whereas AC7 is rapidly activated and diminished by phorbol ester treatment.

In addition, AC5 was also reported to be activated by PKC (64). Yet a marked enhancement in catalytic AC activity by PKC was only found *in vitro* and not or only modestly *in vivo* (59,64,65). Recently, Kawabe and co-workers showed a fivefold increase of forskolin-stimulated cAMP accumulation in HEK-293 cells stably expressing AC5, with only a 60% increase of basal activity upon phorbol ester treatment (65). Using purified AC5 and PKC isoenzymes, Kawabe and co-workers demonstrated that PKCα and PKCζ directly phosphorylate AC5, which was accompanied by a significant increase in catalytic AC activity (64). The pattern of phosphorylation exhibited by the two PKC isoenzymes as demonstrated by phosphopeptide mapping was different, suggesting that AC5 is regulated in a PKC isoenzyme-specific manner. Recently, PKC was also found to directly phosphorylate AC6 and was found to be responsible for the observed A2a adenosine receptor–mediated desensitization of cAMP responses (66,67).

The other isoforms of ACs (AC1, AC3, AC6) that have been tested are either not stimulated or only modestly stimulated (59,68). No information is available as to whether the newly cloned AC9 is affected by PKC.

Protein Kinase A

During studies on glucagon-induced desensitization in chick hepatocytes, one component of desensitization was cAMP dependent. Reconstitution with G_s did not restore full activity to membranes from cells treated either with glucagon or 8-Br-cAMP. Treatment of membranes from naive but not desensitized cells *in vitro* with PKA resulted in decreased AC activity (69). Studies on the kin⁻ variant of S49 cells also indicate that PKA treatment results in decreased forskolin-stimulated AC activity (69,70). Since G_s is not a target for PKA regulation (71), it appeared that AC is the most likely target for the PKA-mediated phosphorylation. Polymerase chain reaction (PCR) analysis had shown that chick hepatocytes contain AC5 and AC6 (69). We have also cloned the mouse AC6 from S49 cells (69). Since both chick hepatocytes and murine S49 cells have AC6 in common, we had proposed that this may be one of the isoforms inhibited by PKA-dependent phosphorylation (4). Recent studies in our laboratory have confirmed this. Treatment of Sf9 cell membranes containing AC6 with PKA results in decreased stimulation by $G_{s\alpha}$ (27). The effect is specific for the AC6 family. AC1 and AC2 are not affected by PKA treatment. Ishikawa and co-workers have shown that PKA treatment of AC5 results in decreased activity as well (72).

There is a single predicted PKA phosphorylation site in AC6 (see Fig. 2). We have mutated the Ser at this site (Ser-674) to an Ala and tested to see if effects of PKA are lost (27). AC6 can be phosphorylated by PKA; however, the S674A mutant AC6 is not phosphorylated by PKA. Furthermore, when S674A mutant AC6 is treated with PKA, there is no effect on $G_{s\alpha}$ stimulation. These observations indicate that PKA regulates G_s interactions with AC6 by phosphorylation at Ser-674. Such regulation plays an important role in heterologous desensitization. AC6 may be a locus of heterologous desensitization in several cell types.

CaM Kinase

Storm and co-workers have reported that elevation of Ca^{2+} suppresses activity of AC3 expressed in HEK-293 cells (73). An inhibitor of CaM kinase relieves this suppression, suggesting that CaM kinase may be involved. Co-expression of activated fragment of CaM kinase also results in suppression of AC3 activity (73). In a recent study, Storm and co-workers have shown that CaM kinase II phosphorylates and inhibits AC3 in the intact cell (74). AC1 is also inhibited by CaM kinase (75). Inhibition of about 50% results from a suppression of maximal activity, rather than a change in sensitivity to Ca^{2+}. This inhibition is specific for the isoforms of CaM kinases. CaM kinase IV inhibits AC1 whereas CaM kinase II does not. Mutation of Ser-545 or Ser-552 in AC1 to Ala results in abolishing CaM kinase-mediated inhibition of AC1. In contrast to AC1 and AC3, AC8 is not inhibited by either CaM kinase II or IV. It is thought that such differential regulation could represent a mechanism of attenuation of receptor-regulated cAMP increases and could be a mechanism for connection between the cAMP and Ca^{2+} signaling pathways. It is also not known if other ACs are regulated by CaM kinases.

Other Protein Kinases

It has been reported that AC9 can be activated by the inhibition of the phosphatase calcineurin (76). These observations suggest that AC9 needs to be phosphorylated for it to be active. The identity of the protein kinases that can phosphorylate and activate AC9 remains currently unknown. Patel and co-workers have reported that when AC5 is expressed in HEK-293 cells, epidermal growth factor (EGF) stimulates AC activity in *in vitro* assays. Such stimulation is not observed when other isoforms are expressed (77). EGF stimulation appears to involve activation of $G_{s\alpha}$, though it is unclear whether this is because the EGF receptor activates G_s. It is currently not known if AC5 or other ACs are substrates for tyrosine kinases.

Regulation of Adenylyl Cyclases by P-Site Ligands

ACs are potently inhibited by purine nucleotides. These compounds are referred to as P-site inhibitors because their inhibition is strictly dependent on the presence of an intact purine ring (78). The most potent P-site inhibitors are the dideoxy adenosine analogs (79). The stimulated forms of ACs, such as activity in the presence of Mn^{2+} or hormone-stimulated activity, are most susceptible to inhibition by P-site ligands. The exact location of P-site inhibition is currently unknown. Inhibition is thought to occur because of direct interactions with a domain distinct from the catalytic site, although some parts of the P-site appear also to be part of the catalytic site. Some mutations in AC1 were shown to alter ATP and P-site ligand binding to the same extent, whereas for other mutations the effects were found to be different (20). Mutation of Lys 923 to Ala in AC1 resulted in a greater than 90% inhibition of basal, G_s, and CaM activity. The mutant enzyme showed substantial (40% of wild type) activity in the presence of Mn^{2+} and forskolin but showed a hundredfold decrease in sensitivity to P-site inhibition. This implies that Lys 923 is important both for catalytic activity in the presence of Mg-ATP and for P-site ligands. Besides, the configuration of the catalytic site appears to be different for Mg-ATP and Mn-ATP. Recent studies have shown that the different AC isoenzymes display distinct sensitivity toward P-site-mediated inhibition (80). The general order of sensitivity to P-site inhibition is AC1 > AC6 = AC8 > AC2.

FUNCTIONAL CONSEQUENCES OF DIFFERENTIAL REGULATION

Adenylyl Cyclase Isoforms and Proliferation

Intracellular cAMP has long been known as a modulator of cell proliferation in many cell types. Depending on the cell type, cAMP has been reported to induce growth stimulatory and inhibitory effects. Though these results may appear to conflict, they demonstrate that cAMP, dependent on the physiologic status of the cell, exerts dual controls, either positively or negatively, with regard to cell proliferation. Growth factors are known to stimulate proliferation through interaction with receptor tyrosine kinases, thereby activating the Ras–Raf–MeK–MAP (mitogen-activated protein) kinase signaling pathway (81,82). Several groups, including this laboratory, have shown that elevation of intracellular levels of cAMP by G_s, addition of 8-Br-cAMP or forskolin, suppresses signaling through the MAP kinase pathways (83–86). Studies in our laboratory have shown that modest increases (50–100%) in intracellular levels of cAMP induced by expression of mutant-activated $G_{s\alpha}$ was sufficient to suppress Ras-induced transformation in NIH-3T3 cells (86). Suppression of the MAP kinase signaling pathway by elevation of cAMP was found to result from phosphorylation of Raf by PKA, disrupting the binding of Raf to Ras and thereby inhibiting the subsequent signaling (83,84,87). In addition, lowering of cAMP levels in Rat-1 fibroblasts was shown to trigger proliferation (88). As such, cAMP can function as a negative modulator of proliferative signals, from growth factors and Ras that activate the MAP kinase pathway.

Various pathophysiologic disorders, including cancer, are ascribed to unregulated proliferation. Cho-Chung and co-workers showed the inhibitory effects of cAMP on a broad spectrum of human cancer cell lines in culture as well as in human xenograft in athymic mice by the use of cAMP analogs (89). Elevation of cAMP levels in most cells has little or no effect on cell proliferation. Thus, the interaction between the cAMP and MAP kinase signaling pathway may provide a physiologic means to inhibit unregulated proliferation.

One way to modulate levels of cAMP that affect proliferation is by the differential regulation of the AC isoforms. The basal activity of some ACs may contribute to the steady-state intracellular levels of cAMP. In cells where elevation of cAMP induces growth inhibitory effects, high basal activity may attenuate proliferation or processes that lead to neoplastic transformation, whereas low basal activity may permit proliferation. As stated earlier, AC1 and AC2 show high basal activity, whereas AC6 displays low basal activity (20,22,23). Interestingly, AC1 and AC2 are most abundantly expressed in the brain, an organ that displays high levels of cAMP and contains nerve cells that persist throughout life without dividing and without being replaced. In contrast, AC6 is more abundantly expressed in peripheral tissues, where cAMP levels are often low, and cells need to differentiate and proliferate to maintain proper organ function. As such, the basal activity displayed by the different AC isoforms may regulate ambient intracellular cAMP levels, thereby contributing to the differentiation status of the cell.

The regulatory properties of the ACs may also be used to control unregulated proliferation. Elevation of cAMP simultaneously with activation of the MAP kinase signaling pathways could inhibit mitogenic signaling. Research in our laboratory and others has demonstrated that AC2 is activated by PKC, thereby raising intracellular levels of cAMP (58–61). Activation of growth factor receptors such as platelet-derived growth factor (PDGF) and EGF receptors results in the stimulation of several downstream effectors, including Ras, phospholipase C-γ (PLCγ) and phosphatidylinositol 3 kinase. It is important to realize that growth factors induce activation of PKC through activation of PLCγ. Thus, if growth factor receptors are activated and AC2 is present, activation of PKC should give rise to an elevation of cAMP. This elevation in cAMP should result in the inhibition of signal transmission from Ras and Raf to MAP kinase, leading to a reduction in mitogenic signaling. The ectopic expression of AC2 therefore could be used therapeutically to regulate proliferation. AC2 can raise intracellular levels of cAMP only when activated by growth factors and may thus inhibit proliferation in a conditional manner. Such conditional regulation of proliferation could be useful in cells that are predisposed to aberrant proliferation such as precancerous cells. If the proposed studies are successful, they could lay the groundwork for novel gene therapy–based approaches for the treatment of proliferative disorders. Preliminary results in our laboratory have indicated that ectopic expression of AC2 attenuates PDGF-induced signaling, suggesting that AC2 may function as a conditional modulator of regulation. The presence of AC2 can serve to integrate signals between different signaling pathways and as such modulate proliferative responses.

It is also noteworthy that cells that are most capable of proliferating, such as fibroblasts, are abundant in AC6 as assessed by G_i inhibition. Since receptor tyrosine kinases often activate phospholipase C and raise intracellular Ca^{2+}, it is possible that this may lead to a transient inhibition of cAMP production as well as transient activation of phosphodiesterases. As a result, PKA would be inhibited, thereby facilitating growth factor–initiated signal transmission through the Raf/ERK-1,2 pathway and thus promoting a proliferative response. Thus, the complement of ACs in a cell could regulate its proliferative status. The possible role of AC isoforms in regulating cell proliferation is summarized in Fig. 3.

Adenylyl Cyclase Isoforms and Regulation of Synaptic Plasticity

The cAMP signaling pathway is important for learning and memory in a variety of species. The molecular mechanisms underlying the role of cAMP in memory have been the focus of intense studies. Particular attention has been directed toward the understanding of the role of the cAMP pathway in long-term potentiation (LTP) of synaptic responses in the hippocampus. LTP is widely viewed as an electrophysiologic paradigm for learning and memory (90). It is generally accepted that the primary event in the induction of LTP is an elevation of intracellular Ca^{2+} in the postsynaptic neuron. This elevation results from Ca^{2+} influx through the N-methyl-

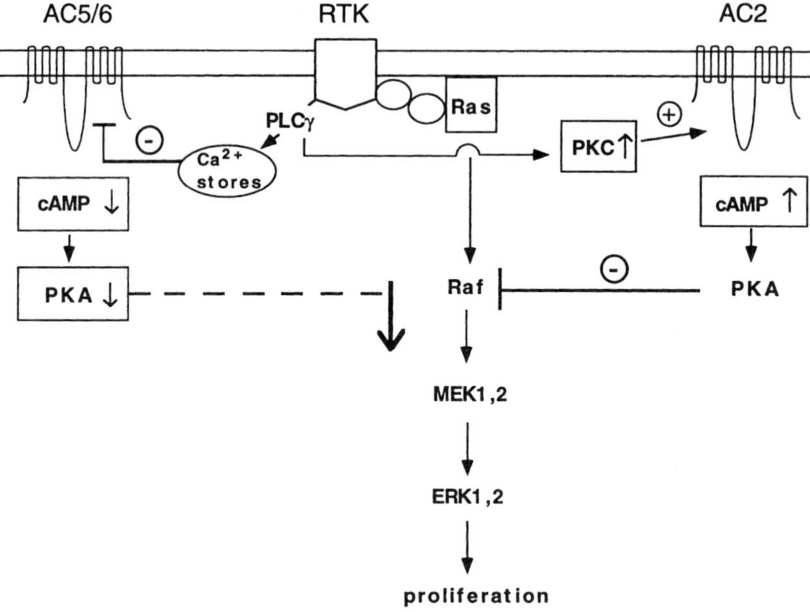

FIG. 3. Role of ACs in proliferation. A hypothetical scheme is shown in which receptor tyrosine kinase signaling to both cAMP and MAP kinase pathways may result in altered MAP kinase-triggered proliferation. Stimulation of growth factor receptors results in activation of PKC, which would give rise to an elevation of cAMP in the presence of AC2. This elevation in cAMP would result in the inhibition of signal transmission from Ras and Raf to MAP kinase, leading to a reduction in mitogenic signaling. In addition, stimulation of receptor tyrosine kinases leads to activation of phospholipase C and to an increase in intracellular Ca^{2+}. In the presence of AC5/6, this may lead to a transient inhibition of cAMP. As a result, protein kinase A would be inhibited, thereby facilitating growth factor–initiated signal transmission through the Raf/ERK-1, -2 pathway and thus promoting a proliferative response. RTK, receptor tyrosine kinases; MAPK, MAP kinase. Other abbreviations are described in the text.

D-aspartate (NMDA) receptor. NMDA has been shown to elevate cAMP through a Ca^{2+}/CaM-dependent mechanism (91). Ca^{2+}/CaM-dependent activation of ACs plays two distinct roles in LTP.

The early phase of LTP, which involves the first hour of potentiated responses, utilizes biochemical signaling reactions but does not involve protein synthesis or gene expression (90). In this phase of LTP, the cAMP pathway plays a regulatory role. Stimulation that produces LTP also elevates cAMP levels in the postsynaptic neuron through CaM, indicating a role for AC1 and AC8. This elevation of cAMP is essential for LTP because inhibitors of PKA when injected into the postsynaptic neuron block LTP. However, postsynaptic elevation of cAMP by itself does not enhance synaptic responses (92). In contrast, expression of constitutively activated CaM kinase by itself appears to be sufficient to enhance postsynaptic responses (93,94); hence, the CaM kinase pathway is thought to be the primary pathway for LTP. The cAMP pathway appears to enhance signal flow through the CaM kinase pathway by inhibiting phosphoprotein phosphatases that deactivate the CaM kinase

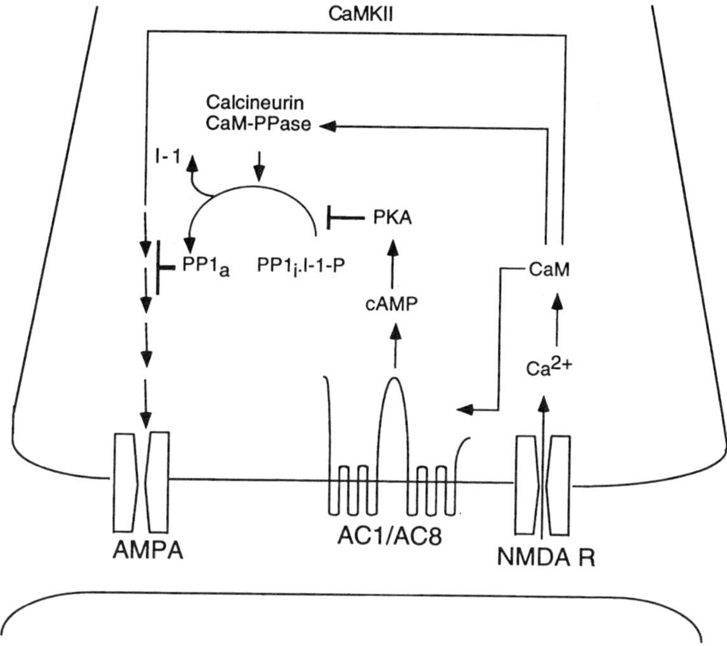

FIG. 4. Role of AC1 and/or AC8 in the early phase of LTP of synaptic responses. A schematic representation of some of the signaling pathways in early LTP is shown. Activation of NMDA receptors leads to a rise in Ca^{2+}. The Ca^{2+}/CaM in turn activates AC1 and or AC8, resulting in elevation of cAMP. Ca^{2+}/CaM also activates CaM kinase II, which appears to be the primary pathway in signaling LTP. The cAMP pathway can enhance signal flow through the CaM kinase pathway by inhibiting phosphoprotein phosphatases that deactivate the CaM kinase (92). PP1a, activated protein phosphatase 1; PP1i, inactive PP1; I-1-P, phosphorylated (activated) inhibitor-1. Other abbreviations are described in the text.

(92). This presumably results in sustained CaM kinase signaling, which in turn results in potentiated postsynaptic responses. Interactions between the cAMP and CaM kinase pathways in early LTP are summarized in Fig. 4. It is noteworthy that in this system, initial Ca^{2+}/CaM signal can be routed through multiple pathways and that interactions between these pathways appear to be essential for the physiological response. Such interactions between these signaling pathways gives the cell an efficient mechanism for regulating physiologic responses.

In addition to playing a regulatory role in the early LTP, the cAMP pathway plays a crucial role in late LTP. This phase of LTP is entirely dependent on cAMP and involves protein synthesis and regulation of gene expression (95). Elevation of cAMP has been shown to be sufficient to evoke late LTP (96). Hence, it appears likely that cAMP-regulated gene expression will play a key role in late LTP. In transgenic mice where PKA is selectively inhibited in the hippocampus and other forebrain areas, there is a selective loss of the late phase of LTP along with a decrease in spatial and long-term memory (97). cAMP-regulated gene expression is mediated through the transcription factor cAMP-responsive element binding protein (CREB), and elevation of

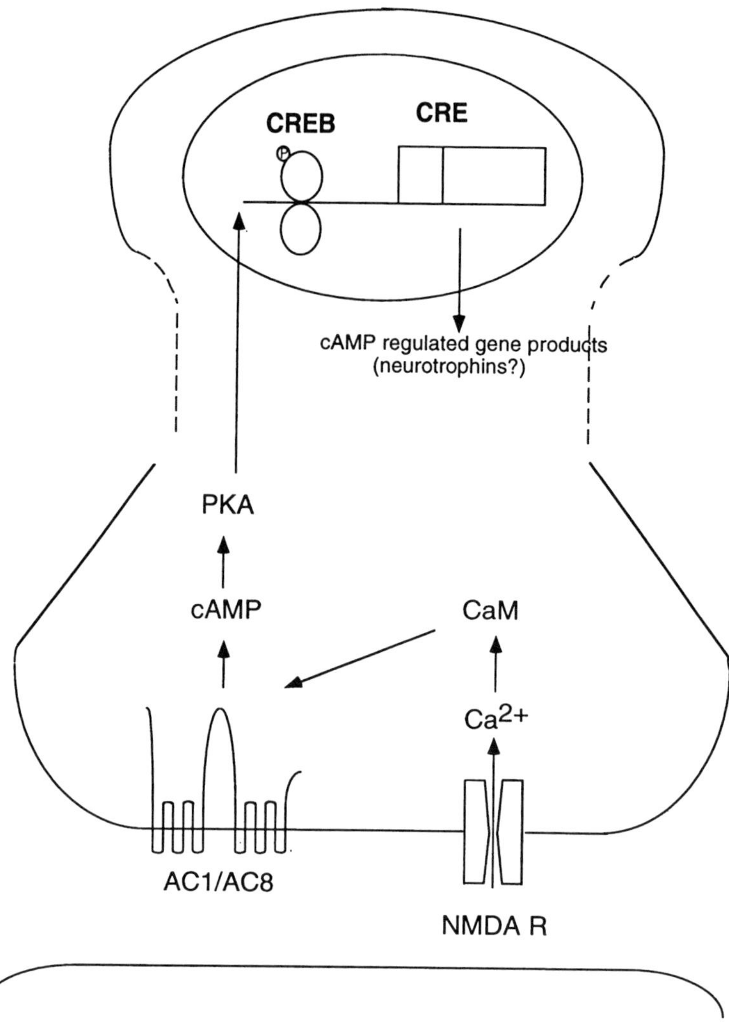

FIG. 5. Role of AC1 and/or AC8 in the late phase of LTP. The late phase of LTP requires gene expression and protein kinase A stimulation. Since the initial intracellular signal for LTP is Ca^{2+}, a schematic representation of how Ca^{2+} can regulate gene expression through protein kinase A is shown. Activation of NMDA receptors leads to a rise in Ca^{2+}, which in turn results in an increase of cAMP in the presence of AC1 and/or AC8. Elevation of cAMP and Ca^{2+} activates CREB, thereby regulating gene expression and contributing to late LTP. Currently, the genes regulated by this process have not been identified. One speculative possibility is that genes that encode neurotrophins are induced in a cAMP-dependent manner. CRE, cAMP-responsive elements. Other abbreviations are described in the text.

cAMP and Ca^{2+} is known to activate CREB in the neuronal cells (98,99). Thus, regulated gene expression, which is crucial for the late phase of LTP and hippocampus-based memory, may be triggered by activation of the cAMP pathway. With respect to the relevant AC isoforms, the key observation is that the Ca^{2+}/CaM sensitivity of AC1 and AC8 makes it possible for a linkage between the early postsynaptic events involving NMDA receptor–mediated Ca^{2+} entry and the later nuclear events. Thus AC1 and AC8 could be key components of the primary biochemical pathways required for learning and memory. The roles of ACs in early and late LTP are summarized in Figs. 4 and 5.

CONCLUSION

cAMP has been long recognized as an important intracellular messenger capable of regulating such diverse functions as sugar metabolism in the liver, steroidogenesis in the ovary, and cardiac contractility. Recently, it has been recognized that in addition to these crucial physiologic functions where the cAMP pathway functions as the primary signal transmittal pathway there are a number of other key functions, such as differentiation, proliferation, and synaptic plasticity, where the cAMP pathway plays an important regulatory role. The varied signal-receiving capability of the different isoforms of ACs appears to be critical in determining whether or not the cAMP pathway regulates these processes. The molecular mechanisms by which the cAMP pathway interacts with other signal pathways are only beginning to be elucidated. Future studies will have to ascertain the role of individual ACs and their signal-receiving capabilities in integrated signaling networks that underlie complex biologic responses.

ACKNOWLEDGMENTS

Research in our laboratory is supported by NIH grants DK-38761 and GM-54508. M.J.S. was supported by a fellowship from the Netherlands Organization for Scientific Research (NWO).

REFERENCES

1. Iyengar R, Birnbaumer L: Signal transduction by G-proteins. *ISI Atlas Sci* 1987;1:213–221.
2. Gilman AG: G proteins: transducers of receptor-generated signals. *Annu Rev Biol Chem* 1987;56: 651–693.
3. Neer EJ: Heterotrimeric G proteins: organizers of transmembrane signals. *Cell* 1995;80:249–257.
4. Iyengar R: Molecular and functional diversity of mammalian G_s-stimulated adenylyl cyclases. *Faseb J* 1993;768–775.
5. Premont RT, Matsuoka I, Mattei M-G et al: Identification and characterization of a widely expressed form of adenylyl cyclase. *J Biol Chem* 1996;271:13900–13907.
6. Mons N, Harry A, Dubourg P et al: Immunohistochemical localization of adenylyl cyclase in rat brain indicates a highly selective concentration at synapses. *Proc Natl Acad Sci U S A* 1995;92:8473–8477.
7. Premont RT, Chen J, Ma HW et al: Two members of a widely expressed subfamily of hormone-stimulated adenylyl cyclases. *Proc Natl Acad Sci U S A* 1992;89:9809–9813.
8. Ishikawa Y, Katsushika S, Chen L et al: Isolation and characterization of a novel cardiac adenylyl cyclase cDNA. *J Biol Chem* 1992;268:13553–13557.

9. Yoshimura M, Cooper DMF: Cloning and expression of a Ca^{2+} inhibitable adenylyl cyclase from NCB-20 cells. *Proc Natl Acad Sci U S A* 1992;89:6710–6720.

10. Cali JJ, Parekh RS, Krupinski J: Splice variants of type VIII adenylyl cyclase. Differences in glycosylation and regulation by Ca^{2+}/calmodulin. *J Biol Chem* 1996;271:1089–1095.

11. Krupinski J, Coussen F, Bakalyar HA et al: Adenylyl cyclase amino acid sequence: possible channel- or transporter-like structure. *Science* 1989;244:1558–1564.

12. Levin LR, Han P-Y, Hwang PM et al: The *Drosophila* learning and memory gene rutabaga encodes a Ca^{2+}/calmodulin responsive adenylyl cyclase. *Cell* 1992;68:479–489.

13. Livingstone MS, Sziber PP, Quinn WG: Loss of calcium/calmodulin responsiveness in adenylate cyclase of rutabaga, a *Drosophila* learning mutant. *Cell* 1984;37:205–215.

14. Tang WJ, Gilman AG: Construction of a soluble adenylyl cyclase activated by G_s alpha and forskolin. *Science* 1995;268:1769–1772.

15. Whisnant RE, Gilman AG, Dessauer CW: Interaction of the two cytosolic domains of mammalian adenylyl cyclase. *Proc Natl Acad Sci U S A* 1996;93:6621–6625.

16. Yan S-Z, Hahn D, Huang Z-H, Tang W-J: Two cytoplasmic domains of mammalian adenylyl cyclase form a $G_{s\alpha}$- and forskolin-activated enzyme *in vitro*. *J Biol Chem* 1996;271:10941–10945.

17. Scholich K, Barbier AJ, Mullenix JB, Patel TB: Characterization of soluble forms of non-chimeric type 5 adenylyl cyclase. *Proc Natl Acad Sci U S A* 1997;94:2915–2920.

18. Zhang G, Liu Y, Ruoho AE, Hurley JH: Structure of the adenylyl cyclase catalytic core. *Nature* 1997;386:247–253.

19. Dessauer CW, Gilman AG: Purification and characterization of a soluble adenylyl cyclase activated by $G_{s\alpha}$ and forskolin. *J Biol Chem* 1996;271:16967–16974.

20. Tang WJ, Stanzel M, Gilman AG: Truncation and alanine-scanning mutants of type I adenylyl cyclase. *Biochemistry* 1995;34:14563–14572.

21. Droste M, Mollner S, Pfeuffer T: Localisation of an ATP-binding site on adenylyl cyclase type I after chemical and enzymatic fragmentation. *FEBS Lett* 1996;391:209–214.

22. Bakalyar HA, Reed RR: Identification of a specialized adenylyl cyclase that may mediate odorant detection. *Science* 1990;250:1403–1406.

23. Pieroni JP, Harry A, Chen J et al: Distinct characteristics of the basal activities of adenylyl cyclases 2 and 6. *J Biol Chem* 1995;270:21368–21373.

24. Katsushika S, Chen L, Kawabe J-I et al: Cloning and characterization of a sixth adenylyl cyclase isoform types V and VI constitute a subgroup within the mammalian adenylyl cyclase family. *Proc Natl Acad Sci U S A* 1992;89:8774–8778.

25. Somkuti SG, Hildebrandt JD, Herberg JT, Iyengar R: Divalent cation regulation of adenylyl cyclase: an allosteric site on the catalytic component. *J Biol Chem* 1982;257:6387–6393.

26. Harry A, Chen Y, Magnusson R et al: Differential regulation of adenylyl cyclases by $G_{\alpha s}$*. *J Biol Chem* 1997; 19017–19021.

26a. Yan S-Z, Huang Z-H, Rao VD et al. Three distinct regions of mammalian adenylyl cyclase forms a site for $G_{s\alpha}$ activation. *J Biol Chem* 1997; 272:18849–18854.

27. Chen Y, Harry A, Li J et al: A $G_{\alpha s}$ interaction region in adenylyl cyclase: selective involvement in protein kinase A regulation of adenylyl cyclase 6. (submitted)

28. Chen J-Q, Iyengar R: Inhibition of cloned adenylyl cyclases by mutant activated G_i-alpha and specific suppression of type 2 adenylyl cyclase inhibition by phorbol ester treatment. *J Biol Chem* 1993;268:12253–12256.

29. Taussig R, Iniguez Lluhi JA, Gilman AG: Inhibition of adenylyl cyclase by G_i alpha. *Science* 1993;261:218–221.

30. Taussig R, Tang WJ, Hepler JR, Gilman AG: Distinct patterns of bidirectional regulation of mammalian adenylyl cyclases. *J Biol Chem* 1994;269:6093–6100.

31. Wong YH, Conklin BR, Bourne HR: G_z-mediated hormonal inhibition of cyclic AMP accumulation. *Science* 1992;255:339–342.

32. Kozasa T, Hepler JR, Smrcka AV et al: Purification and characterization of recombinant G16 alpha from Sf9 cells: activation of purified phospholipase C isozymes by G-protein alpha subunits. *Proc Natl Acad Sci U S A* 1993;90:9176–9180.

33. Hildebrandt JD, Codina J, Birnbaumer L: Interaction of the stimulatory and inhibitory regulatory proteins of the adenylyl cyclase system with the catalytic component of cyc-S49 cell membranes. *J Biol Chem* 1984;259:13178–13185.

34. Tang WJ, Gilman AG: Type-specific regulation of adenylyl cyclase by G protein beta gamma subunits. *Science* 1991;254:1500–1503.

35. Tang WJ, Gilman AG: Adenylyl cyclases. *Cell* 1992;70:869–872.

36. Taussig R, Quarmby LM, Gilman AG: Regulation of purified type I and type II adenylyl cyclases by G protein beta gamma subunits. *J Biol Chem* 1993;268:9–12.
37. Chen J, DeVivo M, Dingus J et al: A region of adenylyl cyclase 2 critical for regulation by G protein beta gamma subunits. *Science* 1995;268:1166–1169.
38. Yan K, Gautam N: A domain on the G protein β subunit interacts with both adenylyl cyclase 2 and the muscarinic atrial potassium channel. *J Biol Chem* 1996;271:17597–17600.
39. Chen Y, Weng G, Li J et al: A surface on the G-protein beta-subunit involved in interactions with adenylyl cyclases. *Proc Natl Acad Sci U S A* 1997;94:2711–2714.
40. De Waard M, Liu H, Walker D et al: Direct binding of G-protein βγ complex to voltage-dependent calcium channels. *Nature* 1997;385:446–450.
41. Zamponi GW, Bourinet E, Nelson D et al: Cross-talk between G proteins and protein kinase C mediated by the calcium channel α1 subunit. *Nature* 1997;385:442–446.
42. Katada T, Kasukabe K, Oinuma M, Ui M: A novel mechanism for the inhibition of adenylyl cyclase via GTP binding proteins: calmodulin dependent inhibition of the cyclase catalyst by the βγ-subunits of GTP binding proteins. *J Biol Chem* 1987;262:11897–11900.
43. Tang WJ, Krupinski J, Gilman AG: Expression and characterization of calmodulin-activated (type I) adenylyl cyclase. *J Biol Chem* 1991;266:8595–8603.
44. Cali JJ, Zwaagstra JC, Mons N et al: Type VIII adenylyl cyclase: a Ca^{2+}/calmodulin-stimulated enzyme expressed in discrete regions of rat brain. *J Biol Chem* 1994;269:12190–12195.
45. Fagan KA, Mahey R, Cooper DMF: Functional co-localization of transfected Ca^{2+}-stimulated adenylyl cyclases with capacitive Ca^{2+} entry sites. *J Biol Chem* 1996;271:12438–12444.
46. Wayman GA, Impey S, Wu Z et al: Synergistic activation of the type I adenylyl cyclase by Ca^{2+} and G_s-coupled receptors *in vivo*. *J Biol Chem* 1994;269:25400–25405.
47. Vorherr T, Knopfel L, Hofmann F et al: The calmodulin binding domain of nitric oxide synthase and adenylyl cyclase. *Biochemistry* 1993;32:6081–6088.
48. Wu Z, Wong ST, Storm DR: Modification of the calcium and calmodulin sensitivity of the type I adenylyl cyclase by mutagenesis of its calmodulin binding domain. *J Biol Chem* 1993;268:23766–23768.
49. Steer ML, Levitzki A: The control of adenylate cyclase by calcium in erythrocyte ghosts. *J Biol Chem* 1975;250:2080–2084.
50. Chiono M, Mahey R, Tate G, Cooper DM: Capacitative Ca^{2+} entry exclusively inhibits cAMP synthesis in C6–2B glioma cells; evidence that physiologically evoked Ca^{2+} entry regulates Ca^{2+}-inhibitable adenylyl cyclase in non-excitable cells. *J Biol Chem* 1995;270:1149–1155.
51. Ishikawa Y, Shuichi K, Chen L et al: Isolation and characterization of a novel cardiac adenylyl cyclase cDNA. *J Biol Chem* 1992;267:13553–13557.
52. Watson PA, Krupinski J, Kempinski AM, Frankenfield CD: Molecular cloning and characterization of the type VII isoform of mammalian adenylyl cyclase expressed widely in mouse tissues and in S49 mouse lymphoma cells. *J Biol Chem* 1994;269:28893–28898.
53. Hellevuo K, Yoshimura M, Mons N et al: The characterization of a novel human adenylyl cyclase which is present in brain and other tissues. *J Biol Chem* 1995;270:11581–11589.
54. Bushfield M, Murphy GJ, Lavan BE et al: Hormonal regulation of Gi_2 alpha-subunit phosphorylation in intact hepatocytes. *Biochem J* 1990;268:449–457.
55. Katada T, Gilman AG, Watanabe Y et al: Protein kinase C phosphorylates the inhibitory guanine-nucleotide binding regulatory component and apparently suppresses its function in hormonal inhibition of adenylate cyclase. *Eur J Biochem* 1985;151:431–437.
56. Yoshimasa T, Sibley DR, Bouvier M et al: Cross-talk between cellular signalling pathways suggested by phorbol ester induced adenylate cyclase phosphorylation. *Nature* 1987;327:67–70.
57. Simmoteit R, Schulzki H-D, Palm D et al: Chemical and functional analysis of components of adenylyl cyclase from human platelets treated with phorbol esters. *FEBS Lett* 1991;285:99–103.
58. Lustig KD, Conklin BR, Herzmark P et al: Type II adenylyl cyclase integrates coincident signals from G_s, G_i, and G_q. *J Biol Chem* 1993;268:13900–13905.
59. Jacobowitz O, Chen J, Premont RT, Iyengar R: Stimulation of specific types of G_s-stimulated adenylyl cyclases by phorbol ester treatment. *J Biol Chem* 1993;268:3829–3832.
60. Yoshimura M, Cooper DMF: Type-specific stimulation of adenylyl cyclase by protein kinase C. *J Biol Chem* 1993;268:4604–4607.
61. Jacobowitz O, Iyengar R: Phorbol ester–induced stimulation and phosphorylation of adenylyl cyclase 2. *Proc Natl Acad Sci U S A* 1994;91:10630–10634.
62. Zimmermann G, Taussig R: Protein kinase C alters responsiveness of adenylyl cyclases to G protein α and βγ subunits. *J Biol Chem* 1996;271:27161–27166.

63. Tsu RC, Wong YH: G_i-mediated stimulation of type II adenylyl cyclase is augmented by G_q-coupled receptor activation and phorbol ester treatment. *J Neurosci* 1996;16:1317–1323.

64. Kawabe J, Iwami G, Ebina T et al: Differential activation of adenylyl cyclase by protein kinase C isoenzymes. *J Biol Chem* 1994;269:16554–16558.

65. Kawabe J, Ebina T, Toya Y et al: Regulation of type V adenylyl cyclase by PMA-sensitive and insensitive protein kinase C isoenzymes in intact cells. *FEBS Lett* 1996;384:273–276.

66. Chern Y, Chiou JY, Lai HL, Tsai MH: Regulation of adenylyl cyclase VI activity during desensitization of the A2a adenosine receptor mediated cyclic AMP response: role for protein phosphatase 2A. *Mol Pharmacol* 1995;48:1–8.

67. Lai H-L, Yang T-H, Messing RO et al: Protein kinase C inhibits adenylyl cyclase type VI activity during desensitization of the A2a-adenosine receptor–mediated cAMP response. *J Biol Chem* 1997;272:4970–4977.

68. Choi EJ, Wong ST, Dittman AH, Storm DR: Phorbol ester stimulation of the type I and type III adenylyl cyclases in whole cells. *Biochemistry* 1993;32:1891–1894.

69. Premont RT, Jacobowitz O, Iyengar R: Lowered responsiveness of the catalyst of adenylyl cyclase to stimulation by G_s in heterologous desensitization: a role for adenosine 3′,5′-monophosphate-dependent phosphorylation. *Endocrinology* 1992;131:2774–2784.

70. Kunkel M, Freidman WJ, Clark RB: Cell free heterologous desensitization of adenylyl cyclase in S49 cell membranes mediated by cAMP dependent kinase. *FASEB J* 1989;3:2067–2074.

71. Premont RT, Iyengar R: Heterologous desensitization of the liver adenylyl cyclase: analysis of the role of G-proteins. *Endocrinology* 1989;125:1151–1160.

72. Iwami G, Kawabe J, Ebina T et al: Regulation of adenylyl cyclase by protein kinase A. *J Biol Chem* 1995;270:12481–12484.

73. Wayman GA, Impey S, Storm DR: Ca^{2+} inhibition of type III adenylyl cyclase *in vivo. J Biol Chem* 1995;270:21480–21486.

74. Wei J, Wayman G, Storm DR: Phosphorylation and inhibition of type III adenylyl cyclase by calmodulin-dependent protein kinase II *in vivo. J Biol Chem* 1996;271:24231–24235.

75. Wayman GA, Wei J, Wong S, Storm DR: Regulation of type I adenylyl cyclase by calmodulin kinase IV *in vivo. Mol Cell Biol* 1996;16:6075–6082.

76. Paterson JM, Smith SM, Harmar AJ, Antoni FA: Control of a novel adenylyl cyclase by calcineurin. *Biochem Biophys Res Commun* 1995;214:1000–1008.

77. Chen Z, Nield HS, Sun H et al: Expression of type V adenylyl cyclase is required for epidermal growth factor-mediated stimulation of cAMP accumulation. *J Biol Chem* 1995;270:27525–27530.

78. Johnson RA, Yeung SM, Stubner D et al: Cation and structural requirements for P site-mediated inhibition of adenylate cyclase. *Mol Pharmacol* 1989;35:681–688.

79. Desaubry L, Shoshani I, Johnson RA: 2′,5′-Dideoxyadenosine 3′-polyphosphates are potent inhibitors of adenylyl cyclases. *J Biol Chem* 1996;271:2380–2382.

80. Johnson RA, Desaubry L, Bianchi G et al: Isoenzyme-dependent sensitivity of adenylyl cyclases to P-site mediated inhibition by adenine nucleosides and nucleoside 3′-polyphosphates. *J Biol Chem* 1997;272:8962–8966.

81. Schlessinger J: How tyrosine kinases activate Ras. *Trends Biochem Sci* 1993;18:273–275.

82. Egan SE, Weinberg RA: The pathway to signal achievement. *Nature* 1993;365:781–783.

83. Wu J, Dent P, Jelinek T et al: Inhibition of the EGF-activated MAP kinase signaling pathway by adenosine 3′-5′-monophosphate. *Science* 1993;262:1065–1068.

84. Häfner S, Adler HS, Mischak H et al: Mechanism of inhibition of Raf-1 by protein kinase A. *Mol Cell Biol* 1994;14:6696–6703.

85. Cook SJ, McCormick F: Inhibition of cAMP-Ras–dependent activation of Raf. *Science* 1993;262:1069–1072.

86. Chen J, Iyengar R: Suppression of Ras-induced transformation of NIH 3T3 cells by activated G alpha s. *Science* 1994;263:1278–1281.

87. Chuang E, Barnard D, Hettich L et al: Critical binding and regulatory interactions between Ras and Raf occur through a small, stable N-terminal domain of Raf specific Ras effector residues. *Mol Cell Biol* 1994;14:5318–5325.

88. Van Corven EJ, Groenink A, Jalink K et al: Lysophosphatidate-induced cell proliferation: identification and dissection of signaling pathways mediated by G proteins. *Cell* 1989;59:45–54.

89. Cho-Chung YS: Suppression of malignancy targeting cyclic AMP signal transducing proteins. *Biochem Soc Trans* 1992;20:425–430.

90. Bliss TVP, Collingridge GL: A synaptic model of memory: long-term potentiation in the hippocampus. *Nature* 1993;361:31–39.

91. Chetkovich DM, Sweatt JD: NMDA receptor activation increases cyclic AMP in area CA1 of the hippocampus via calcium/calmodulin stimulation of adenylyl cyclase. *J Neurochem* 1993;61: 1933–1942.
92. Blitzer RD, Wong T, Nouranifar R et al: Postsynaptic cAMP pathway gates early LTP in hippocampal CA1 region. *Neuron* 1995;15:1403–1414.
93. Pettit DL, Perlman S, Malinow R: Potentiated transmission and prevention of further LTP by increased CaMKII activity in postsynaptic hippocampal slice neurons. *Science* 1994;266:1881–1885.
94. Lledo P-M, Hjelmstad GO, Mukherji S et al: Calcium/calmodulin-dependent kinase II and long-term potentiation enhance synaptic transmission by the same mechanism. *Proc Natl Acad Sci U S A* 1995;92:11175–11179.
95. Nguyen PV, Abel T, Kandel ER: Requirement of a critical period of transcription for induction of a late phase of LTP. *Science* 1994;265:1104–1107.
96. Frey U, Huang YY, Kandel ER: Effects of cAMP stimulate a late stage of LTP in hippocampal CA1 neurons. *Science* 1993;260:1661–1664.
97. Abel T, Nguyen PV, Barad M et al: Genetic demonstration of a role for PKA in the late phase of LTP and in the hippocampus-based long-term memory. *Cell* 1997;88:615–626.
98. Impey S, Mark M, Villacres EC et al: Induction of CRE-mediated gene expression by stimuli that generate long-lasting LTP in area CA1 of the hippocampus. *Neuron* 1996;16:973–982.
99. Bito H, Deisseroth K, Tsien RW: CREB phosphorylation and dephosphorylation: a Ca^{2+} and stimulus duration-dependent switch for hippocampal gene expression. *Cell* 1996;87:1203–1214.

*Advances in Second Messenger and
Phosphoprotein Research*, Vol. 32,
edited by Dermot M. F. Cooper
Lippincott–Raven Publishers, Philadelphia © 1998

2

Ca^{2+}-Sensitive Adenylyl Cyclases

Dermot M. F. Cooper,[*][‡] Jeffrey W. Karpen,[†][‡] Kent A. Fagan,[‡] and
Nicole E. Mons[§]

*Departments of [*]Pharmacology and [†]Physiology, and [‡]Neuroscience Program,
University of Colorado Health Sciences Center, Denver, Colorado, 80262, and
[§]Laboratory of Behavioural Neurosciences, URA-CNRS 339, University of Bordeaux I,
Talence, France*

Cloning and expression studies of the adenylyl cyclase (AC) family have established
that the Ca^{2+}-signaling pathway can play a major role in influencing cellular cyclic
adenosine monophosphate (cAMP) production. Thus, all of the nine cloned ACs can
be regulated either directly or indirectly by Ca^{2+} or protein kinase C (PKC; Table 1).
It is particularly striking that, when grouped by their structural relatedness, the ACs
comprise four subfamilies, which reflect this susceptibility to regulation by Ca^{2+}-
signaling pathways (see Table 1). Ca^{2+} in submicromolar concentrations directly
stimulates or inhibits two of the subfamilies; PKC stimulates another subfamily, and
Ca^{2+}, via calcineurin, indirectly inhibits the fourth subfamily. Moreover, in some
cases, the classic regulators of AC (i.e., G protein α subunits) do not exert a dominant regulatory influence.[1] Thus whereas $G_{i\alpha}$ inhibits AC5 and AC6 very effectively
(although no more efficaciously than Ca^{2+} inhibits these species), $G_{i\alpha}$ is incapable of
regulating the AC2 subgroup (indeed, it is unclear whether any mechanism exists to
inhibit AC2). Furthermore, $G_{i\alpha}$ regulates the AC1 subfamily most effectively when
activity is stimulated by Ca^{2+} (1). The effects of G_s are also rather variable; it is
poorly able to stimulate AC1 compared with Ca^{2+}, whereas it is much more effective
than Ca^{2+} at stimulating AC3 and similar to Ca^{2+} at stimulating AC8.

This chapter focuses on the most acute regulation by Ca^{2+}—signaling pathways of
the ACs; that is, direct effects of Ca^{2+} on the Ca^{2+}-stimulable AC1, AC3, and AC8
and the Ca^{2+}-inhibitable AC5 and AC6.[2] The *in vitro* and *in vivo* sensitivity to Ca^{2+}
is discussed, as are possible mechanisms for the strict requirement for localized Ca^{2+}

---

[1]A detailed discussion of the regulation of ACs by G protein α and $\beta\gamma$ subunits is presented in Chapters
1, 3, and 4.

[2]Chapter 5 deals fully with the regulation of ACs by protein kinase C and Chapter 8 discusses Ca^{2+} acting indirectly via calcineurin on AC9.

TABLE 1. *Adenylyl cyclases grouped by their structural relatedness*

	Ca^{2+}	PKC	$\beta\gamma$	$G_{s\alpha}$	$G_{i\alpha}$
AC1	Stim	No	Inhib	Mild	Ca^{2+}-dep[‡]
AC3	?[*]	No	Inhib	Yes	?
AC8	Stim	No	Inhib	Yes	Ca^{2+}-dep[‡]
AC2	No	Stim	Stim	Yes	No
AC4	No	Stim	Stim	Yes	No
AC7	No	Stim	Stim	Yes	No
AC5	Inhib	No	No	Yes	Yes
AC6	Inhib	No	No	Yes	Yes
AC9	Inhib[†]	?	No	Yes	?

The regulatory properties summarized emanate from cells transfected with appropriate cDNAs, generally followed by *in vitro* exposure to either EGTA-buffered Ca^{2+} solutions in the submicromolar range, or various G protein subunits. (The *in vitro* Ca^{2+} effects are mirrored when intact transfected cells are subjected to capacitative Ca^{2+}-entry, [2,3].) The PKC data summarized come from treatment of intact cells with phorbol esters. Phosphorylation experiments carried out in broken cells give a range of additional effects, as reviewed by Ishikawa (Chapter 5). Stim, stimulation; inhib, inhibition.

[*]The effects of Ca^{2+} on AC3 are not clearcut. Apparently, in *in vitro* assays only extremely high concentrations of Ca^{2+} (>50 μM) stimulate the enzyme (4). Capacitative Ca^{2+} entry does not regulate AC3 at all in short (1-minute) assays (3), but in extended (30-minute) assays, ionophore-mediated $[Ca^{2+}]_i$-rises inhibit cAMP accumulation in cells transfected with AC3 (5; see later discussion).

[†]The effect of Ca^{2+} on AC9 is likely mediated by calcineurin and is not observed *in vitro* (see Chapter 8).

[‡]AC1 is inhibited by $G_{i\alpha}$ in *in vitro* assays when activity is stimulated by Ca^{2+}/calmodulin (1); AC8 is inhibited by endogenous somatostatin receptors in HEK-293 cells when activity is stimulated by capacitative Ca^{2+}-rises (Fagan and Cooper, unpublished).

entry to regulate these cyclases in the intact cell. The expression of cyclases, on both a tissue and a subcellular scale, is presented, along with the potential for oscillations in both [cAMP] and $[Ca^{2+}]_i$ to be generated by Ca^{2+}-sensitive ACs.

STRUCTURAL ISSUES

This section speculates on some issues that may bear on the sensitivity of cyclases to Ca^{2+} entry, leaving detailed discussions of the structural requirement for G protein subunit regulation of ACs to Chapters 1 and 3.

The Role of the 12 Membrane-Spanning Domains

The overall structure of ACs remains one of the most tantalizing puzzles of this enzyme. The two six-membrane spanning motifs were probably the greatest surprise to emerge when the first cyclase was cloned in 1989 (6). At that time, Krupinski and co-workers noted the gross similarity of the structure to channels, and the even closer similarity to transporters (Fig. 1). All the cyclases can be considered to be members of the ABC-binding cassette superfamily, which includes the cystic fibrosis transmembrane conductance regulator (CFTR) gene, the P-glycoprotein, and

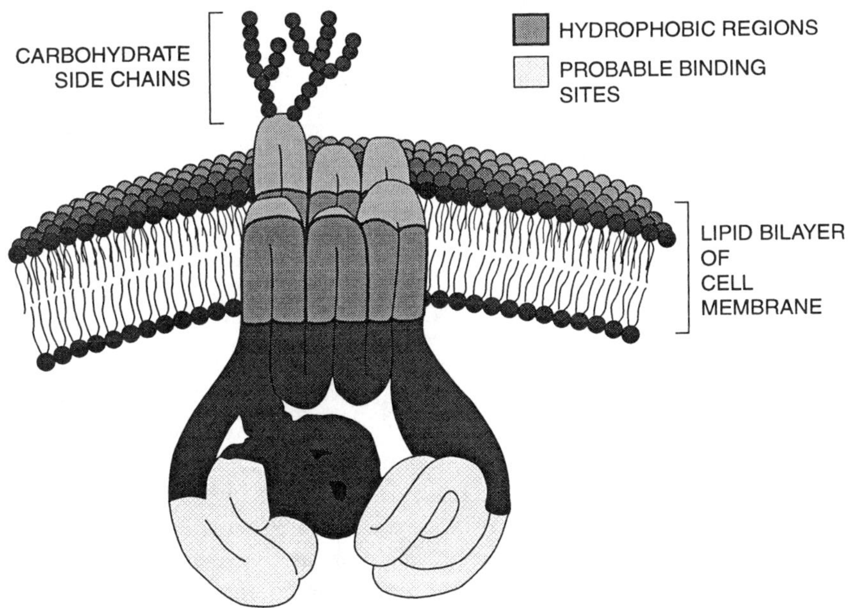

FIG. 1. A speculative structure for the cystic fibrosis transmembrane conductance regulator (CFTR), based on hydropathy plots and the known ion-conducting activity of the channel. It is presented here to draw attention to the possibly similar structure that may be adopted by the closely related ACs. (Adapted with permission from Gelehrter TD, Collins FS: *Principles of medical genetics*. Baltimore: Williams & Wilkins, 1990:216[10].)

bacterial permeases. These proteins have the same putative membrane structure, along with two cytosolic adenosine triphosphate (ATP)–binding domains (7), which are known in the cyclases as the C_{1a} and C_{2a} domains (the C_{1b} and C_{2b} domains are the remainder of the respective cytosolic loops). Krupinski and co-workers (6) considered the possibility that the structure might serve to export cAMP, which now seems unlikely, based on studies with *Dictyostelium* mutants (see Chapter 6). Recent data add to the puzzle of the 12 transmembrane-spanning domains, because the first cytosolic loop of AC1 when connected by a soluble stretch of amino acids to the second cytosolic loop of AC2 possesses AC activity that is regulated by G protein α and βγ subunits, forskolin, and adenosine P-site analogs (8). Indeed, these two cytosolic loops when purified separately can be reconstituted to give a fully functional AC (9).

The endogenous cyclase of *Paramecia* was reported to be sensitive to membrane hyperpolarization (11). Indeed, these data indicated not only that the cyclase was stimulated by hyperpolarization but that an enriched preparation of the enzyme could also function as a K^+-channel when incorporated into lipid bilayers. This enzyme has not yet been cloned and so whether it shares the 12-membrane spanning structure of mammalian enzymes is not known. More recently, a prolonged (30-minute) depolarization of cultured neurons with high K^+ resulted in increased cAMP accumulation (12). The most exciting implication of this result was that the endogenous mammalian AC was (like the *Paramecia* enzyme) sensitive to membrane potential. Of course, whether this was a direct or an indirect effect on AC activity could not be determined in such protracted experiments. Nevertheless, it remains a compelling possibility that the complex membrane-inserted feature of cyclase structure plays an important role in its regulation. It may yet turn out that the exquisite sensitivity of Ca^{2+}-regulated ACs to Ca^{2+} entry (see later) depends on this structural feature of cyclases.

Oligomerization of Adenylyl Cyclases

All the AC complementary deoxyribonucleic acids (cDNAs) cloned so far have protein molecular weights predicted to lie between 115 and 145 kDa (1). Such molecular weights correspond to the lower limits estimated from earlier studies. A number of lines of evidence from this older literature had suggested that ACs might exist as dimers or even as higher-order structures. Target size analysis led Rodbell and co-workers to propose that the minimal size of the activated state of AC activity was a dimer, whereas in the unactivated, resting condition, the enzyme was in even higher-order assemblies of receptors, G proteins, and catalytic units (13). Hydrodynamic analyses of detergent-solubilized catalytic activity from bovine cerebral cortex by Neer and co-workers (14) also suggested that monomer/dimer transitions could occur in the catalytic unit.

The possession by G protein β subunits of so-called WD repeats (i.e., trp-asp repeats) is postulated to permit G proteins to organize macromolecular assemblies of signaling molecules (15). This provides a formal basis for the original proposal of the Rodbell group on the oligomerization of G proteins and raises the possibility that the cyclases might be organized in similar higher-order arrays by such G proteins. An interesting protein recently purified and cloned by the Monneron group, *striatin*, occurs in regions where AC5 is expressed and possesses eight WD repeats (16). This protein co-purified with AC activity and initially was believed to confer some stabilizing activity on the cyclase. Whether such a function develops for this protein remains to be seen, although it is certainly true that proteins are to be anticipated that retain cyclases in appropriate cellular domains (see later).

More recently, studies conducted by Tang and co-workers (17) using truncation mutants of AC1 and AC2 have suggested that AC might dimerize. An antibody directed to the C-terminus of AC1 immunoprecipitated AC activity from detergent extracts of Sf9 cells expressing AC1. As expected, this antibody would not immunoprecipitate AC activity from cells expressing AC1 truncated at the C-terminal, which lacked the epitope, although it was enzymatically active. When this truncated AC1 was expressed along with a mutant AC1 that possessed no enzymatic activity but that did contain the C-terminal epitope, a significant amount of the enzymatic activity could be immunoprecipitated. These data suggest that the functional C-terminal truncation mutant and the inactive (epitope-containing) mutant associated. Thus the active cyclase could comprise associations of more than one molecule.

Obviously, dimerization requires putative dimerization domains. One such possible domain for ACs is a leucine-zipper motif that is conserved, in both sequence and location, in several of the AC subtypes. AC types 3, 5, 6, and 8 all contain a leucine-zipper motif located on the N-terminal side of the absolutely conserved "riki" sequence.

AC1 **L** VKLLNE **L** FGKFDE **L** ATENHC *R* **RIKIL** (327–352) (Bovine)

AC2 **L** VHMLNE **L** FGKFDQ *I* AKENEC *M* **RIKIL** (310–336) (Rat)

AC3 **L** VKLLNE **L** FARFDK **L** AAKYHQ **L** **RIKIL** (340–366) (Rat)

AC4 **L** VLMLNE **L** FGKFDQ *I* AKEHEC *M* **RIKIL** (294–320) (Rat)

AC5 **L** VMTLNE **L** FARFDK **L** AAENHC **L** **RIKIL** (425–454) (Canine)

AC6 **L** VMTLNE **L** FARFDK **L** AAENHC **L** **RIKIL** (398–424) (Mouse)

AC7 **L** VVVLNE **L** FGKFDQ *I* AKANEC *M* **RIKIL** (300–326) (Human)

AC8 **L** VRMLNE **L** FARFDR **L** AHEHHC **L** **RIKIL** (432–458) (Rat)

The AC types that do not express the canonical leucine-zipper motif still preserve the amphipathic nature of this region, as judged by helical-wheel presentations.[3]

A study from the Ishikawa laboratory (19) underscores the importance of this region in catalytic activity. Using synthetic peptides derived from regions of canine AC5, they found that only one peptide (425–444), which overlaps with the leucine-zipper motif, significantly inhibited activity in *in vitro* studies. This peptide also inhibited purified AC2 activity. Interestingly, if the two leucine residues present in this peptide (which comprise the third and fourth leucines of the leucine-zipper motif) were changed (conservatively) to isoleucine, the peptide was still effective in inhibiting AC6 activity, whereas nonconservative substitutions of leucine to proline abolished the inhibitory effect of the peptide. This peptide does not appear to block G_s stimulation of AC6, and the inhibition is also seen with forskolin stimulation. The inhibition is also not dependent on competition for ATP, because the Michaelis constant (K_m) is unaffected in kinetic analyses. The ability of the isoleucine-substituted peptide to inhibit activity also suggests that the ACs that do not have the complete leucine-zipper motif but that still possess the overall amphipathic helical structure have similar properties.

In vitro Sensitivity to Ca²⁺ of Adenylyl Cyclases

Stimulation by Ca²⁺

The stimulation of ACs (AC1, AC3, and AC8) by Ca²⁺ is mediated by loosely bound calmodulin (CaM), which can be dissociated experimentally by washing plasma membranes in EGTA-containing buffers (see ref. 20 and references therein).

---

[3]Leucine zippers have recently been shown to stabilize the pentameric structure of phospholamban (18).

Putative CaM-binding domains were identified in AC1 by Pfeuffer's group (21). Peptides corresponding to these sites were used to identify a region close to the plasma membrane in C_{1b} as the likely site that participated in the binding of CaM. Mutagenesis studies of AC1 (22) and chimeras of AC1 and AC2 (23) strongly support this assignation. Furthermore, the soluble AC1/AC2 chimera described earlier, which does not extend to this domain of AC1, is also not sensitive to Ca^{2+} *in vitro*. In a recent review (1), Gilman's group cautions that a number of mutations in other regions of AC1 also resulted in impaired Ca^{2+}/CaM regulation. Their caution underscores the fact that the tertiary structure of an "active domain" may comprise either contiguous or distant parts of the linear amino acid sequence. No information is yet available on the possible domains that mediate the Ca^{2+}/CaM responsiveness on AC8 or AC3. For instance, the region implicated in AC1 is not present on either AC3 or AC8 (see also Chapter 3 for further discussion).

Inhibition by Ca^{2+}

Unlike the Ca^{2+}-stimulable ACs, extensive washing of plasma membranes with EGTA does not result in the loss of susceptibility of Ca^{2+}-inhibitable ACs to regulation by submicromolar Ca^{2+} (20,24,25). There is no obvious (contiguous) E-F hand or other Ca^{2+}-binding motif in AC5 or AC6. Consequently, it is really not clear whether CaM or another Ca^{2+}-binding protein mediates the effect, or if Ca^{2+} acts directly on the cyclase. The studies performed to date do not absolutely preclude the involvement of CaM, because although Ca^{2+} generally promotes the binding of CaM to its target (hence, the efficacy of washing with EGTA to remove CaM), examples are encountered where CaM is not dissociable from its target (e.g., phosphorylase kinase [26] and nitric oxide synthase [27]). A very recent report indicates that Ca^{2+} inhibition of AC5 is mediated directly within a 112-amino-acid stretch of the C_{1b} domain, independently of CaM (28). Whether this represents the inhibition by submicromolar Ca^{2+} discussed here or an inhibitory site that is sensitive to supramicromolar Ca^{2+}, and that is shared by all ACs (20,29) regardless of their response to submicromolar Ca^{2+}, is not clear in these early studies.

It is also worth considering the possibility that other proteins could mediate the effect, including the plethora of both putative and definite Ca^{2+}-binding proteins that have been discovered. The family of proteins related to recoverin (30) attracted our attention. These proteins all display four E-F hands and seemed like potential candidates to mediate the inhibitory effect of Ca^{2+} on AC. We therefore performed reverse-transcription-PCR on various clones of C6–2B glioma cells, which express predominantly AC6 AC (31) with degenerate oligonucleotides for Ca^{2+}-binding proteins of the recoverin family, including neurocalcin, hippocalcin, and neural visinin-like Ca^{2+}-binding protein types 1, 2, and 3 (NVP-1, NVP-2, and NVP-3) (30,32–35). Sequencing of cloned products revealed both hippocalcin and NVP-3. Attempts to knock out the Ca^{2+}-binding proteins by antisense strategies were without effect on AC inhibition (Nakahashi and Cooper, unpublished). Thus, whether CaM or another

Ca^{2+}-binding protein, or even Ca^{2+} directly, mediates Ca^{2+}-inhibition remains unresolved.

Regulation of Cloned Adenylyl Cyclases by Ca^{2+} in the Intact Cell

Direct Effects of Ca^{2+} on Adenylyl Cyclase

The physiologic relevance of putatively Ca^{2+}-sensitive ACs rests on whether or not they are regulated by cellular transitions in [Ca^{2+}]$_i$. Reports abound in the literature indicating that in a variety of cell lines and tissues, hormones or other agents that elevate [Ca^{2+}]$_i$ alter cAMP accumulation. In certain of these cases, a Ca^{2+}-regulatable AC has been identified in plasma membranes from these sources, so that it seems reasonable to interpret the intact cell results as indicating direct effects of Ca^{2+} on the cyclase activity (see refs. 20, 24, and 25 and references therein). However, hormone receptor occupancy can result in numerous effects, including liberation of βγ subunits of G proteins, activation of PKC and other signaling pathways (e.g., triggering of arachidonate release), any of which could have secondary effects on ACs. Therefore, we believed it important to develop a method that used the cellular capacity to elevate [Ca^{2+}]$_i$, without additional consequences, so that we could at least interpret our own results fairly unambiguously. This method exploited capacitative Ca^{2+} entry as a physiological means of elevating [Ca^{2+}]$_i$, which was not subject to the uncertainties in interpretation raised earlier. Thus if stores are depleted of Ca^{2+} by the store reuptake pump inhibitor thapsigargin, in the absence of extracellular Ca^{2+}, released Ca^{2+} is extruded and the cell is primed to receive the entry of extracellular Ca^{2+}. Upon reintroduction of Ca^{2+} to the external medium, Ca^{2+} rushes in in a capacitative manner (36). By such means, both endogenous and transfected Ca^{2+}-sensitive ACs have been shown to be regulated by simple [Ca^{2+}]$_i$ rises (2,3,37,38).

More detailed studies have revealed a surprising preference for Ca^{2+} entry rather than for release from internal stores to regulate these cyclases. The first studies showed that an endogenous AC6 in C6–2B glioma cells responded only to entry rather than to release mediated by thapsigargin, ionomycin, or the P2u agonist uridine triphosphate (UTP) (37). This requirement was maintained even when the release of Ca^{2+} (as elicited by ionomycin) was far greater than the Ca^{2+} entry triggered by UTP (37). These conclusions were later also shown to apply to transfected Ca^{2+}-stimulable ACs (AC1 and AC8) in HEK-293 cells (3).[4] In this case, a dramatic release of Ca^{2+} from internal stores triggered by ionomycin was incapable of regulating the cyclases, whereas again a much more modest entry in response to thapsigargin-mediated depletion of stores stimulated the enzymes. (These results

---

[4]In this study, AC3 was also examined and was found to be quite unresponsive to [Ca^{2+}]$_i$-rises in short (1-minute) assays. By contrast, in extended assays (30 minutes), cAMP accumulation in cells transfected with AC3 was inhibited as a consequence of the action of CaM kinase II, which was activated by 10 μM A23187 and 1.8 mM CaCl$_2$ (39).

also demonstrate that heterologously-expressed cyclases possess the information to be targeted to regions of Ca^{2+} entry.)

One interpretation for the selective efficacy of entry over release at regulating ACs was that a plasma membrane–oriented enzyme like AC was simply closer to entry sites than to potentially remote release sites (3). To address this latter possibility, a nonspecific method was devised to allow discrete entry of Ca^{2+} through the plasma membrane. Thus in the absence of extracellular Ca^{2+}, cells were first depleted of Ca^{2+} by thapsigargin, which was followed by the introduction of a high concentration of ionomycin. This treatment allowed the concentration-driven transport of Ca^{2+} through the plasma membrane, in addition to capacitative Ca^{2+} entry, which would have been triggered by the depletion of the stores. Under these circumstances, very low concentrations of external Ca^{2+} gave rise to extremely elevated $[Ca^{2+}]_i$. Some capacitative Ca^{2+} entry still occurred, but this is a low-affinity process, which is initiated at higher $[Ca^{2+}]_{ex}$. When the effects of ionomycin-mediated entry were compared with capacitative Ca^{2+} entry, it was clear that only the latter entry mode could modulate transfected AC8. The capacitative entry component that underlay the ionomycin-mediated entry was fully responsible for any effects of the ionophore on AC8 (3). These results strongly suggest an absolute requirement for capacitative entry rather than a simple $[Ca^{2+}]_i$-rise—even at the plasma membrane—for the regulation of Ca^{2+}-sensitive ACs,[5] and reinforce the concept of an intimate association between capacitative entry sites and the cyclases.

This intimate functional association between ACs and capacitative Ca^{2+} entry channels (I_{CRAC})[6] suggests a number of possible, and not necessarily mutually exclusive, explanations. (1) The two molecular species could be in intimate contact; (2) the AC itself could be an I_{CRAC}; (3) the two entities could be present in the same lipid microenvironment of the plasma membrane; (4) the two species could be confined to the same subcellular organelle; (5) they might be held in the same plasma membrane domain by cellular cytoskeletal elements. Some notes follow on what we can find to discriminate between these possibilities.

1. In an attempt to put numbers on the intimacy of the relationship between the I_{CRAC} and AC, an approach introduced by Zweifach and Lewis (41) was modified for the more challenging situation of cell population experiments. These authors had probed the relationship between a feedback inhibitory Ca^{2+} site on the I_{CRAC} of Jurkat T-lymphocytes. They compared the efficacy of equilibrating cells with 12 mM EGTA or 12mM BAPTA at blocking the Ca^{2+} feedback and found that, even though both chelators had equal affinity for Ca^{2+} (in fact, EGTA has a slightly higher affinity for Ca^{2+}), only BAPTA could preclude the feedback effect. This disparity reflected the approximately 100-fold greater on-rate of BAPTA for Ca^{2+} compared

---

[5]Similar findings have been encountered with the AC6 endogenously expressed in C6–2B cells (Fagan and Cooper, in preparation).

[6]I_{CRAC} is used in this chapter not in the strict sense of the mast cell channel first characterized by Hoth and Penner (40) but in a generic sense to indicate any channels that are activated by store depletion.

with EGTA, which allowed the authors to calculate that the site at which Ca^{2+} acted must have been <4 nm from the entry pore (meaning that the site was probably part of the channel). Against such a background, we compared the effects of loading cells with either EGTA-AM or BAPTA-AM, to determine whether there would be any selective difference between the two chelators. (As far as we could tell, based on the quenching of intracellular fura-2 responses, equivalent concentrations of both deesterified chelators were achieved.) In fact, although BAPTA blocked the effect of Ca^{2+} entry on AC, EGTA-AM, even at 10 times higher starting concentrations, was unable to fully reverse the effect (Fagan and Cooper, in preparation). This finding strongly supports the notion that Ca^{2+} entering the cell binds rapidly to its regulatory site on AC.

It may even be worth raising the possibility that the intimacy of the interaction between I_{CRAC} and AC could reflect juxtapositioning of the two proteins and sensing of the changes in activity of the channel through allosteric interactions with the cyclase. This is to say that it is not the amount of Ca^{2+} entering the cell per se that regulates the ACs but the activity states of the channel that regulate the enzyme. The latter possibility could be addressed by substituting a cation that could be carried by the channel but that was poorly able to regulate the cyclase. Sr^{2+} might be appropriate given that it is approximately 100 times less potent than Ca^{2+} at stimulating brain cyclase (42).

2. The membrane topography of ACs, the apparent ion-transporting activity of *Paramecia* cyclase, and the intimate dependence of cyclases on Ca^{2+} entry are circumstantial reasons for considering the possibility that the cyclases might themselves act as I_{CRAC}s. Already *Trp* channels with some but not all (43) the properties of I_{CRAC} have been cloned (44,45). These cloned channels are not cyclases, but what of the real I_{CRAC}? This issue is somewhat intractable experimentally, for the following reasons. First, I_{CRAC}s are only active when stores are depleted. Thus, expression of purified cyclase in black lipid membranes in the hopes of detecting a current (which may well have been attempted) would not be expected to be successful. Second, it might be expected that if cyclases were I_{CRAC}s, their expression in HEK-293 cells could change capacitative Ca^{2+} entry. However, cAMP modifies both InsP$_3$ receptor activity and release from stores (46), and because of the interdependence of store status and capacitative entry, it would be hard to distinguish indirect effects of increased basal cAMP levels from effects of increased I_{CRAC} expression.

3. A reason for considering the lipid microenvironment of ACs is that ionomycin is only poorly able to penetrate cholesterol-rich phospholipid micelles (47). It is thus conceivable that the reason that ionomycin-mediated entry of Ca^{2+} into cells does not regulate AC is that the [Ca^{2+}]$_i$-rise would not occur near a cyclase that was concentrated in cholesterol-rich domains. One way of addressing this issue is to use the detergent digitonin. Digitonin, at low concentrations, selectively partitions in cholesterol-rich lipids (48). Thus, at low concentrations, digitonin could be expected to provide access to regions of the plasma membrane that were inaccessible to ionomycin. When very low concentrations of digitonin were used in C6–2B cells along with low concentrations of [Ca^{2+}]$_{ex}$ to generate a very modest rise in [Ca^{2+}]$_i$, it was indeed possible to inhibit the endogenous AC6 activity (Fagan and Cooper, in preparation). This finding suggests that AC might well be localized in cholesterol-rich do-

mains and that active capacitative Ca^{2+} entry need not occur to regulate AC. However, two additional possibilities preserve the need for capacitative entry in the normal regulation of Ca^{2+}-sensitive ACs: (1) placing sufficient $[Ca^{2+}]$ in the appropriate microdomain may merely simulate an *in vitro* assay and not actually a whole cell situation, and (2) it is possible that low concentrations of digitonin trigger capacitative Ca^{2+} entry mechanisms, even though no release that can be measured in population experiments occurs.

4. Caveolae are functional domains in which signal-transducing elements (such as AC and I_{CRAC}s) might be concentrated. These flask-shaped structures, which are approximately 50 nm in diameter, were first reported in endothelial cells, where they were first implicated in the transcytosis of macromolecules (49). These structures have now been found in many more cell types (50,51). Accumulating evidence has suggested that caveolae, by creating a distinct subcompartment of the plasma membrane, could play an important physiologic role in nonneuronal cells by coordinating cytosolic Ca^{2+} concentration through a combination of channel-mediated influx and energy-requiring Ca^{2+} extrusion (52). Recent morphologic and biochemical studies provide support that caveolae contain a variety of signaling molecules, including two proteins related to transmembrane Ca^{2+} transport, the plasmalemma Ca^{2+} pump and an $InsP_3$ receptor-like protein (53,54). In addition, caveolae appear to be enriched in many molecules that are implicated in different cellular signaling cascade pathways, such as $G_{s\alpha}$ subunits and glycosyl-phosphatidyl-inositol (GPI)-linked proteins (55–57). The caveolae are commonly identified by the marker protein caveolin, although apparently this 21-kDa protein is not always present (58). In addition, cholesterol is reported to be enriched in caveolae (55,59,60), an observation that fits with the efficacy of Ca^{2+} entry mediated by digitonin over that mediated by ionomycin at regulating AC, which was discussed earlier. In that context, C6–2B cells express high levels of caveolin immunoreactivity and indeed caveolar structures at the ultrastructural level (Gonzales and co-workers, in preparation). Whether or not there would be any functional significance to the detection of AC in caveolae remains to be determined. For instance, it is conceivable that caveolae represent a device for delivering or retrieving AC from the plasma membrane rather than from the functional microdomain where AC and I_{CRAC} interact. After all, well-defined, closed morphologic domains are not strictly required for two regulatory elements to interact that are juxtaposed by other forces.

5. Attempts have been made to determine whether the cytoskeleton might play a role at co-localizing the AC and I_{CRAC}. In C6–2B cells, disruption of the cytoskeleton with cytochalasin D (verified by phalloidin labeling) had no effect on the ability of I_{CRAC} to regulate the endogenous AC6 (Fagan and Cooper, unpublished).

Thus, although the foregoing studies reinforce the intimate association between AC and I_{CRAC}, the precise mechanism is still not clear. To gain a different perspective on the details of this interaction, a strategy has been developed to study the $[Ca^{2+}]_i$-transients sensed by a Ca^{2+}-regulatable AC. We adapted an approach pioneered by Rizzuto and co-workers for the measurement of $[Ca^{2+}]_i$ in discrete cellular subdomains (61,62). We created a chimera of AC6 and aequorin. Aequorin, with its cell-permeant co-factor, coelenterazine, luminesces upon binding Ca^{2+} (63). The affinity of aequorin for Ca^{2+} is also lower than that of fura-2, which renders it more capable of measuring higher concentrations of Ca^{2+}. The AC-aequorin cDNA when transfected in HEK-293 cells is expressed at the plasma membrane as assessed by immunohistochemical criteria. Both the AC activity and the aequorin functionality are unaffected by the placement of the aequorin at the C-terminal of the AC; the AC also remains sensitive to capacitative Ca^{2+} entry. Thus, it seems as though a tool is available that will measure I_{CRAC} under a variety of circumstances. In addition, the dynamics of $[Ca^{2+}]_i$ changes in the domain of a cyclase can be compared with the Ca^{2+}-mediated changes in AC activity.

Indirect Effects of Ca²⁺ on Adenylyl Cyclase

Calcineurin-mediated inhibition of AC9 is discussed in detail in Chapter 8. Another indirect mechanism is suggested by recent data, indicating that in prolonged incubations (0.5 hour) CaM kinase II inhibits cAMP accumulation in AC3-expressing HEK-293 cells (39). This result is a little paradoxical at first because in *in vitro* experiments AC3 was shown to be mildly stimulable by high (>50 μM) Ca^{2+} concentrations in the presence of G_s stimulation (4). AC3 was cloned initially from rat olfactory neuroepithelium (64) and because very high concentrations of Ca^{2+} are achieved only transiently around the mouth of Ca^{2+} channels, and odorants via G_s are also elevated cAMP in neuroepithelium, the attractive hypothesis was assembled proposing AC3 as a coincidence detector in the discrimination of odorant stimuli (65). The more recent report of AC3 *inhibition* as a consequence of $[Ca^{2+}]_i$-elevation via CaM kinase II activation in prolonged assays suggests that a careful series of experiments in a system in which AC3 naturally predominates is required to compare the effects of short-term and long-term physiologic elevations of Ca^{2+} on AC3 (if indeed long-term elevations of Ca^{2+} are achieved in tissues where AC3 is expressed). If regulation by CaM kinase II is shown to occur in cells where AC3 is expressed naturally, this may be a further means for fine-tuning the long-term effects of the cAMP-generating system on gene expression, for instance. These effects are of course not to be confused with the short-term, direct effects of physiologic elevation of $[Ca^{2+}]_i$ on ACs.

OSCILLATIONS IN cAMP

The presence of Ca^{2+}-sensitive ACs in cells that exhibit Ca^{2+} oscillations, or cells in which cAMP regulates Ca^{2+} entry, leads to a number of possible scenarios by which cAMP concentrations might oscillate. Ca^{2+} oscillations that are independent of cAMP and that are widely encountered (66–68) could cause cAMP oscillations in cells that contain either Ca^{2+}-stimulable or Ca^{2+}-inhibitable cyclases. The requirements are that the Ca^{2+} concentration change near the surface membrane (69), that Ca^{2+} regulation of cyclase occur more rapidly than the oscillations, and that phosphodiesterase (PDE) activity be rapid enough to hydrolyze cAMP while cyclase activity is regulated. In addition, if a Ca²⁺-inhibitable cyclase is involved, it must be stimulated in some way in the absence of Ca²⁺. Because these mechanisms are relatively easy to visualize, this section focuses instead on negative feedback loops between cAMP and Ca²⁺ that are capable, under certain conditions, of generating oscillations in the concentrations of both messengers. A minimum mechanism that can give rise to sustained oscillations is presented later. This system is comprised of elements known to coexist in a variety of cell types. A brief description of this system has appeared previously (70).

In Fig. 2, cAMP is generated by a Ca²⁺-inhibitable cyclase. cAMP opens channels in the surface membrane, causing an increase in Ca²⁺ entry. This is known to occur either by a direct action of cAMP on cyclic nucleotide-gated channels (71) that are permeable to Ca²⁺ or through the phosphorylation of L-type Ca²⁺ channels by cAMP-dependent protein kinase A (PKA; 72). The rise in Ca²⁺ inhibits cAMP synthesis. Phosphodiesterase activity reduces the concentration of cAMP and channels close. Ca²⁺ is then pumped out of the cell, allowing another round of cAMP synthesis to begin. The model is described by the following system of nonlinear differential equations:

$$dx/dt = aw - bx \quad (1)$$

$$dy/dt = c_m x_i^{N_1}/(x^{N_1} + K_1^{N_1} - dy) \quad (2)$$

$$dz/dt = ey - fz \quad (3)$$

$$dw/dt = h - g_m wz^{N_2}/(z^{N_2} + K_2^{N_2}) \quad (4)$$

There are four time-dependent variables in the scheme: x is the concentration of cAMP, y is the concentration of active channels, z is the concentration of free Ca²⁺, and w is the concentration of active cyclase. There are eight rate constants or con-

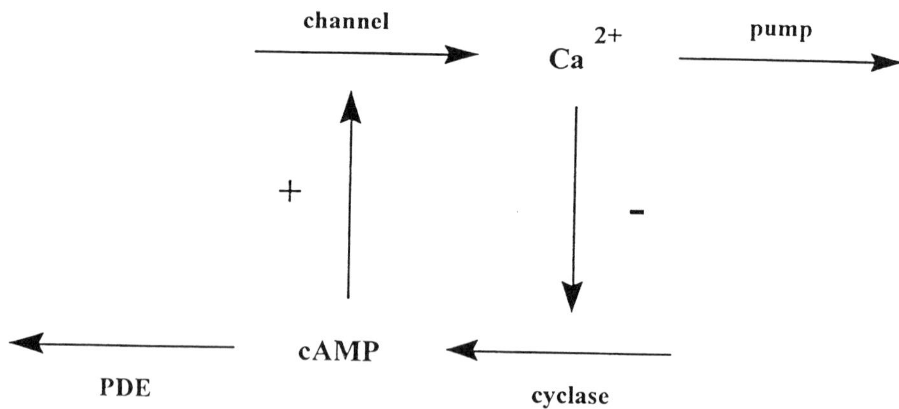

FIG. 2. Elements governing the control of [cAMP] and [Ca²⁺].

centration-independent rates: a is the rate constant of cAMP synthesis by AC, b is the rate constant of cAMP hydrolysis by PDE, c_m is the maximal rate of channel activation at saturating cAMP, d is the rate constant for channel deactivation, e is the rate constant of Ca^{2+} influx, f is the rate constant of Ca^{2+} pumping or exchange, g_m is the maximal rate constant for cyclase deactivation at saturating Ca^{2+}, and h is the rate of reversal of Ca^{2+} inhibition of cyclase. Channel activation by cAMP and cyclase inhibition by Ca^{2+} are both modeled as cooperative processes, dependent on the binding of several molecules of messenger. N_1 and N_2 are the Hill coefficients (an index of cooperativity) for those processes. K_1 is the concentration at which cAMP stimulation of channel activation is half-maximal, and K_2 is the concentration at which Ca^{2+} inhibition of cyclase is half-maximal.

For clarity and to illustrate the minimum requirements for sustained oscillations in cAMP and Ca^{2+}, we have made several simplifying (but plausible) assumptions. (1) The concentrations in the cytosol are uniform (i.e., the effects of diffusion and spatial gradients are not considered). For small cellular compartments, such as dendritic spines in neurons (see discussion of cyclase localization to sites of Ca^{2+} entry), spaces between caveolae in nonexcitable cells, or other regions of restricted diffusion near the surface membrane, it is reasonable to assume that diffusional equilibrium is achieved rapidly on the timescale of the preceding reactions. A recent striking demonstration of second messenger compartmentalization near the surface membrane is in cardiac myocytes, where β-adrenergic receptors have been shown to regulate nearby Ca^{2+} channels via local elevations of cAMP (73). If diffusion of Ca^{2+} and cAMP away from their sites of entry or synthesis is appreciable, the effect could be to reduce the amplitude of oscillations or eliminate them altogether. On the other hand, under the right conditions, diffusion could enhance the possibility that oscillations occur. Lledo and co-workers (74) showed that calbindin overexpression in GH3 cells muted Ca^{2+}-transients evoked by voltage-depolarization. Removal of Ca^{2+} and cAMP is necessary for oscillations, and, as described later, the balance of rates in the feedback loop is critical. Diffusion could serve to remove the messengers at an appropriate rate, in conjunction with Ca^{2+} pumping and cAMP hydrolysis. (2) The precursors from which the four variable species are formed are present in large excess, and are therefore taken to be constant. This is certainly true for ATP and extracellular Ca^{2+}, although it is not necessarily the case for inactive channels or inactive cyclase. If Ca^{2+} and cAMP levels oscillate, however, active channel and active cyclase will also oscillate between discrete levels, making it possible that the supply of channels and cyclase will not be significantly depleted. (3) The rates of decay of the variable species are directly proportional to concentration. This will be true if the decay occurs spontaneously. If the decay is enzyme-catalyzed, the assumption is that the variable species are present at concentrations less than the K_m values of the enzymes. When the variable species oscillate between discrete levels, these levels are likely to remain within the linear range of the enzymes. In general, relaxing these assumptions will alter the quantitative behavior of the system somewhat but not the ability of the system to exhibit oscillations.

Figure 3A shows a simulation that predicts sustained or stable oscillations in the levels of both cAMP and Ca^{2+}. The two messengers oscillate at the same frequency but somewhat out of phase. The values of the parameters, which are based on realistic estimates for the molecules involved, were as follows: $a = 2$ s^{-1}, $b = 0.5$ s^{-1}, $c_m = 0.45$ μM s^{-1}, $d = 1$ s^{-1}, $e = 1$ s^{-1}, $f = 0.5$ s^{-1}, $g_m = 2$ s^{-1}, $h = 0.11$ μM s^{-1}, $K_1 = 2$ μM, $K_2 = 0.2$ μM, $N_1 = 3$, and $N_2 = 3$. The initial concentrations of the four variables were $x_0 = 1$ μM, $y_0 = 0.05$ μM, $z_0 = 0.1$ μM, and $w_0 = 0.5$ μM. At $t = 0$, active channel, Ca^{2+}, and Ca^{2+}-regulated cyclase were all at steady-state (dy/dt, dz/dt, $dw/dt = 0$). The system was perturbed by a step change in the AC rate constant, a, to the value of 2 s^{-1} ($dx/dt = 0.5$ μM s^{-1} at $t = 0$). A perturbation of this type, although not instantaneous, would occur upon G protein stimulation. The differential equations were solved numerically by a fourth-/fifth-order Runge-Kutta method, using MATLAB software with the extension SIMULINK (MathWorks, Natick, MA).

There are two basic requirements for sustained oscillations. The first is a certain degree of cumulative cooperativity in the processes regulated by cAMP and Ca^{2+}. Assuming other conditions are met, the product of the Hill coefficients for cAMP activation of the channel and Ca^{2+} inhibition of the cyclase must exceed a certain value. Figure 3B shows that damped oscillations are predicted when the Hill coefficients for both processes are assumed to have a value of 2 instead of 3 (with similar or identical values for all the rates in the scheme; see the figure legend). If the Hill coefficient for one of the processes is 2 and the other is 3, damped oscillations that persist for a longer period of time are predicted. The second requirement for sustained oscillations is a minimum series of reaction steps that each take a certain amount of time. If one assumes that either activation of the channel by cAMP or inhibition of cyclase activity by Ca^{2+} is very rapid compared with other steps in the loop (the synthesis and breakdown of cAMP or the entry and pumping of Ca^{2+}), then either damped oscillations or no oscillations are predicted. With a minimum number of time-consuming reaction steps, an overall balance of rates in the scheme is required. As an illustration, if the rate constant of cAMP hydrolysis by PDE is assumed to be 0.1 s^{-1} (fivefold lower than in the simulation in Fig. 3A), the concentration of each messenger is predicted to rise and fall with little secondary oscillation (Fig. 3C). As the PDE rate constant is raised (with all the other rates held constant), the system exhibits damped oscillations and then sustained oscillations for values between 0.3 and 1.5 s^{-1}. At higher values, damped oscillations and then no oscillations are predicted. Although we have just considered the effects of changing a particular rate constant, the overall reaction rates are the critical determinants of whether oscillations occur. In the case of the channel and the cyclase, higher turnover numbers (for Ca^{2+} entry and cAMP synthesis) and lower protein concentrations, which may be more realistic in some cells, give the same behavior. The low effective turnover number for Ca^{2+} entry in the simulation in Fig. 3A reflects the assumption that a high percentage of entering Ca^{2+} is rapidly buffered.

How likely is it that these requirements will be met? At the moment, it is difficult to answer the question because we do not know the detailed kinetic properties of all the proteins in any one cell type. In different cells, there are alternative molecules that could serve a role in this feedback loop, and in some cases slightly different schemes could give rise to oscillating levels of the two messengers. In terms of the

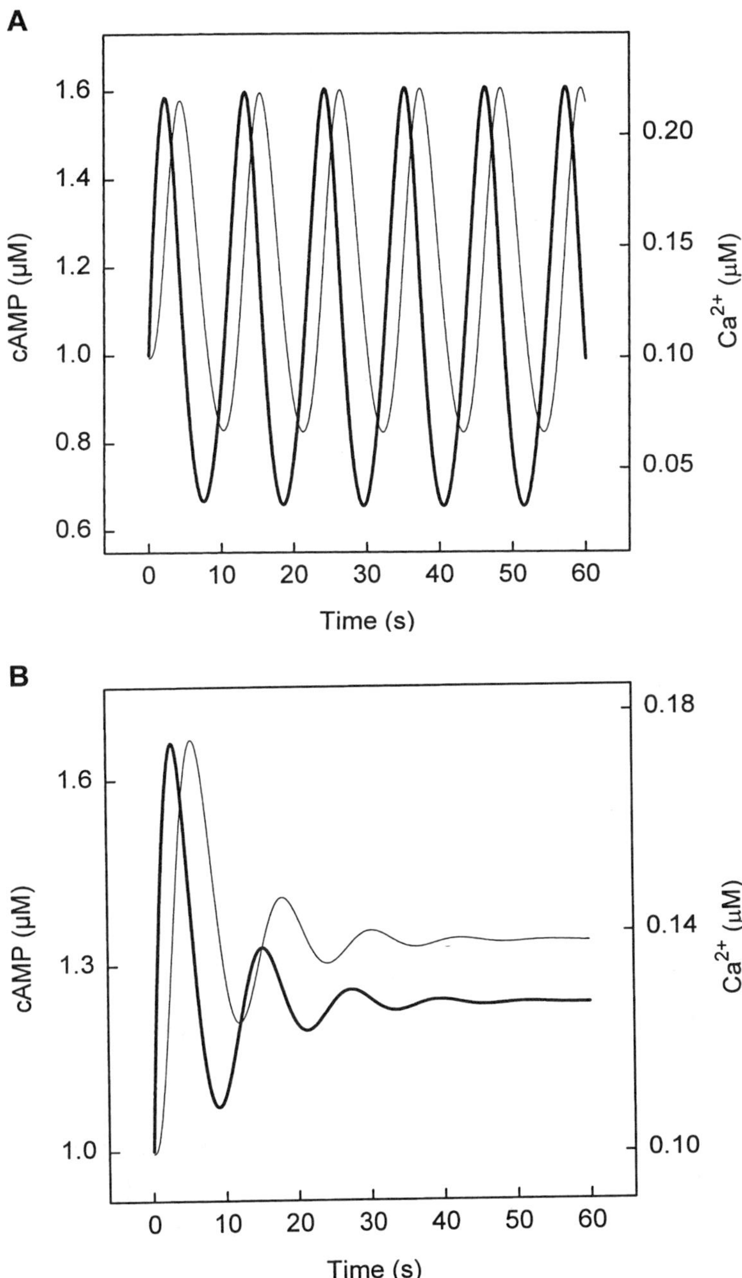

FIG. 3. (A) Sustained oscillations in cAMP (*thick line*) and Ca^{2+} levels (*thin line*) predicted by the negative feedback loop described in the text. **(B)** Damped oscillations in cAMP and Ca^{2+} levels predicted with the following differences from the simulation in panel A: $N_1 = N_2 = 2$, $c_m = 0.25$ μM s⁻¹, $h = 0.2$ μM s⁻¹.

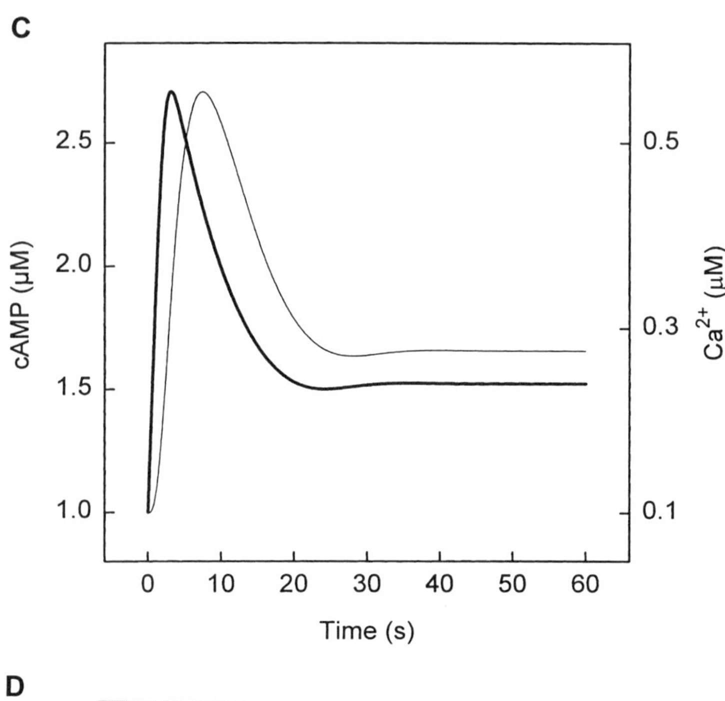

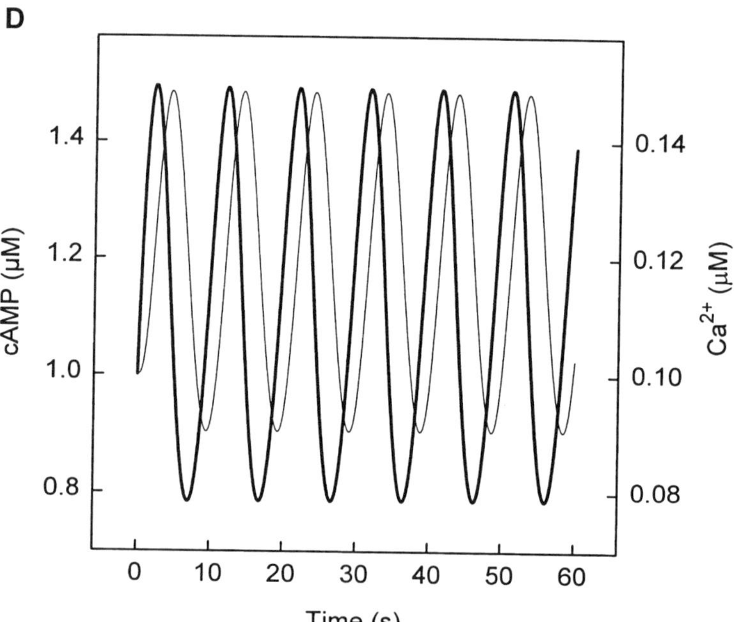

FIG. 3. *(Continued)* (**C**) Rise and fall in cAMP and Ca^{2+} levels, with little secondary oscillation, predicted with only one difference from panel A: $b = 0.1$ s^{-1}. (**D**) Sustained oscillations in cAMP and Ca^{2+} levels predicted with lower Hill coefficients for channel regulation by cAMP and cyclase inhibition by Ca^{2+} ($N_1 = N_2 = 2$), with the necessary additional cooperativity being provided by Ca^{2+} stimulation of PDE activity ($N_3 = 2$). See text for other parameters and initial conditions.

cooperativity requirements, Hill coefficients of 3 for both channel activation by cAMP and cyclase inhibition by Ca^{2+} probably represent steeper concentration dependencies than exist in most cells. However, the necessary cooperativity could be distributed between other steps in the loop. For example, Ca^{2+} stimulation of PDE is a cooperative process mediated by CaM in many cell types (75). Figure 3D shows the effect that this process can have on oscillations. In this simulation, the Hill coefficients for channel activation by cAMP, cyclase inhibition by Ca^{2+}, and PDE activation by Ca^{2+} were all assumed to be 2. The following equation was substituted for equation 1:

$$dx/dt = aw - bvx \quad (5)$$

A fifth time-dependent variable, v, represents the active PDE concentration. The following equation was added to describe PDE activation and deactivation:

$$dv/dt = i_m z^{N_3}/(z^{N_3} + K_3^{N_3}) - jv \quad (6)$$

The maximal rate of PDE activation at saturating Ca^{2+} is i_m, N_3 is the Hill coefficient for the activation process, K_3 is the Ca^{2+} concentration at which PDE activation is half-maximal, and j is the rate constant for conversion of active to inactive PDE. In the simulation shown in Fig. 3D, the values of these parameters and others that differed from the simulation in Fig. 3A were as follows: $i_m = 0.22$ μM s⁻¹, $N_1 = N_2 = N_3 = 2$, $K_3 = 0.2$ μM, $j = 0.44$ s⁻¹, $b = 5.6$ μM⁻¹ s⁻¹, $c_m = 0.25$ μM s⁻¹, $h = 0.2$ μM s⁻¹. The initial concentration of PDE, v_0, was 0.1 μM. As in the other simulations, at $t = 0$, all variable species other than cAMP were at steady-state, and the system was perturbed by a step change in the AC rate constant. Clearly, in comparing this simulation to the one in Fig. 3B, the additional cooperativity in PDE activation causes the system to exhibit stable oscillations. The Hill coefficients in this case are all realistic. It should be noted that Ca^{2+} stimulation of PDE can substitute for Ca^{2+} inhibition of cyclase in the preceding scheme in generating sustained oscillations (not shown). It was necessary, with only two cooperative processes in the mechanism (channel and PDE activation), to once again assume a value of 3 for the Hill coefficients. Other cooperative processes could similarly enhance the possibility of oscillations, including inhibition of channel activity by Ca^{2+} (76).

Regarding the requirement for a minimum series of time-consuming reaction steps, phosphorylation of channels and cyclase could provide the necessary delays. The association of binding proteins might also serve this purpose. In the case of cyclic nucleotide-gated channels, the activation is expected to be more rapid, although this may depend on the particular channel subtype. Also, higher-frequency oscillations might be expected in systems in which all of the steps are rapid, but comparable in their kinetics.

Measurements of Ca^{2+} in living cells have identified widespread oscillatory behaviors, and have revolutionized how signaling by Ca^{2+} is considered. On the other hand, what evidence exists to suggest that cAMP levels might oscillate? It is currently impossible to detect cAMP in living cells with either the temporal or spatial resolution of the expected oscillations, even though a single cell method using fluorescent energy transfer between the subunits of microinjected PKA subunits has been developed (77; and see Chapter 10). Notwithstanding the less than optimal temporal resolution of this system, some very attractive observations have been made. For instance, Backsai and co-workers showed the establishment in cultured *Aplysia* neurons of very large spatial gradients in cAMP accumulation between the cell body and distal processes in response to serotonin (77). DeBernardi and Brooker (78) have shown an inverse correlation between $[Ca^{2+}]_i$ and [cAMP] within the same single C6–2B glioma cell (which expresses AC6; see Chapter 10). A need to improve this methodology is suggested by the likelihood that rapid diffusion of cAMP away from its site of synthesis will tend to mute oscillations over greater cellular distances. Furthermore, the use of PDE inhibitors, as are commonly employed in [cAMP] determinations, would totally compromise any oscillations in cAMP for reasons that are obvious from the foregoing discussion. Indeed, PDEs, when active, are probably critical determinants of spatial and temporal patterns of cAMP signaling. Cyclic AMP has been shown to oscillate in a synchronized preparation of frog ventricular strips as a function of the normal contraction/relaxation cycle (79). Given that cardiac myocytes express predominantly AC5 and AC6, their participation in the model described seems easy to imagine.[7] It has also been suggested for some years that oscillations in AC activity underlie oscillations in extracellular cAMP levels observed in *Dictyostelium* (80). Several mechanisms have been proposed to explain this behavior (68).

In terms of underlining the importance of all the elements required for the maintenance of oscillations in cAMP and Ca^{2+}, some observations from the ethereal world of learning and memory may be informative. It is widely accepted that at some level Ca^{2+}-stimulated ACs are involved in learning and memory. Some of the most compelling support comes from *Drosophila* mutants, in which both the *dunce* and *rutabaga* mutants are impaired in associative learning. It is worth noting, particularly in view of the oscillation model discussed earlier, that changes in any of the participating elements would be expected to destroy the fidelity of the system. Thus, it is particularly interesting that both the *dunce* and *rutabaga* mutations result in a similarly impaired associative learning disability, even though in one case AC is inactivated, resulting in 50% less cAMP production, whereas in the other case, PDE is inactivated, resulting in eight times more cAMP production (81). However, in both cases, the fidelity of the cAMP signals is destroyed. These considerations reinforce the possibility that it is the oscillatory potential of the system that is actually critical to the entraining regimes. Similarly, knockout mice rendered deficient in the Ca^{2+}-binding protein, calbindin D28 (an intracellular Ca^{2+}-buffer), also exhibited deficits in learning and memory (82). Thus, perturbations in the fidelity of the system in any direction compromise the target of the oscillations.

---

[7]Long-term inhibition of AC3 by CaM kinase II has been proposed to contribute to Ca^{2+} oscillations in cardiac tissue (39). It is difficult to see how this might occur, given that AC3 is a very minor species in cardiac tissue at both the messenger ribonucleic acid (mRNA) level and the functional level, where the activity is prominently directly inhibitable by Ca^{2+}

The potential importance of cAMP oscillations, of course, is that rather than the classical and familiar amplitude mode, cAMP might actually signal as a function of the frequency of these oscillations, as has also been proposed for $[Ca^{2+}]_i$ oscillations (67). It is therefore perhaps worth speculating a little on the requirements for cAMP to act as a frequency encoder. For cAMP oscillations to be signals, mechanisms must exist for their decoding. In the case of Ca^{2+} oscillations, a detailed model has been presented whereby CaM kinase II, because of its arrangement in multimeric arrays and its Ca^{2+}-dependent autophosphorylation, can incrementally respond to sequential Ca^{2+}-transients (83). Frequency detection by CaM kinase II depends on the organization of identical subunits in oligomeric arrays. In the case of cAMP, then, are analogous features available? PKAs apparently do not form large oligomeric complexes, although they do occur as dimers. However, PKAs can be organized in multimeric arrays through the so-called PKA-anchoring proteins (AKAPs) (84). Whether such arrangements facilitate communication among PKA subunits and whether such a property affects their susceptibility to regulation by cAMP remain to be determined.

REGIONAL DISTRIBUTION OF ADENYLYL CYCLASE ISOFORMS

Adenylyl cyclase is highly enriched in the brain, which is the richest source of ACs in terms of the number of isoforms expressed. Evidence for specific regional distribution of AC isoforms in the central nervous system (CNS) comes from *in situ* hybridization studies, since isoform-specific antibodies have not yet been described. This distribution has been largely described elsewhere (85), but some more recent data are incorporated into the following summary. The strikingly distinct localization of AC isoforms in different brain areas is consistent with a highly regulated role of these cyclases in many neuronal functions. Although all nine isoforms can be detected in the CNS, AC3 and AC6 occur at very low levels in most brain areas. All Ca^{2+}-sensitive ACs are remarkably discretely expressed, whereas Ca^{2+}-insensitive cyclases (except AC7) are widely expressed both in the CNS (Table 2) and peripherally. Although the expression patterns displayed by ACs do not correlate directly with any specific neurotransmitter or neuromodulator pathways, in some cases, compelling correlations can be inferred from the expression of Ca^{2+}-sensitive ACs and the specific functions associated with particular regions.

Ca²⁺-Stimulable Adenylyl Cyclases

At present, the Ca^{2+}-stimulable ACs (AC1, AC3, AC8) seem to be strictly neural-specific. It had been proposed, based on functional assays, that Ca^{2+}-stimulated activity was strictly neural-specific and that any such activity that had been claimed in nonneuronal tissues (e.g., heart or adrenal) actually correlated with the degree of innervation of the tissue and that nonneuronal tissue or cell lines did not express this activity (20,86). Subsequent northern blot analysis of AC1 in various bovine tissues

TABLE 2. *Distribution of adenylyl cyclase isoforms in selected adult rat brain areas*

	Hippocampus dentate gyrus	CA1–CA3	Cerebellum Purkinje cells	Granular cells	Striatum	Hypothalamus	Cortex
Ca^{2+}-stimulable adenylyl cyclase	AC1 AC8	AC1 AC8	AC8	AC1 AC3^{low} AC8	AC8	AC8	AC1 AC8
Ca^{2+}-insensitive adenylyl cyclase	AC2 AC7^{low} AC9	AC2 AC7^{low} AC9	AC2 AC7	AC2 AC7 AC9	AC2 AC9	AC2 AC2^{low}	AC2 AC9
Ca^{2+}-inhibitable adenylyl cyclase					AC5		

Localization data were compiled from *in situ* hybridization data (see reference 83 and references therein).

supports this suggestion (87). Whatever peculiar feature of neurons or excitable cells that results in their garnering all the Ca^{2+}-stimulable ACs is far from evident at present.

The pattern of expression of AC1 mRNA provides a particularly good example of both regional and cell-type selective expression of an individual AC isoform in the CNS (88,89). *In situ* hybridization analysis shows a preponderance of AC1 mRNA in hippocampus, cerebral cortex, and cerebellum, with a particularly high level of expression in glutamatergic neurons. Recent developmental studies showed interesting differences in the ontogenetic expression of cyclase mRNAs (90). A transitory high expression of AC1 mRNA was seen in areas such as striatum, brain stem, and thalamus, which express quite low levels in the adult. Within hippocampus, the strongest expression is observed in pyramidal cells in CA1–CA2 fields and granular cells in the dentate gyrus (89). AC1 mRNA was also prominently expressed in cortical pyramidal neurons and cerebellar granular cells, with very low levels of expression in GABAergic neurons, such as Purkinje cells or cortical interneurons. Within hippocampus, the enrichment of AC1 mRNA in the CA1–CA2 field and dentate gyrus juxtaposes with Ca^{2+} entry through glutamatergic N-methyl-D-aspartate (NMDA) receptor activation, supporting the involvement of AC1 in Ca^{2+}-dependent synaptic plasticity mechanisms, which are popularly associated with learning and memory processes. Evidence for involvement of AC1 in hippocampal learning and memory has been provided by gene disruption in transgenic mice. These mice showed impaired hippocampal-dependent learning abilities and an altered long-term potentiation (LTP) in the CA1 region (91).[8]

AC3 was first reported to be restricted to the olfactory system (64). However, northern blot analysis also detected AC3 mRNA in brain, spinal cord, and adrenal and other tissues (92). The virtual restriction of AC3 to olfactory neuroepithelia neurons suggested a unique role for AC3 in olfactory signal transduction. It has been postulated that AC3 may be involved in the rapid but transient increase of cAMP caused by many odorants which, after binding to a cyclic nucleotide-dependent cation channel in the membrane of cilia, causes a depolarization and an initiation of neuronal signals (93).

AC8 is also restricted to the brain (94–96) in a pattern that is distinct from that of AC1. The highest level of expression of AC8 mRNA is in hippocampus, cortex, piriform and entorhinal cortex, amygdaloid complex, and hypothalamus (95,96). Although AC8 mRNA is expressed at lower levels than AC1 in many brain areas, AC8 mRNA is found in both excitatory glutamatergic and inhibitory GABAergic neuronal populations. Within hippocampus, AC8 mRNA is expressed in the CA1–CA3 layers and dentate gyrus as well as in GABAergic interneurons present in the molecular layers. The distinct pattern of AC1 and AC8 messages in the CA fields suggests that these two ACs may have specific but complementary physiologic functions in hippocampus. Within hypothalamus, a robust Ca^{2+}-stimulated AC is found

---

[8]Although cynics might comment that few signaling molecules have *not* been implicated in hippocampal LTP, the truth of that observation may be telling us that cellular regulation is a remarkably interdigitated process with few truly independent components.

in plasma membranes (97). A moderate expression of AC8 mRNA (95), coupled with the absence of AC1 mRNA (97) in hypothalamic nuclei, suggests that AC8 may be the prime integrator of $[Ca^{2+}]_i$ and cAMP-signaling in hypothalamus.

Ca²⁺-Inhibitable Adenylyl Cyclases

Both of the Ca^{2+}-inhibitable ACs (AC5, AC6) occur in brain, as well as in peripheral tissues. Northern analyses detected AC5 abundantly expressed in heart and neural tissues (98). However, PCR analysis also detected AC5 in other peripheral tissues, such as kidney, liver, testis, and skeletal muscle (99). AC6 is widely detected, with its strongest expression in heart and brain (94,99–101). Recent PCR studies have demonstrated that in renal tubule, AC5 and AC6 displayed a differential pattern of mRNA expression. This study provided evidence that in epithelial cells, AC5 and AC6 not only are cell-type specific but also display different patterns of regulation induced by $[Ca^{2+}]_i$ on either glucagon or vasopressin-dependent cAMP levels (102). Within the brain, strong and selective expression of AC5 mRNA occurs in the striatum, with no significant expression of AC6 mRNA in any discrete region (38,103). AC5 mRNA was predominantly expressed in striatal medium-sized GABAergic neurons, but it was also found in the limbic system, including the islands of Calleja, nucleus accumbens, and olfactory tubercle. AC5 may therefore participate in cognition and emotional behavior originating in these areas. Interestingly, in the ontogenetic study cited earlier (90), AC5 mRNA steadily increased during synaptic maturation in the striatum, matching neuronal differentiation in this area. However, in neostriatal GABAergic neurons, most of the known effects of elevation of cAMP are mediated by activation of dopaminergic D1-receptors that regulate the activity of PKA and protein phosphatase-1 (PrP1) (104). The D1-receptor pathways also trigger processes that modulate Ca^{2+} entry through activation of glutamatergic-inotropic receptors. It seems likely therefore, because AC5 is directly regulated by both Ca^{2+} and G_s pathways, that AC5 is a central integrator of dopaminergic and glutamatergic inputs (85).

Ca²⁺-Insensitive Adenylyl Cyclases

The Ca^{2+}-insensitive isoforms (AC2, AC4, AC7, and AC9) are all widely distributed in mammalian tissues. AC2 is abundant in lung and brain (105), whereas AC4, AC7, and AC9 are found in most tissues examined (94,106–109). *In situ* hybridization showed that AC2 and AC9 were abundant and widely distributed in rat brain (89,109,110), suggesting roles in numerous signaling systems. Although AC2 mRNA was widely expressed, the strongest expression was found in hippocampus, cortex, and cerebellum. Within hippocampus, AC2 mRNA was strongly expressed in glutamatergic cells of pyramidal and granular layers as well as in GABAergic cells within the molecular layers. In addition, AC2 mRNA was also moderately expressed in several areas particularly rich in monoaminergic and catecholaminergic neurons, such as hypothalamus, substantia nigra, and locus coeruleus (85,97). Interestingly, AC7 mRNA is largely confined to cerebellum (107). In rat cerebellum, AC7 mRNA is prominently expressed in Purkinje cells, whereas granular cells express predominantly AC2 mRNA. Both Purkinje and granular cells also express distinct PKC isoforms, which suggests a possible cell-type association of AC- and PKC-dependent pathways. Recent analyses showed high to moderate levels of mRNA for AC9 in hippocampus, cerebellum, cortex, striatum, brain stem, and amygdala (109). As yet, there are no *in situ* hybridization data on the precise neuronal expression of AC9 in these areas.

Overall, it is clear that ACs are expressed very specifically in mammalian brain and are therefore poised to respond differently and very specifically to local increases in $[Ca^{2+}]_i$ and/or PKC in various brain regions. However, given that most brain areas—such as hippocampus, striatum, and cerebellum—express multiple AC isoforms, which can display differing responses to Ca^{2+}, G proteins, or PKC (70), unraveling the role of particular ACs in precise neuronal functions will be a major challenge for the future.

MICRODISTRIBUTION OF ADENYLYL CYCLASES

Recent studies have shown distinct subcellular targeting of protein phosphatase and kinase isoforms, which are major determinants in the specific phosphorylation events that occur following activation of second messenger signaling pathways (84,111). Best characterized are the AKAPs, which occur in the postsynaptic densities of neuronal cells. Immunohistochemical studies using a pan-specific AC antibody detected a preferential post- or presynaptic distribution of AC immunoreactivity by electron microscopy (112). Within the hippocampal molecular layer of the CA1 field, immunoreactivity was concentrated near the postsynaptic densities in dendritic spines, which confirms that AC protein is selectively transported from cell bodies to predetermined sites to modulate synaptic information (Fig. 4). It is already established that the dendritic spines are areas of high concentration of Ca^{2+} channels and pumps, as well as PKA, CaM kinase II, and PDE (113–115). Indeed, dendritic spines are increasingly being viewed as areas of independent neuronal activity and $[Ca^{2+}]_i$ homeostasis (116,117). Therefore, an enrichment of both AC and protein kinases in dendritic spines, in juxtaposition with Ca^{2+} entry through NMDA or AMPA receptors, points to a dramatic facilitation in reception and propagation of signals carried by cAMP and Ca^{2+}, which have been implicated during the induction of LTP (see Chapter 1).

SUMMARY AND FUTURE DIRECTIONS

It is now clear that all the ACs described to date are potentially subject to regulation by the Ca^{2+}-signaling pathway, whether by Ca^{2+} or by PKC, directly or indi-

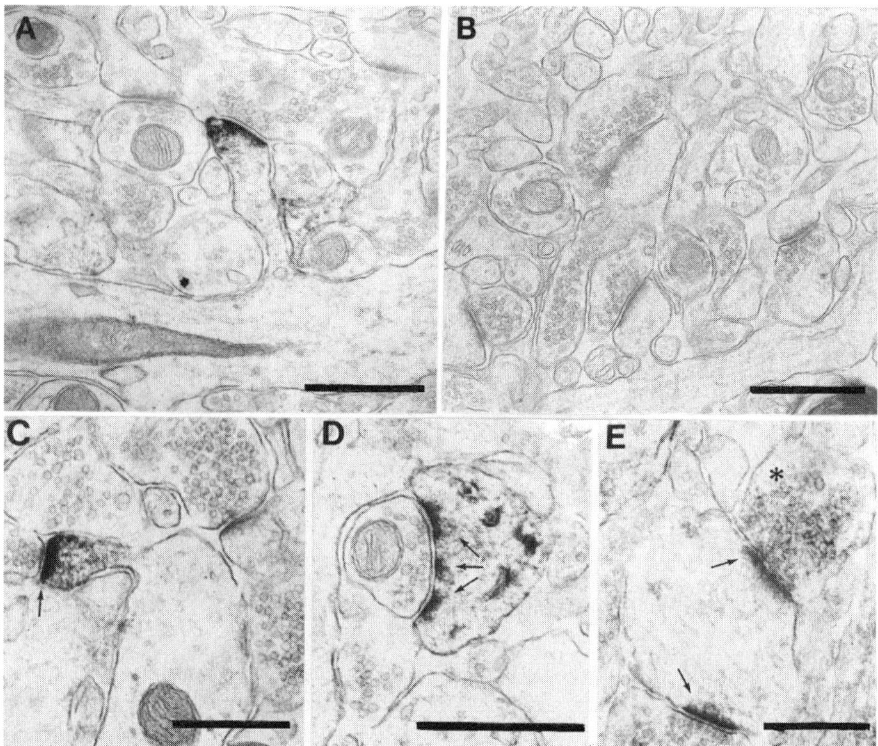

FIG. 4. Preferential association of AC immunoreactivity with the postsynaptic compartment in the molecular layer of the hippocampal CA1 region. Immunocytochemical localization was with the AC-comm antiserum (112). (**A**) Strongly stained postsynaptic density (psd) with a weak diffusion of immunoreactivity on the spinehead and no staining on the dendritic shaft. (**B**) Control section. No labeling is seen on psd's after preadsorption of antibodies. (**C**) One or (**D–E**) multiple-labeled psd's (*arrows*) were in contact with labeled (*asterisk*) or unlabeled terminals. (**A–C**, × 47,300; **D**, × 33,800; **E**, × 28,200; bars = 0.5 μM). (Modified with permission from Mons N, Harry A, Dubourg P et al: Immunohistochemical localization of adenylyl cyclase in rat brain indicates a highly selective concentration at synapses. *Proc Nat Acad Sci U S A* 1995;92: 8473–8477.)

rectly. Whether particular species will always be regulated as a consequence of activation of Ca²⁺ signaling in every cellular context remains to be determined, but the fact that this potential exists should bring these two major signaling pathways into the same frame of reference when considering physiologic regulation. Some physiologic roles for Ca²⁺-sensitive ACs have been raised elsewhere in this volume and examples are emerging from the literature, based on what is now known about the sensitivity of various AC species. Examples can be anticipated where the systems will interact synergistically (as in the hippocampus), antagonistically (predicted for platelets), and harmonically (as in cardiac tissue). However, we are just on the threshold of understanding how important these interactions are in general cellular regulation.

Many structure-functional questions remain to be resolved over the next few years—not least of which is the domains of cyclases that confer not only *in vitro* sensitivity to Ca^{2+} but also *in vivo* sensitivity to Ca^{2+} entry. Whether ACs expressed in neurons will display the same sensitivity to Ca^{2+} entry and by what means—whether voltage-gated, receptor-operated, or capacitative—has yet to be addressed. Now that some x-ray crystallographic structure has been determined for part of a cyclase (118), it may be anticipated that precise information will begin to be gathered on the conformational changes associated with various regulatory events.

The vast majority of experiments performed in cAMP signaling treat the second messenger as though it is purely an amplitude-encoding signal. However, a careful search for cAMP oscillations may yet revolutionize how signaling by this messenger is viewed, because the elements required for the production for cAMP oscillations appear to be available, particularly by virtue of the sensitivity of ACs to Ca^{2+}. All in all, the next few years promise to bring us closer to understanding how the two "classic" second messengers govern function and the steps taken by nature to ensure their interdigitation.

ACKNOWLEDGMENTS

Research described from the laboratory of D.M.F.C. is supported by NIH grants GM324843 and NS 28389.

REFERENCES

1. Sunahara RK, Dessauer CW, Gilman AG: Complexity and diversity of mammalian adenylyl cyclases. *Annu Rev Pharmacol Toxicol* 1996;36:461–480.
2. Cooper DMF, Yoshimura M, Zhang Y et al: Capacitative Ca^{2+} entry regulates Ca^{2+}-sensitive adenylyl cyclases. *Biochem J* 1994;297:437–440.
3. Fagan KA, Mahey R, Cooper DMF: Functional co-localization of transfected Ca^{2+}-stimulable adenylyl cyclases with capacitative Ca^{2+}-entry sites. *J Biol Chem* 1996;271:12438–12444.
4. Choi, EJ, Xia ZG, Storm DR: Stimulation of the type III olfactory adenylyl cyclase by calcium and calmodulin. *Biochemistry* 1992; 31:6492–6498.
5. Nielsen MD, Chan GCK, Poser SW, Storm DR: Differential regulation of type I and type VIII Ca^{2+}-stimulated adenylyl cyclases by Gi-coupled receptors *in vivo. J Biol Chem* 1996;271:33308–33316.
6. Krupinski J, Coussen F, Bakalyar HA et al: Adenylyl cyclase amino acid sequence: possible channel or transporter-like structure. *Science* 1989;244:1558–1564.
7. Hyde SC, Emsley P, Hartshorn MJ et al: Structural model of ATP-binding proteins associated with cystic fibrosis, multidrug resistance and bacterial transport. *Nature* 1990;346:363–365.
8. Dessauer CW, Gilman AG: Purification and characterization of a soluble form of mammalian adenylyl cyclase. *J Biol Chem* 1996;271:16967–16974.
9. Yan S-Z, Hahn D, Huang Z-H, Tang W-J: Two cytoplasmic domains of mammalian adenylyl cyclase form a Gsα- and forskolin-activated enzyme *in vitro. J Biol Chem* 1996;271:10941–10945.
10. Gelehrter TD, Collins FS: *Principles of medical genetics*. Baltimore: Williams & Wilkins, 1990;216.
11. Schultz JE, Klumpp S, Benz R et al: Regulation of adenylyl cyclase from *Paramecium* by an intrinsic potassium conductance. *Science* 1992;255:600–603.

12. Reddy R, Smith D, Sayman G et al: Voltage-sensitive adenylyl cyclase activity in cultured neurons. *J Biol Chem* 1995;255:14340–14346.

13. Rodbell M, Lad PM, Nielsen TB et al: The structure of adenylate cyclase systems. *Adv Cyc Nuc Res* 1981;14:3–14.

14. Neer EJ, Echeverria D, Knox S: Increase in the size of soluble brain adenylate cyclase with activation by guanosine 5′-(βγ-imido)triphosphate. *J Biol Chem* 1980;255:9782–9789.

15. Neer, EJ, Schmidt CJ, Nambudripad R, Smith TF: The ancient regulatory-protein family of WD-repeat proteins. *Nature* 1994;371:297–300.

16. Castets F, Bartoli M, Barnier JV et al: A novel calmodulin-binding protein, belonging to the WD-repeat family, is localized in dendrites of a subset of CNS neurons. *J Cell Biol* 1996;134:1051–1062.

17. Tang W-J, Stanzel M, Gilman AG: Truncation and alanine-scanning mutants of type I adenylyl cyclase. *Biochemistry* 1995;34:14563–14572.

18. Simmerman HKB, Kobayashi YM, Autry JM, Jones LR: A leucine zipper stabilizes the pentameric membrane domain of phospholamban and forms a coiled-coil pore structure. *J Biol Chem* 1996;271:5941–5946.

19. Kawabe J-I, Ebina T, Ismail S et al: A novel peptide inhibitor of adenylyl cyclase (AC). *J Biol Chem* 1994;269:24906–24911.

20. Caldwell KK, Boyajian CL, Cooper DMF: The effects of Ca²⁺ and calmodulin on adenylyl cyclase activity in plasma membranes derived from neural and non-neural cells. *Cell Calcium* 1992;13:107–121.

21. Vorherr T, Knopfel L, Hofmann et al: The calmodulin binding domain of nitric oxide synthase and adenylyl cyclase. *Biochemistry* 1993;31:6081–6088.

22. Wu ZL, Wong ST, Storm DR: Modification of the calcium and calmodulin sensitivity of the type I adenylyl cyclase by mutagenesis of its calmodulin binding domains. *J Biol Chem* 1993;268:23766–23768.

23. Levin LR, Reed RR: Identification of functional domains of adenylyl cyclase using in vivo chimeras. *J Biol Chem* 1995;270:7573–7539.

24. Boyajian CL, Garritsen A, Cooper DMF: Bradykinin stimulates Ca²⁺ mobilization in NCB-20 cells leading to direct inhibition of adenylylcyclase: a novel mechanism for inhibition of cAMP production. *J Biol Chem* 1991;266:4995–5003.

25. Cooper DMF, Mons N, Fagan K: Ca²⁺-sensitive adenylyl cyclases. *Cell Signalling* 1994;8:823–840.

26. Klee CB, Crouch TB, Richman PG: Calmodulin. *Annu Rev Biochem* 1980;49:489–515.

27. Cho HJ, Xie Q, Calacay J et al: Calmodulin is a subunit of nitric oxide synthase from macrophages. *J Exp Med* 1992;176:599–604.

28. Scholich K, Barbier AJ, Mullenix JB, Patel TB: Characterization of soluble forms of non-chimeric type 5 adenylyl cyclase. *Proc Natl Acad Sci U S A* 1997;94:2915–2920.

29. Steer ML, Levitzki A: The control of adenylate cyclase by calcium in erythrocyte ghosts. *J Biol Chem* 1975;268:23766–23768.

30. Dizhoor AM, Ray S, Kumar S et al: Recoverin: a calcium sensitive activator of retinal rod guanylate cyclase. *Science* 1991;251:915–918.

31. DeBernardi MA, Munshi R, Yoshimura M, Brooker G: Predominant expression of type-VI adenylate cyclase in C6–2B rat glioma cells may account for inhibition of cyclic AMP accumulation by calcium. *Biochem J* 1993;293:325–328.

32. Yamagata K, Goto K, Kuo C-H et al: Visinin: a novel calcium binding protein expressed in retinal cone cells. *Neuron* 1991;2:469–476.

33. Kawamura S, Murakami M: Calcium-dependent regulation of cyclic GMP phosphodiesterase by a protein from frog retinal rods. *Nature* 1991;349:420–423.

34. Kuno T, Kajimoto Y, Hashimoto T et al: cDNA cloning of a neural visinin-like Ca²⁺-binding protein. *Biochem Biophys Res Comm* 1992;184:1219–1225.

35. Kajimoto Y, Shirai Y, Mukai H et al: Molecular cloning of two additional members of the neural visinin-like Ca²⁺-binding protein gene family. *J Neurochem* 1993;61:1091–1096.

36. Putney JW Jr: The capacitative model for receptor-activated calcium entry. *Adv Pharmacol* 1991;22:251–269.

37. Chiono M, Mahey R, Tate G, Cooper DMF: Capacitative Ca²⁺-entry exclusively inhibits cAMP synthesis in C6–2B glioma cells. *J Biol Chem* 1995;270:1149–1155.

38. Mons N, Cooper DMF: Selective expression of one Ca²⁺-inhibitable adenylyl cyclase in dopaminergically innervated rat brain regions. *Brain Res Mol Brain Res* 1994;22:236–244.

39. Wei J, Wayman G, Storm DR: Phosphorylation and inhibition of type III adenylyl cyclase by calmodulin-dependent protein kinase II *in vivo*. *J Biol Chem* 1996;271:24231–24235.

40. Hoth M, Penner R: Calcium release-activated calcium current in rat mast cells. *J Physiol* (Lond) 1993;465:359–86.
41. Zweifach A, Lewis RS: Rapid inactivation of depletion-activated calcium current (I_{CRAC}) due to local calcium feedback. *J Gen Physiol* 1995;105:209–226.
42. Piascik MT, Wisler PL, Johnson CL, Potter JD: Ca^{2+}-dependent regulation of guinea pig brain adenylyl cyclase. *J Biol Chem* 1980;255:4176–4181
43. Clapham DE: Trp is Cracked but is CRAC Trp? *Neuron* 1996;16:1069–1072.
44. Zitt C, Zobel A, Obukhov AG et al: Cloning and functional expression of a human Ca^{2+}-permeable cation channel activated by calcium store depletion. *Neuron* 1996;16:1189–1196.
45. Zhu X, Jiang M, Peyton M et al: trp, a novel mammalian gene family essential for agonist-activated capacitative Ca^{2+} entry. *Cell* 1996;85;661–671.
46. Burgess GM, Bird GStJ, Obie JF, Putney JW Jr: The mechanism for synergism between phospholipase C- and adenylylcyclase-linked hormones in liver: cyclic AMP–dependent kinase augments inositol trisphosphate-mediated Ca^{2+} mobilization without increasing the cellular levels of inositol polyphosphates. *J Biol Chem* 1991;266:4772–4781.
47. Blau L, Weissmann G: Transmembrane calcium movements mediated by ionomycin and phoshatidate in liposomes with fura 2 entrapped. *Biochemistry* 1988;27:5661–5666.
48. Gogelein H, Huby A: Interaction of Saponin and digitonin with black lipid membranes and lipid monolayers. *Biochim Biophys Acta* 1984;773:32–38.
49. Schnitzer JE, Oh P, Pinney E, Allard J: Filipin sensitive caveolae-mediated transport in endothelium: reduced trancytosis, scavenger endocytosis, and capillary permeability of select macromolecules. *J Cell Biol* 1994;127:1217–1232.
50. Lisanti MP, Scherer PE, Vidugiriene J et al: Characterization of caveolin-rich membrane domains isolated from an endothelial-rich source: implications for human disease. *J Cell Biol* 1994;126:111–126.
51. Parton RG: Caveolae and caveolins. *Curr Opin Cell Biol* 1996;8:542–548.
52. Crone CE: Modulation of solute permeability in microvascular endothelium. *Fed Am Soc Exp Biol* 1986;45:77–83.
53. Fujimoto T, Miyamashi A, Mikoshiba K: Inositol 1,4,5-trisphosphate receptor-like protein in plasmalemmal caveolae is linked to actin filaments. *J Cell Sci* 1995;108:7–15.
54. Fujimoto T: Calcium pump of the plasma membrane is localized in caveolae. *J Cell Biol* 1993;120:1147–1157.
55. Rothberg KG, Ying YS, Kolhouse JF et al: Cholesterol controls the clustering of the glycophospholipid-anchored membrane receptor for 5-methyltetrahydrofolate. *J Cell Biol* 1990;111:2931–2938.
56. Ying YS, Anderson RGW, Rothberg KG: Each caveola contains multiple glycosyl-phosphatidylinositol-anchored membrane proteins. *Cold Spring Harb Symp Quant Biol* 1992;57:593–604.
57. Chun M, Liyanage UK, Lisanti MP, Lodish HF: Signal transduction of a G protein–coupled receptor in caveolae: colocalization of endothelin and its receptor with caveolin. *Proc Natl Acad Sci U S A* 1994;91:11728–11732.
58. Conrad PA, Smart EJ, Ying Y-S et al: Caveolin cycles between plasma membrane caveolae and the Golgi complex by microtubule-dependent and microtubule-independent steps. *J Cell Biol* 1995;131:1421–1433.
59. Parton RG, Simons K: Digging into caveolae. *Science* 1995;269:1398–1399.
60. Fielding PE, Fielding CJ: Plasma membrane caveolae mediate the efflux of cellular free cholesterol. *Biochemistry* 1995;34:14288–14292.
61. Rizzuto R, Simpson AW, Brini M, Pozzan T: Rapid changes of mitochondrial Ca^{2+} revealed by specifically targeted recombinant aequorin. *Nature* 1992;358:325–327.
62. Rizzuto R, Brini M, Pozzan T: Targeting recombinant aequorin to specific intracellular organelles. *Meth Cell Biol* 1994;40:339–358.
63. Cobbold PH, Rink TJ: Fluorescence and bioluminescence measurement of cytoplasmic free calcium. *Biochem J* 1987;248:313–328.
64. Bakalyar HA, Reed RR: A specialized adenylyl cyclase may mediate odorant detection. *Science* 1990;250:1403–1406.
65. Anholt RR: Signal integration in the nervous system: adenylyl cyclases as molecular coincidence detectors. *TINS* 1994;17:37–41.
66. Cuthbertson KS, Cobbold PH: Oscillations in cell calcium. *Cell Calcium* 1991;12:61–62.
67. Meyer T, Stryer L: Calcium spiking. *Annu Rev Biophys Biophys Chem* 1991;20:153–174.

68. Goldbeter A: *Biochemical oscillations and cellular rhythms.* Cambridge, UK: Cambridge University Press, 1990.
69. Dolmetsch RE, Lewis RS: Signaling between intracellular stores and depletion-activated Ca^{2+}-channels generates $[Ca^{2+}]_i$ oscillations in T lymphocytes. *J Gen Physiol* 1994;103:365–388.
70. Cooper DMF, Mons N, Karpen JW: Adenylyl cyclases and the interaction between calcium and cAMP signalling. *Nature* 1995;374:421–424.
71. Finn JT, Grunwald ME, Yau K-W: Cyclic nucleotide-gated ion channels: an extended family with diverse functions. *Annu Rev Physiol* 1996;58:395–426.
72. Trautwein W, Hescheler J: Regulation of cardiac L-type calcium current by phosphorylation and G proteins. *Annu Rev Physiol* 1990;52:257–274.
73. Jurevicius J, Fischmeister R: cAMP compartmentation is responsible for a local activation of cardiac Ca^{2+} channels by β-adrenergic agonists. *Proc Natl Acad Sci U S A* 1996;93:295–299.
74. Lledo PM, Somasundaram B, Morton AJ et al: Stable transfection of calbindin-D28k into the GH3 cell line alters calcium currents and intracellular calcium homeostasis. *Neuron* 1992;9:943–954.
75. Rybalkin SD, Beavo JA: Multiplicity within cyclic nucleotide phosphodiesterases. *Biochem Soc Trans* 1996;24:1005–1009.
76. Chen T-Y, Yau K-W: Direct modulation by Ca^{2+}-calmodulin of cyclic nucleotide-activated channel of rat olfactory receptor neurons. *Nature* 1994;368:545–548.
77. Backsai BJ, Hochner B, Mahaut-Smith M et al: Spatially resolved dynamics of cAMP and protein kinase A subunits in Aplysia sensory neurons. *Science* 1993;260:222–226.
78. DeBernardi MA, Brooker G: Single cell Ca^{2+}/cAMP cross-talk monitored by simultaneous Ca^{2+}/cAMP fluorescence ratio imaging. *Proc Nat Acad Sci U S A* 1996;93:4577–4582.
79. Brooker G: Oscillation of cyclic adensosine monophosphate concentration during the myocardial contraction cycle. *Science* 1973;182:933–934.
80. Roos W, Scheidegger C, Gerisch G: Adenylate cyclase activity oscillations as signals for cell aggregation in *Dictyostelium discoideum. Nature* 1977;266:259–261.
81. Livingstone MS, Sziber PP, Quinn WG: Loss of calcium/calmodulin responsiveness in adenylyl cyclase of *rutabaga,* a *Drosophila* learning mutant. *Cell* 1984;37:205–215.
82. Molinari S, Battini R, Ferrari S et al: Deficits in memory and hippocampal long-term potentiation in mice with reduced calbindin D28K expression. *Proc Nat Acad Sci U S A* 1996;93:8028–8033.
83. Hanson PJ, Meyer T, Stryer L, Schulman H: Dual role of calmodulin in autophosphorylation of multifunctional CaM kinase may underlie decoding of calcium signals. *Neuron* 1994;12:943–956.
84. Faux MC, Scott JD: More on target with protein phosphorylation: conferring specificity by location. *TIBS* 1996;21:312–315.
85. Mons N, Cooper DMF: Adenylate cyclases: critical foci in neuronal signaling. *TINS* 1995;18:536–542.
86. Cooper DMF, Caldwell KK: Does calmodulin regulate non-neuronal adenylate cyclase? *Biochem J* 1988;254:935–936.
87. Xia ZG, Choi EJ, Fan W et al: Type-I calmodulin-sensitive adenylyl cyclase is neural specific. *J Neurochem* 1993;60:305–311.
88. Xia Z, Refsdal CD, Merchant KN et al: Distribution of mRNA for the calmodulin-sensitive adenylate cyclase in rat brain: expression in areas associated with learning and memory. *Neuron* 1991;6:431–443.
89. Mons N, Yoshimura M, Cooper DMF: Discrete expression of Ca^{2+}/calmodulin-sensitive and Ca^{2+}-insensitive adenylyl cyclases in the rat brain. *Synapse* 1993;14:51–59.
90. Matsuoka I, Suzuki Y, Defer N et al: Differential expression of type I, II and V adenylyl cyclase gene in the postnatal developing rat brain. *J Neurochem* 1997;68:498–506.
91. Wu ZL, Thomas SA, Villacres EC et al: Altered behavior and long-term potentiation in type I adenylyl cyclase mutant mice. *Proc Nat Acad Sci U S A* 1995;92:220–224.
92. Xia Z, Choi EJ, Wang F, Storm DR: The type III calcium/calmodulin-sensitive adenylyl cyclase is not specific to olfactory sensory neurons. *Neurosci Lett* 1992;144:169–173.
93. Breer H: Odor recognition and second messenger signaling in olfactory receptor neurons. *Semin Cell Biol* 1994;5:25–32.
94. Krupinski J, Lehman TC, Frankenfield CD et al: Molecular diversity in the adenylyl cyclase family: evidence for 8 forms of the enzyme and cloning of type VI. *J Biol Chem* 1992;267:24858–24862.
95. Cali JJ, Zwaagstra JC, Mons N et al: Type VIII adenylyl cyclase: a Ca^{2+}/calmodulin-stimulated enzyme expressed in discrete regions of rat brain. *J Biol Chem* 1994;269:12190–12195.

96. Parma J, Stengel D, Gannage MH et al: Sequence of a human brain adenylyl cyclase partial cDNA: evidence for a consensus cyclase specific domain. *Biochem Biophys Res Comm* 1991;179:455–462.

97. Mons N, Cooper DMF: Adenylyl cyclase messenger RNA expression does not reflect the predominant Ca^{2+}/calmodulin-stimulated activity in the hypothalamus. *J Neuroendocrinol* 1994;6:665–671.

98. Ishikawa Y, Katsushika S, Chen L et al: Isolation and characterization of a novel cardiac adenylyl cyclase cDNA. *J Biol Chem* 1992;267:13553–13557.

99. Premont RT, Chen JQ, Ma HW et al: Two members of a widely expressed subfamily of hormone-stimulated adenylyl cyclase. *Proc Nat Acad Sci U S A* 1992;89:9809–9813.

100. Yoshimura M, Cooper DMF: Cloning and expression of a Ca^{2+}-inhibitable adenylyl cyclase from NCB-20 cells. *Proc Nat Acad Sci U S A* 1992;89:6716–6720.

101. Katsushika S, Chen L, Kawabe JI et al: Cloning and characterization of a 6th adenylyl cyclase isoform: type V and type VI constitute a subgroup within the mammalian adenylyl cyclase family. *Proc Nat Acad Sci U S A* 1992;89:8774–8778.

102. Chabardes D, Firsov D, Aarab L et al: Localization of mRNAs encoding Ca^{2+}-inhibitable adenylyl cyclases along the renal tubule: functional consequences for regulation of the cAMP content. *J Biol Chem* 1996;271:19264–19271.

103. Glatt CF, Snyder SH: Cloning and expression of an adenylyl cyclase localized to the corpus striatum. *Nature* 1993;361:536–538.

104. Kötter R: Postsynaptic integration of glutamatergic and dopaminergic signals in the striatum. *Prog Neurobiol* 1994;44:163–196.

105. Feinstein PG, Schrader KA, Bakalyar HA et al: Molecular cloning and characterization of a Ca^{2+}/calmodulin insensitive adenylyl cyclase from rat brain. *Proc Nat Acad Sci U S A* 1991;88: 10173–10177.

106. Watson PA, Krupinski J, Kempinski AM, Frankenfield CD: Molecular cloning and characterization of the type VII isoform of mammalian adenylyl cyclase expressed widely in mouse tissues and in S49 lymphoma cells. *J Biol Chem* 1994;269:28893–28898.

107. Hellevuo K, Yoshimura M, Mons N et al: The characterization of a novel human adenylyl cyclase which is present in brain and other tissues. *J Biol Chem* 1995;270:11581–11589.

108. Paterson JM, Smith SM, Harmar AJ, Antoni FA: Control of a novel adenylyl cyclase by calcineurin. *Biochem Biophys Res Comm* 1995;214:1000–1008.

109. Premont R, Matsuoka I, Mattei MG et al: Identification and characterization of a widely expressed form of adenylyl cyclase. *J Biol Chem* 1996;271:13900–13907.

110. Furuyama T, Inagaki S, Tagaki H: Distribution of type II adenylyl cyclase mRNA in the rat brain. *Brain Res Mol Brain Res* 1993;19:165–170.

111. Scott JD, Mc Cartney S: Localization of A-kinase through anchoring proteins. *Mol Endocrinol* 1994;8:5–13.

112. Mons N, Harry A, Dubourg P et al: Immunohistochemical localization of adenylyl cyclase in rat brain indicates a highly selective concentration at synapses. *Proc Nat Acad Sci U S A* 1995; 92:8473–8477.

113. Chetkovich DM, Gray R, Johnston D, Sweatt JD: N-methyl-D-aspartate receptor activation increases cAMP levels and voltage-gated Ca^{2+} channel activity in area CA1 of hippocampus. *Proc Natl Acad Sci U S A* 1991;88:6467–6471.

114. Ludvig N, Burmeister V, Jobe PC, Kincaid RL: Electron microscopic immunocytochemical evidence that the calmodulin-dependent cyclic nucleotide phosphodiesterase is localized predominantly at postsynaptic sites in the rat brain. *Neuroscience* 1991;44:491–500.

115. Klauck TM, Scott JD: The postsynaptic density: a subcellular anchor for signal transduction enzymes. *Cell Signal* 1995;7:747–757.

116. Muller W, Connor JA: Dendritic spines as individual neuronal compartments for synaptic Ca^{2+} responses. *Nature* 1991;354:73–76.

117. Yuste R, Denk W: Dendritic spines as basic functional units of neuronal integration. *Nature* 1995;375:682–684.

118. Zhang G, Liu Y, Ruoho AE, Hurley JH: Structure of the adenylyl cyclase catalytic core. *Nature* 1997;386:247–253.

Advances in Second Messenger and Phosphoprotein Research, Vol. 32, edited by Dermot M. F. Cooper
Lippincott–Raven Publishers, Philadelphia © 1998

3

Molecular Diversity of the Adenylyl Cyclases

John Krupinski* and James J. Cali†

*Bristol-Myers Squibb, Princeton, New Jersey 08543, and †Weis Center for Research,
Pennsylvania State University College of Medicine,
Danville, Pennsylvania 17822-2610*

Hormone-sensitive adenylyl cyclase (AC) is a paradigm for the study of receptor-mediated signal transduction. Three classes of interacting components were initially identified in this system: (1) receptors for hormones that regulate cyclic adenosine monophosphate (cAMP) synthesis, (2) regulatory guanosine triphosphate (GTP) binding proteins (G proteins), and (3) the family of enzymes, the ACs (1). The binding of hormone to its receptor regulates catalytic activity of the AC, resulting in the formation of cAMP from adenosine triphosphate (ATP) at the intracellular face of the plasma membrane. Concentrations of cAMP are altered by hundreds of stimulatory or inhibitory hormones, neurotransmitters, and olfactants. The second messenger then propagates the hormone signal by directly activating cyclic nucleotide-gated channels (2) or by activating cAMP-dependent protein kinase, which alters the activity of a variety of intracellular proteins through protein phosphorylation (3). Physiologic effects as diverse as memory and heart rate are, at least in part, regulated by this signaling pathway.

Molecular biologic investigations have revealed that the ACs are actually a large family of enzymes encoded by at least nine different genes, and additional variants arise from alternatively spliced messages. The various subtypes are typically expressed in discrete patterns in only a limited number of tissues. Types I–IX AC (AC1–9) are predicted to share the same overall topology, but significant amino acid conservation is largely confined to two cytoplasmic domains that participate in catalysis. The extensive sequence diversity first suggested that distinct AC isoforms would be differentially regulated through cross-talk with other signal transduction systems in the cell, and this has been borne out by numerous studies in the field. It is the ability of a single type of AC to elicit a unique integrated response to multiple stimuli that begins to provide a physiologic rationale for the molecular diversity of this enzyme family.

THE ADENYLYL CYCLASE FAMILY

G protein–coupled signal transduction systems are the most widely utilized signaling pathways in eukaryotes. These three component systems (receptor, G protein, and effector) were originally defined through their involvement in the hormonal regulation of AC, and the study of the latter enzyme is representative of the entire G protein–linked family of effectors (for other reviews see refs. 1 and 4–7). The β-receptor, G_s, AC system was one of the first examples in which highly purified protein preparations were combined in phospholipid vesicles to reconstitute second messenger synthesis from just three components (8,9). This served as the most rigorous proof that no additional proteins are required for functional activity of these systems. However, if activation by the stimulatory G protein, G_s, were the only critical mechanism that could regulate AC activity, then a single ubiquitous form of the enzyme could serve this fundamental role.

Before any molecular biologic studies of this enzyme family, biochemists had purified three distinct classes of AC: (1) a ubiquitous membrane-associated enzyme stimulated by G_s; (2) a calmodulin (CaM)- and G_s-stimulated, membrane-associated enzyme abundant in brain; and (3) a CaM-stimulated, soluble enzyme from testes (10–15). The compound forskolin activates all the membrane-associated forms of the enzyme, and the development of a forskolin–Sepharose affinity support (16,17) facilitated the isolation of sufficient quantities of bovine brain AC to allow the direct determination of the sequence of 14 tryptic peptides (18). Serendipitously, the longest tryptic peptide included amino acids from one of the few well-conserved regions in all ACs, and an oligonucleotide "guessmer" based on its sequence was used to isolate complementary deoxyribonucleic acid (cDNA) clones for AC1–4 by screening only three different libraries (18–21). This was the first indication of the extent of the underlying molecular diversity of this enzyme family.

Full-length cDNAs encoding nine distinct mammalian ACs have been isolated by molecular cloning (18–35), and it seems reasonable to expect that additional forms may be identified in specialized cell types. The cDNAs have been obtained from multiple species: bovine and human type I (18,36); rat types II, III, and IV (19–21); canine, rat, and rabbit type V (22,25,26,35); rat, canine, and mouse type VI (23–25,27); mouse and human type VII (30,31); rat and human type VIII (29,32,37); and mouse type IX (33,34). Type VI and type VII messages were both detected in ribonucleic acid (RNA) from a clonal S49 mouse lymphoma cell line, implying that a single cell can express more than one isoform of AC (27). For at least types V, VI, and VIII, additional protein variants arise from alternative splicing of the messages (discussed later). All nine of the ACs have been detected in both humans and rats, indicating that the different isoforms are not simply species variants. The genes are not clustered in the genome, because each of the nine human AC genes has been mapped to a distinct chromosome except for types VII and IX, which are on different arms of chromosome 16 (33,36,38–41).

HOMOLOGY AMONG THE ADENYLYL CYCLASES

Figure 1 shows an alignment of the protein sequences of ACs I–IX with conserved regions highlighted in either gray (similar) or black (identical). A dendrogram derived from the alignment indicates that the ACs can be divided into subfamilies based on relative amino acid similarity between isoforms (Fig. 2). Types V and VI are the most similar of the distinct gene products (69% identity, 78% similarity) and form a subfamily. Types II, IV, and VII form another distinct subfamily (52–57% identity, 65–70% similarity), whereas types I, III, VIII, and IX each constitute a separate branch in the dendrogram. Alternative splicing of the type VIII message generates additional variants that could be included in its subfamily (37). Type IX is at most 34% identical to any other AC (type IV) and is the most distant relative in the enzyme family. When crossing mammalian species, sequence conservation of a given isoform is high (e.g., 97% identity between rat and human type VIII [32,37]), indicating that the source of the cDNA clones has not significantly influenced the division of the ACs into subfamilies.

STRUCTURAL MODEL

Mammalian ACs share a complex topology based on secondary structure analysis and sequence similarity, although experimental evidence supporting the structural model is modest (Fig. 3) (6,18,42). The proteins are predicted to have a variable-length amino-terminal cytoplasmic tail (N) followed by two large alternating sets of hydrophobic and hydrophilic domains. Each large hydrophobic domain (M_1 and M_2) includes six transmembrane spans based on hydropathy plots that assume the spans form α-helices. The two large hydrophilic regions (C_1 and C_2) are approximately 55% similar to one another over a region of approximately 275 amino acids, and each was originally predicted to include a nucleotide-binding domain based on primary sequence similarity to the catalytic domain of the guanylyl cyclases (18,43). This structural model is reminiscent of that proposed for the product of the multidrug resistance gene, the product of the cystic fibrosis gene, various transporters, and specific subunits of voltage-gated channels, although the ACs do not share regions of significant sequence similarity with these other proteins (18). Attempts to demonstrate channel or transport activity have so far failed with the mammalian ACs. Interestingly, AC activity in *Paramecium* is regulated by hyperpolarization. An enzyme that is both an AC and a voltage-independent K^+-channel has been purified from that organism, but its amino acid sequence has not yet been reported (44). A voltage-sensitive AC activity in cultured rat neurons has also been described, although in this case the enzyme is synergistically activated by depolarizing agents and the G_s pathway (45).

The amino-terminal cytoplasmic N domain varies in length and sequence; though short, conserved regions precede the first transmembrane span (see Fig. 1). The

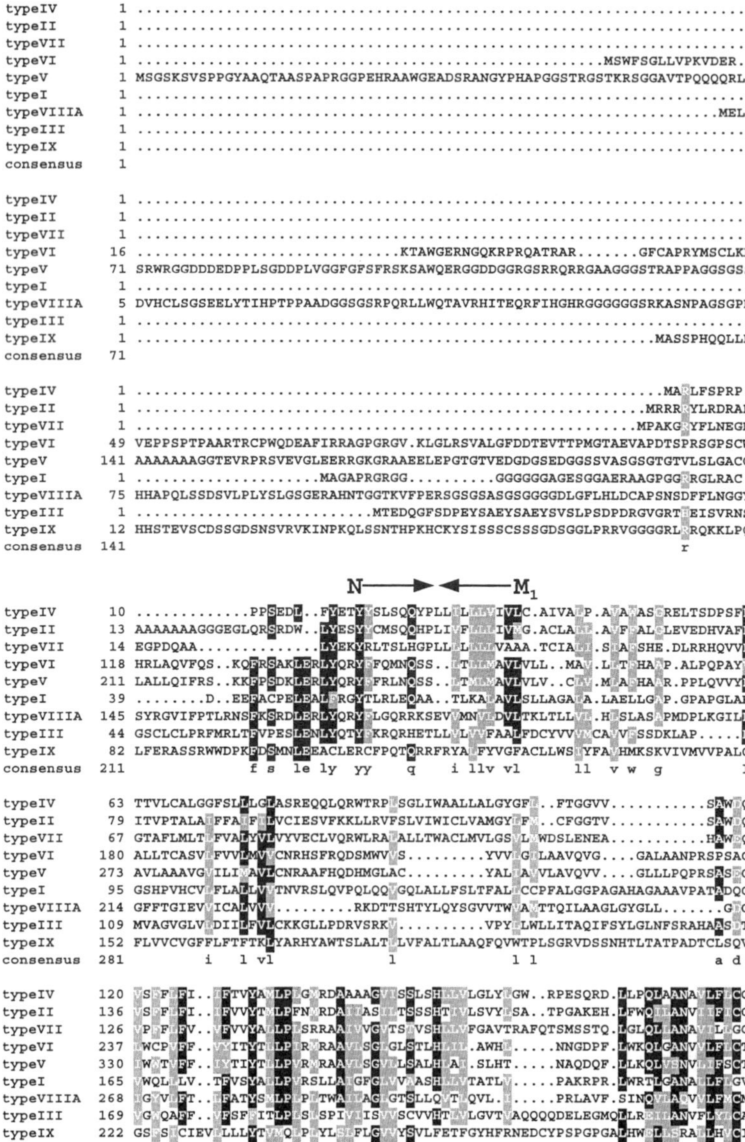

FIG. 1. Amino acid sequence alignment of adenylyl cyclases I–IX. The sequences used are bovine type I (18), rat types II–VI and VIII-A (19–21,27,29,35), and mouse types VII and IX (30,34). The alignment was obtained by using the program Pileup in the Wisconsin Genetics Computer Group's sequence analysis package (Madison, WI) (gap weight = 3.0 and gap length weight = 0.1). The highlighting of conserved regions and the consensus sequence were obtained with the program Boxshade from the Bioinformatics Group at the Swiss Institute for Experimental Research (ISREC, http://ulrec3.unil.ch/software/Box_form.html). At least six of the nine sequences must be either identical (black shading) or similar (gray shading) if the position is highlighted and an entry is included in the consensus sequence. If all nine AC sequences are identical at a given position, then the entry in the consensus is in uppercase. The labeled arrows point to positions in the alignment that form approximate boundaries between the large domains described in Fig. 3. These boundaries have been estimated by inspection based on marked changes in either the chemical nature of the predominant amino acid side chains, or the extent of sequence conservation. (*Figure continues.*)

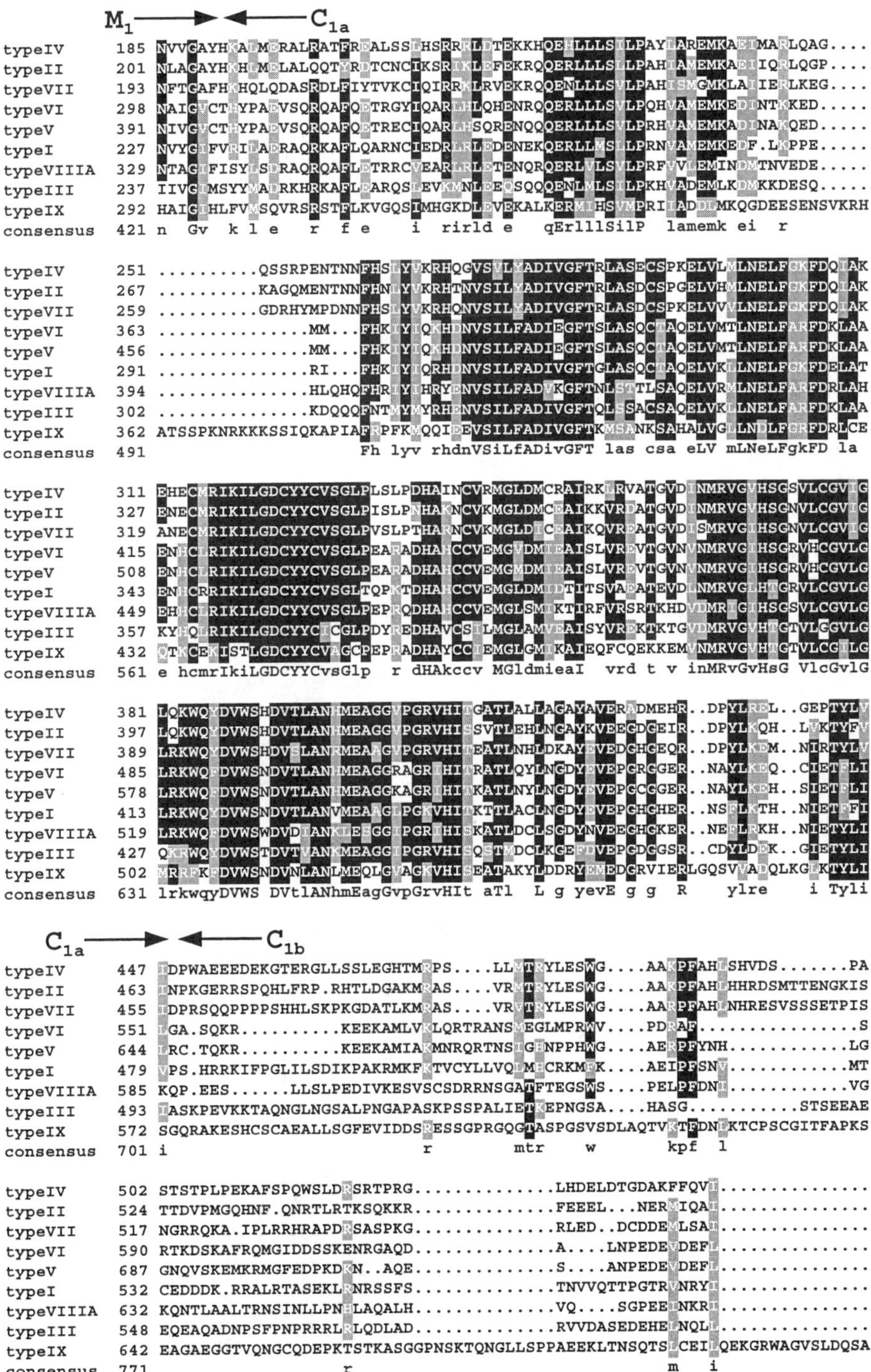
M1 → ← C1a
typeIV   185 NVVGAYHKAIMDRALRATFRDALSSIHSRRFLDTEKKHQEHLLLSILPAYLAREMKAEIMARLQAG....
typeII   201 NLAGAYHKHIMDLALQQTYRDTCNCIKSRIKLDFEKRQQERLLLSILPAHIAMEMKAEIIQRLQGP....
typeVII   193 NFTGAFHKHQLQDASRDLFIYTVKCIQIRRLRVEKRQQENLLLSVLPAHISMGMKLAIIERLKEG....
typeVI   298 NAIGVCTHYPAFVSQRQAFQDTRGYIQARLHLQHENRQQERLLLSVLPQHVAMEMKEDINTKKED.....
typeV   391 NIVGVCTHYPAFVSQRQAFQDTRECIQARLHSQRENQQQERLLLSVLPRHVAMEMKEDINAKQED.....
typeI   227 NVYGIFVRILADRAQRKAFLQARNCIEDRLRLDENEKQERLLMSILPRNVAMEMKEDF.LKPPE.....
typeVIIIA 329 NTAGIFISYLSDRAQRQAFLDTRRCVEARLRLDTENQRQERLVLSVLPRFVVLEMINDMTNVEDE.....
typeIII   237 IIVGIMSYYVADRKHRKAFLDARQSLEVKMNLDEOSQQQENLMLSILPKHVADEMLKDMKKDESQ.....
typeIX   292 HAIGIHLFVVSQVRSRSTFLKVGQSIMHGKDLDVEKALKERMIHSVMPRIIADDLMKQGDEESENSVKRH
consensus 421 n Gv   k l e   r f e   i rirld e   qErlllSilP lamemk ei r

typeIV   251 ...........QSSRPENTNNFHSLYVKRHQGVSVLYADIVGFTRLASECSPKELVLMLNELFGKFDQIAK
typeII   267 ...........KAGQMENTNNFHNLYVKRHTNVSILYADIVGFTRLASDCSPGELVHMLNELFGKFDQIAK
typeVII   259 ...........GDRHYMPDNNFHSLYVKRHQNVSILYADIVGFTRLASDCSPKELVVVLNELFGKFDQIAK
typeVI   363 ....MM...FHKIYIQKHDNVSILFADIEGFTSLASQCTAQELVMTLNELFARFDKLAA
typeV   456 ....MM...FHKIYIQKHDNVSILFADIEGFTSLASQCTAQELVMTLNELFARFDKLAA
typeI   291 ....RI...FHKIYIQRHDNVSILFADIVGFTGLASQCTAQELVKLLNELFGKFDELAT
typeVIIIA 394 ..............HLQHQFHRIYIHRYDNVSILFADVKGFTNLSTTLSAQELVRMLNELFARFDRLAH
typeIII   302 ................KDQQQFNTMYMYRHDNVSILFADIVGFTQLSSACSAQELVKLLNELFARFDKLAA
typeIX   362 ATSSPKNRKKKSSIQKAPIAFRPFKMQQIDEVSILFADIVGFTKMSANKSAHALVGLLNDLFGRFDRLCE
consensus 491   Fh lyv rhdnVSiLfADivGFT las csa eLV mLNeLFgkFD la

typeIV   311 EHECMRIKILGDCYYCVSGLPLSLPDHAINCVRMGLDMCRAIRKLRVATGVDINMRVGVHSGSVLCGVIG
typeII   327 ENECMRIKILGDCYYCVSGLPISLPNHAKNCVKMGLDMCDAIKKVRDATGVDINMRVGVHSGNVLCGVIG
typeVII   319 ANECMRIKILGDCYYCVSGLPVSLPTHARNCVKMGLDICDAIKQVRDATGVDISMRVGIHSGNVLCGVIG
typeVI   415 ENHCLRIKILGDCYYCVSGLPEARADHAHCCVEMGVDMIEAISLVREVTGVNVNMRVGIHSGRVHCGVLG
typeV   508 ENHCLRIKILGDCYYCVSGLPEARADHAHCCVEMGMDMIEAISLVREVTGVNVNMRVGIHSGRVHCGVLG
typeI   343 ENICRRIKILGDCYYCVSGLTQPKTDHAHCCVEMGLDMIDTITSVAEATEVDLNMRVGLHTGRVLCGVLG
typeVIIIA 449 EHHCLRIKILGDCYYCVSGLPEPRQDHAHCCVEMGLSMIKTIRFVRSRTKHDVDMRIGIHSGSVLCGVLG
typeIII   357 KYHQLRIKILGDCYYCVCGLPDYREDHAVCSILMGLAMVEAISYVREKTKTGVDMRVGVHTGTVLGGVLG
typeIX   432 QTKCEKISTLGDCYYCVAGCPEPRADHAYCCIEMGLGMIKAIEQFCQEKKEMVNMRVGVHTGTVLCGILG
consensus 561 e hcmrIkiLGDCYYCvsGlp r dHAkccv MGldmieaI vrd t v inMRvGvHsG VlcGvlG

typeIV   381 LQKWQYDVWSHDVTLANHMEAGGVPGRVHITGATLALLAGAMAVERADMEHR..DPYLREL..GEPTYLV
typeII   397 LQKWQYDVWSHDVTLANHMEAGGVPGRVHISSVTLEHLNGAYKVEEGDGEIR..DPYLKQH..LVKTYFV
typeVII   389 LRKWQYDVWSHDVSLANRMEAAGVPGRVHITEATLNHLDKAYEVEDGHGEQR..DPYLKEM..NIRTYLV
typeVI   485 LRKWQFDVWSNDVTLANHMEAGGRAGRIHITRATLQYLNGDYEVEPGRGGER..NAYLKEQ..CIETFLI
typeV   578 LRKWQFDVWSNDVTLANHMEAGGKAGRIHITKATLNYLNGDYEVEPGCGGER..NAYLKEH..SIETFLI
typeI   413 LRKWQYDVWSNDVTLANVMEAAGLPGKVHITKTTLACLNGDYEVEPGHGHER..NSFLKTH..NIETFFI
typeVIIIA 519 LRKWQFDVWSWDVDIANKLESGGIPGRIHISKATLDCLSGDYNVEEGHGKER..NEFLRKH..NIETYLI
typeIII   427 QKRWQYDVWSTDVTVANKMEAGGIPGRVHISQSTMDCLKGEFDVEPGDGGSR..CDYLDEK..GIETYLI
typeIX   502 MRRFKFDVWSNDVNLANLMEQLGVAGKVHISEATAKYLDDRYEMEDGRVIERLGQSVVADQLKGLKTYLI
consensus 631 lrkwqyDVWS DVtlANhmEagGvpGrvHIt aTl L g yevE g g R   ylre   i Tyli

C1a → ← C1b
typeIV   447 IDPWAEEEDEKGTERGLLSSLEGHTMRPS....LLMTRYLESWG....AAKPFAHLSHVDS.......PA
typeII   463 INPKGERRSPQHLFRP.RHTLDGAKMRAS....VRMTRYLESWG....AAKPFAHLHHRDSMTTENGKIS
typeVII   455 IDPRSQQPPPPSHHLSKPKGDATLKMRAS....VRVTRYLESWG....AARPFAHLNHRESVSSSETPIS
typeVI   551 LGA.SQKR..........KEEKAMLVKLQRTRANSEGLMPRWV....PDRAF................S
typeV   644 IRC.TQKR..........KEEKAMIAKMNRQRTNSIGHNPPHWG....AERPFYNH............LG
typeI   479 VPS.HRRKIFPGLILSDIKPAKRMKFKTVCYLLVQIMHCRKMFK....AEIPFSNV............MT
typeVIIIA 585 KQP.EES......LLSLPEDIVKESVSCSDRRNSGATFTEGSWS....PELPFDNI............VG
typeIII   493 IASKPEVKKTAQNGLNGSALPNGAPASKPSSPALIETKEPNGSA....HASG...........STSEEAE
typeIX   572 SGQRAKESHCSCAEALLSGFEVIDDSRESSGPRGQGTASPGSVSDLAQTVKTFDNLKTCPSCGITFAPKS
consensus 701 i   r   mtr   w   kpf l

typeIV   502 STSTPLPEKAFSPQWSLDRSRTPRG..............LHDELDTGDAKFFQVI...........
typeII   524 TTDVPMGQHNF.QNRTLRTKSQKKR..............FEEEL...NERMIQAI...........
typeVII   517 NGRRQKA.IPLRRHRAPDRSASPKG..............RLED..DCDDEMLSAI...........
typeVI   590 RTKDSKAFRQMGIDDSSKENRGAQD...............A....LNPEDEVDEFI...........
typeV   687 GNQVSKEMKRMGFEDPKDKN..AQE...............S....ANPEDEVDEFI...........
typeI   532 CEDDDK.RRALRTASEKLRNRSSFS..............TNVVQTTPGTRVNRYI...........
typeVIIIA 632 KQNTLAALTRNSINLLPNHLAQALH..............VQ....SGPEEINKRI...........
typeIII   548 EQEAQADNPSFPNPRRRLRLQDLAD..............RVVDASEDEHELNQLL...........
typeIX   642 EAGAEGGTVQNGCQDEPKTSTKASGGPNSKTQNGLLSPPAEEKLTNSQTSLCEILQEKGRWAGVSLDQSA
consensus 771   r   m i

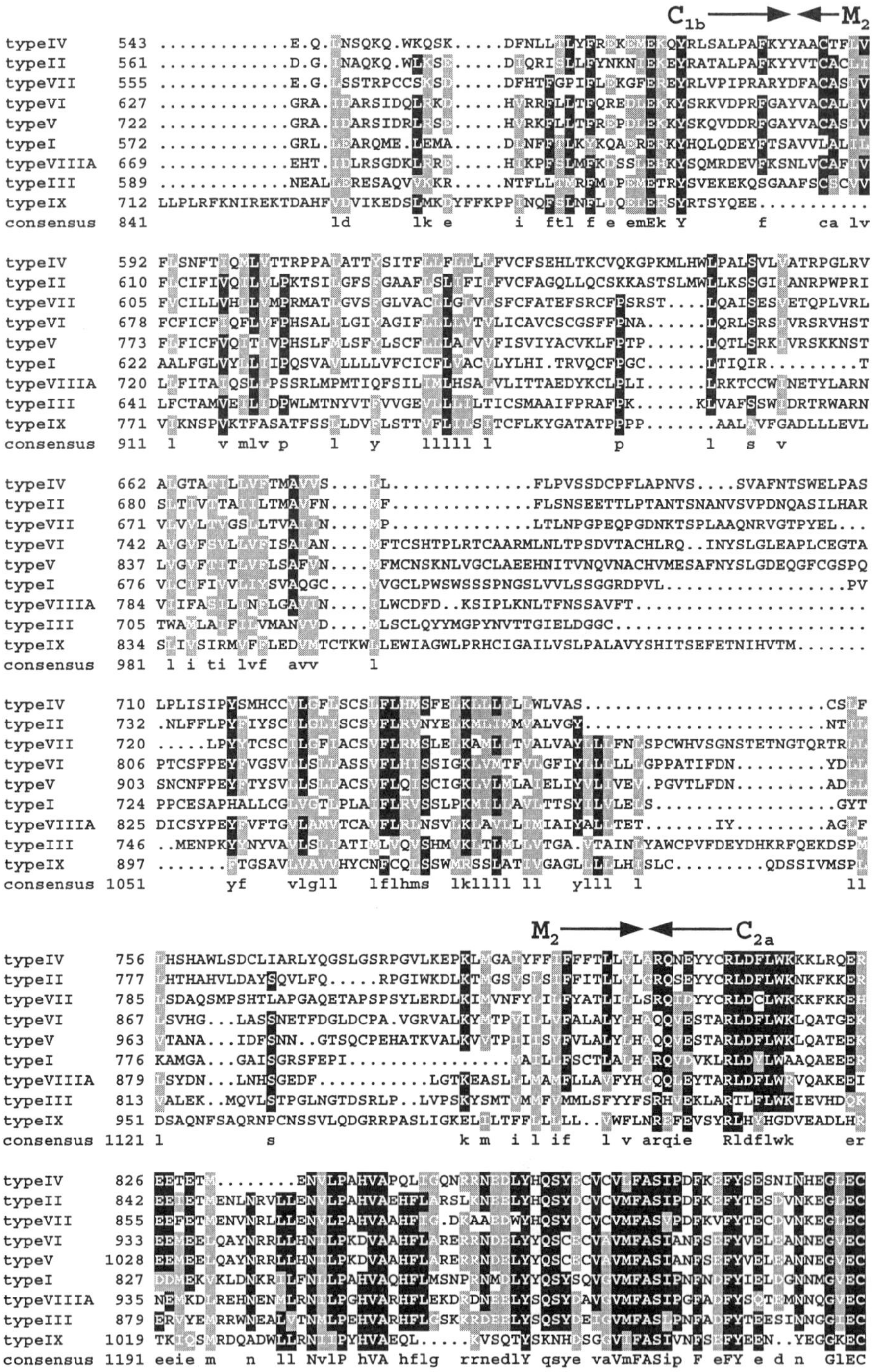

```
                                                          C₁b ─────────►  ◄──── M₂
typeIV     543 ...............E.Q.NSQKQ.WKQSK.....DFNLLLYFREKEMEQYRLSALPAFKYYAACTFIV
typeII     561 ...............D.G.NAQKQ.WLSE.....DIQRISLLFYNKNIEKEYRATALPAFKYYVTCACLI
typeVII    555 ...............E.G.SSTRPCCSKSD....DFHTFGPIELKGFEREYRLVPIPRARYDFACASIV
typeVI     627 ...............GRA.IDARSIDQLRKD....HVRRFLLTFQREDLEKYSRKVDPRFGAYVACALLV
typeV      722 ...............GRA.IDARSIDRLRSE....HVRKFLLTFRPDLEKYSKQVDDRFGAYVACASIV
typeI      572 ...............GRL.IDARQME.LEMA....DINFFLKYKQAREKYHQLQDEYFTSAVVLALIL
typeVIIIA  669 ...............EHT.IDLRSGDKLRRE....HIKPFLMFKDSSLEKYSQMRDEVFKSNLVCAFIV
typeIII    589 ...............NEALIERESAQVVKKR.....NTFLLTMRFMDPEMETRYSVEKEKQSGAAFSCSVV
typeIX     712 LLPLRFKNIREKTDAHFVDVIKEDSLMKDYFFKPPINQFSLNFLDQELERSYRTSYQEE...........
consensus  841                ld       lke e       i  ftl f e emEk Y          f       ca lv

typeIV     592 FISNFTIQMLVTTRPPAATTYSITFIFITLLFVCFSEHLTKCVQKGPKMLHWLPALSVLVATRPGLRV
typeII     610 FICIFIVQILVLPKTSILGFSEGAAFISLIFILFVCFAGQLLQCSKKASTSLMWLLKSSGIIANRPWPRI
typeVII    605 FVCILLVHLLVMPRMATIGVSEGLVACILGIVSFCFATEFSRCFPSRST....LQAISESVETQPLVRL
typeVI     678 FCFICFIQFLVFPHSALILGIYAGIFILLLVTYLICAVCSCGSFFPNA......LQRLSRSIVRSRVHST
typeV      773 FLFICFVQTIVPHSLFMLSFYLSCFILLAIVYFISVIYACVKLFPTP......LQTLSRKIVRSKKNST
typeI      622 AALFGLVYLIIPQSVALLLLVFCICFLVACVLYLHI.TRVQCFPGC......LTIQIR.........T
typeVIIIA  720 LIFITAIQSLIPSSRLMPMTIQFSILIMLHSAIVLITTAEDYKCLPLI......LRKTCCWINETYLARN
typeIII    641 LFCTAMVEILIDPWLMTNYVTVVVGEVLLLILTICSMAAIFPRAFPK......KLVAFSSWIDRTRWARN
typeIX     771 VIKNSPVKTFASATFSSILDVELSTTVFLIISITCFLKYGATATPPPP.......AALIVFGADLLLEVL
consensus  911 l      v mlv p     l    y        lllll l                    p        l   s   v

typeIV     662 AIGTATILLVFTMAVVS.....IL...............FLPVSSDCPFLAPNVS....SVAFNTSWELPAS
typeII     680 SLTIVTTAIILTMAVFN....IF................FLSNSEETTLPTANTSNANVSVPDNQASILHAR
typeVII    671 VLVVLTVGSLLTVAIIN....IP................LTLNPGPEQPGDNKTSPLAAQNRVGTPYEL...
typeVI     742 AVGVFSVLLVFISAIAN....FTCSHTPLRTCAARMLNLTPSDVTACHLRQ..INYSLGLEAPLCEGTA
typeV      837 LVGVFTITLVFLSAFVN....FMCNSKNLVGCLAEEHNITVNQVNACHVMESAFNYSLGDEQGFCGSPQ
typeI      676 VLCIFIVLIYSVAQGC....VVGCLPWSWSSSPNGSLVVLSSGGRDPVL.................PV
typeVIIIA  784 VIIFASILNFLGAVIN....ILWCDFD..KSIPLKNLTFNSSAVFT...............
typeIII    705 TWAMLAIFILVMANVVD....MLSCLQYYMGPYNVTTGIELDGGC...............
typeIX     834 SLIVSIRMVFELEDVMTCTKWLLEWIAGWLPRHCIGAILVSLPALAVYSHITSEFETNIHVTM.......
consensus  981 l  i ti  lvf    avv         l

typeIV     710 LPLISIPYSMHCCVLGFLSCSLFLHMSFELKLLLLLWLVAS........................CSLF
typeII     732 .NLFFLPYFIYSCILGLISCSVFLRVNYELKMLIMMVALVGY.......................NTLL
typeVII    720 .....LPYYTCSCILGFIACSVFLRMSLELKAMLLTVALVAYLLFNLSPCWHVSGNSTETNGTQRTRLL
typeVI     806 PTCSFPEYFVGSVLLSLLASSVFLHISSIGKLVMTFVLGFIYLLLLLGPPATIFDN.........YDLL
typeV      903 SNCNFPEYFTYSVLLSLLACSVFLQISCIGKLVLMLATELIYVLIVEV.PGVTLFDN.........ADLL
typeI      724 PPCESAPHALLCGLVGTLPLAIFLRVSSLPKMILIAVTTSYILVLELS................GYT
typeVIIIA  825 DICSYPEYFVFTGVLAMVTCAVFLRLNSVIKIAVLLIMIAIYALLTET......IY........AGLF
typeIII    746 ..MENPKYYNYVAVLSLIATIMLVQVSHMVKLTLMLLVTGA.VTAINYAWCPVFDEYDHKRFQEKDSPM
typeIX     897 .......FTGSAVLVAVHYCNFCQLSSWMRSSLATVGAGLLLLHLSLC.........QDSSIVMSP
consensus 1051        yf   vlgll  lflhms  lklllll  ll   ylll   l                    ll

                                         M₂ ─────────►  ◄──── C₂a
typeIV     756 IHSHAWLSDCLIARLYQGSLGSRPGVLKEPKLMGATYFFIFFFTLLVLARQNEYYCRLDFLWKKKLRQER
typeII     777 IHTHAHVLDAYSQVLFQ.....RPGIWKDLKTMGSVSLSIFFITLLVLCRQSEYYCRLDFLWKNKFKKER
typeVII    785 ISDAQSMPSHTLAPGAQETAPSPSYLERDLKIMVNFYLILFYATLILLSRQIDYYCRLDCLWKKKFKKEH
typeVI     867 ISVHG...LASSNETFDGLDCPA.VGRVALKYMTPVILLFALALYIHAQQVESTARLDFLWKLQATGEK
typeV      963 VTANA...IDFSNN.GTSQCPEHATKVALKVVTPIIISFVLALYIHAQQVESTARLDFLWKLQATEEK
typeI      776 KAMGA...GAISGRSFEPI...............AILIFSCTLALHARQVDVKLRLDYLWAAQAEEER
typeVIIIA  879 ISYDN...LNHSGEDF.........LGTKEASLILMAIFLLAVFYHCQQLEYTARLDFLWRVQAKEEI
typeIII    813 VALEK..MQVLSTPGLNGTDSRLP..LVPSKYSMTVMIFVMMLSFYYFSRHVEKLARTLFLWKIEVHDQV
typeIX     951 DSAQNFSAQRNPCNSSVLQDGRRPASLIGKELILTFFLLLL..VWFLNRFFEVSYRLHVHGDVEADLH
consensus 1121 l      s         k m  il if  l v arqie Rldflwk       er

typeIV     826 EETETM.......ENVLPAHVAPQLIGQNRRNEDLYHQSYECVCVLFASIPDFKEFYSESNINHEGLEC
typeII     842 EEIETIENLNRVLLENVLPAHVAEHFLARSLKNEELYHQSYDCVCVMFASIPDFKEFYTESDVNKEGLEC
typeVII    855 EEFETMENVNRLLLENVLPAHVAAHFIGDKAAFDWYHQSYDCVCVMFASVPDFKVFYTECDVNKEGLEC
typeVI     933 EEVEEIQAYNRRLLHNILPKDVAAHFLARERRNDELYYQSCECVAVMFASIANFSEFYVELEANNEGVEC
typeV     1028 EEVEEIQAYNRRLLHNILPKDVAAHFLARERRNDELYYQSCECVAVMFASIANFSEFYVELEANNEGVEC
typeI      827 DDMEKVKLDNKRILFNILPAHVAQHFLMSNPRNMDLYYQSYSQVGVMFASIPNFNDFYIELDGNNMGVEC
typeVIIIA  935 NELKDIREHNENVLRNILPGHVARHFLEKDRDNEELYSQSYDAVGVMFASIPGFADFYSQTEMNNQGVEC
typeIII    879 ERVYERRWNEALVTNMLPEHVARHFLGSKKRDEELYSQSYDEIGVMFASIPNFADFYTEESINNGGIEC
typeIX    1019 TKICSMRDQADWLLRNIIPYHVAEQL.....KVSQTYSKNHDSGCVIFASIVNFSEFYEEN..YEGGKEC
consensus 1191 eeie m    n  ll NvlP hVA hflg   rrnedlY qsye vaVmFASip F eFY e d n  GlEC
```

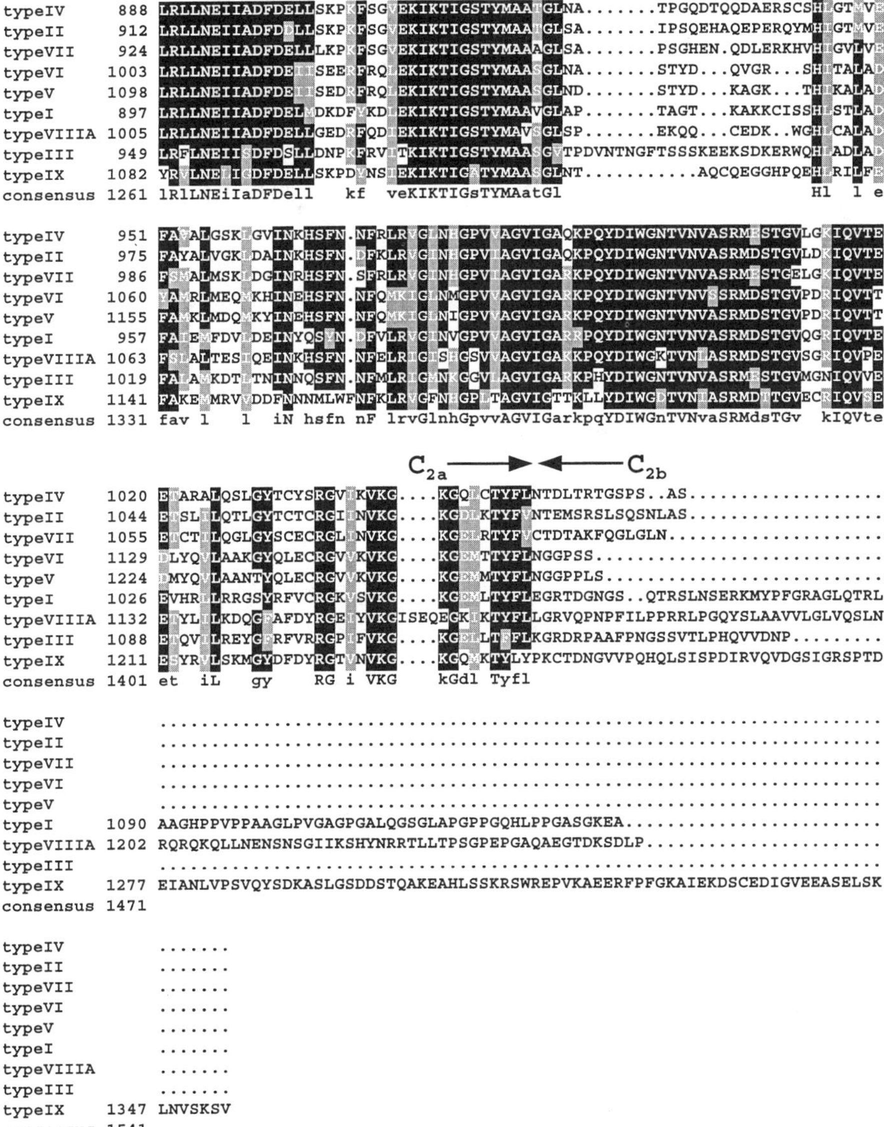

longest N domain is found in rat AC5 where it is 242 amino acids, and the shortest is only 28 amino acids in type IV (20,35). Variability in length and sequence of the N domain even occurs across species in the same AC isoform (22,26,35). Deletion of the N domain of type I dramatically decreases the specific activity of the enzyme (42). The same deletion in rat or canine type V has no effect on G_s- or forskolin-stimulated activity (35,46), although phorbol ester-dependent activation of the rat

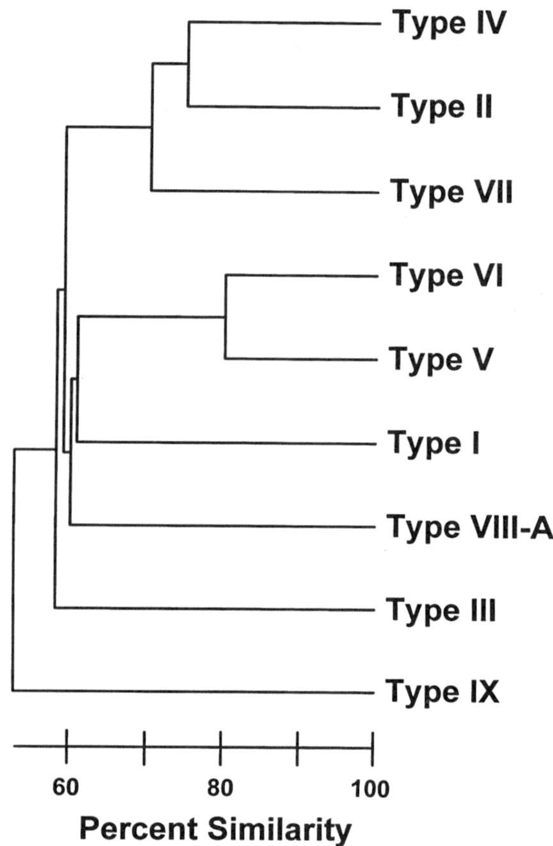

FIG. 2. A dendrogram of the alignment shown in Fig. 1 indicating the percent amino acid similarity found in sequence comparisons of AC types I–IX. The output is again from the program Pileup.

enzyme in intact cells is greatly diminished, implying that the N domain can confer type-specific properties on a given AC isoform (35).

The two large cytoplasmic domains are designated C_1 and C_2, and each can be further divided into "a" and "b" subdomains. The bulk of the sequence conservation resides in the C_{1a} and C_{2a} regions, which span approximately 275 positions in the alignment of Fig. 1. These subdomains each include only one region of five or more consecutive amino acids where there is not significant amino acid conservation shared by at least six of the nine AC isoforms (i.e., regions without shading in Fig. 1). C_{1a} and C_{2a} are homologous to each other (approximately 55% similar), to cognate regions in other isoforms, and to the catalytic domain of the guanylyl cyclases (18,43). Each of these conserved domains participates in catalysis based on studies of soluble ACs (47–50). The intracellular orientation of these catalytic domains within the complete protein (Fig. 3) is consistent with the observation that external stimuli increase intracellular cAMP concentrations. Neither C_{1b} nor C_{2b} is required

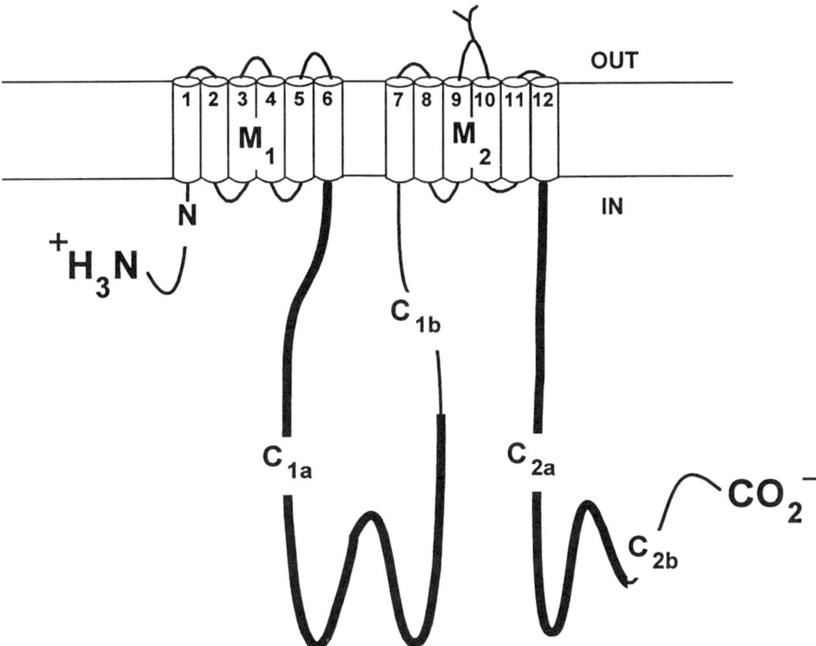

FIG. 3. Schematic representation of the topology shared by the mammalian adenylyl cyclases. The amino-terminal cytoplasmic domain, N, varies widely among the ACs and does not show any extended similarity to other protein families. There are six potential transmembrane-spanning domains, represented as cylinders, in each of the two large hydrophobic domains, M1 and M2. The two large cytoplasmic domains, C_1 and C_2, can each be divided into "a" and "b" subdomains. There is approximately 55% similarity between C_{1a}, C_{2a}, and the catalytic domain of the guanylyl cyclases. The second half of each large cytoplasmic domain, C_{1b} and C_{2b}, varies in a type-specific manner.

for regulation of enzyme activity by the common AC activators forskolin and $G_{s\alpha}$ (51), and each of these regions can vary in length and sequence (see Fig. 1). This suggested that these regions might contribute to type-specific regulation, and consistent with this proposal, the CaM-binding domain of type I has been mapped to C_{1b} (see later) (52–54).

Amino acid identity in M_1 and M_2 is limited, although a preponderance of hydrophobic residues underlies the predicted structural similarity of these regions (see Fig. 1). The essential role of the M domains in membrane association is supported by the observation that full-length ACs are always expressed in membrane fractions, whereas C_1 and C_2 are soluble polypeptides when expressed without M_1 and M_2 (47–50). If the transmembrane spans are simply identified by hydropathy analysis, then some of the connecting loops between helices are predicted to be quite short, and it is unreasonable to expect that they are of sufficient length to make a turn. This suggests that more extended, nonalpha helical conformations may be adopted in the native three-dimensional structure (18). The extracellular loop between transmembrane spans 9 and 10 (see Fig. 3) contains one or two consensus N-linked glycosylation sites (N-x-S/T) in all mammalian ACs except the type VIII-B splice variant and type IX (33,34,37). This is consistent with the results of purification studies that indicated that ACs can be glycoproteins (12,13) and the fact that a truncated type I protein consisting of only M_2 and C_2 is glycosylated when expressed (42). Additional extracellular N-linked glycosylation sites are found between transmembrane spans 11 and 12 in types III, V, VI, VII, VIII, and IX. Treatment with N-glycosidase F increases the electrophoretic mobility of types III, VIII-A, VIII-C, and IX, directly demonstrating that at least some of the consensus glycosylation sites are utilized in the full-length, active proteins (21,33,37). Verification of other aspects of the model, such as the precise boundaries defining the transmembrane spans, will require further study.

ALTERNATIVE SPLICING OF ADENYLYL CYCLASE MESSAGES

The unique type V-α "half molecule" was the first example of alternative splicing of an AC message to be described (55). This splice variant is predicted to encode a truncated form of type V consisting of the amino terminal half of the full-length enzyme. The message extends through what is an exon/intron boundary in the full-length type V precursor message and uses a polyadenylation signal encoded within this "intron" of the type V gene. This message is observed in canine heart RNA, but the protein has not yet been detected (55). Type V-α has no enzymatic activity when expressed alone; however, activity can be reconstituted by co-expression of an artificially generated type V C-terminal half-molecule. Similar experiments had been performed with the two halves of type I, which were also artificially generated (42). Unfortunately, there have not been any reports indicating that the messages encoding these other half-molecules are endogenously expressed, and this raises the question of the normal physiologic role for the V-α splice variant.

Full-length type V cDNAs have been cloned from canine, rabbit, and rat libraries (22,26,35). The rat and rabbit cDNAs encode proteins that are 95% identical overall (26,35). Both of these species variants differ considerably from canine type V in the N domain. For example, the rat enzyme is simply 78 amino acids longer than canine AC5 at the start of the protein. It has been reported that the canine and rat variants of type V are actually splice variants of a single gene and that each is expressed in both species (46). However, the presumably canine-specific probe used in those experiments is based on an incomplete rat cDNA sequence (25). Part of the probe is actually identical in 36 of 37 positions to a sequence in the complete rat cDNA (35), and this would be expected to hybridize across species under the conditions used for the RNA blotting analysis (46). A single genomic clone from the rabbit AC5 gene extends from putative promoter elements through a single exon encoding the N domain and part of the M_1 domain before an intron is found (26). Using either rat heart cDNA or rat genomic DNA as a template, amplification of a fragment including the region that is proposed to be alternatively spliced only generates a single product with the same sequence determined from the rat cDNA clone. If there is alternative splicing of the canine type V message in the region encoding the N-terminal domain, it must be species specific. The distinction is important functionally because the amino terminal cytoplasmic domain of rat type V is required to see phorbol ester–dependent activation of cAMP synthesis in an intact cell (35). cAMP synthesis by canine type V is stimulated less than twofold by phorbol esters when it is expressed in HEK-293 cells, whereas rat AC-V is activated up to ninefold in the same expression system. It is this regulatory mechanism that distinguishes AC5 from the closely related type VI.

Alternative splicing is not unique to the type V gene. A rat type VI cDNA clone has been reported (25) that differs from other canine, mouse, and rat AC-VI cDNAs in that it includes an additional exon immediately 5′ to the sequence that encodes the initiator methionine in the three other clones (23,24,27). The reading frame of the atypical clone remains open to its 5′ end, and the alternative exon encodes an additional methionine, indicating that the protein variants will be distinct. It remains possible that this variant cDNA is incomplete and that the true start site is even further upstream. Thus far, regulatory differences between the two type VI variants have not been described. An additional type VI messenger RNA (mRNA) with an alternatively spliced 3′ untranslated sequence has been discussed; however, its significance is unclear because it does not change the predicted amino acid sequence (27).

Alternative splicing of the type VIII message generates at least three variant proteins (37). Type VIII-A was initially identified simply as type VIII (29). This variant has the same topology predicted for all ACs. The other two messages are derived from the common type VIII-A precursor by an exon-skipping mechanism. The type VIII-B message is missing an exon encoding a 30-amino-acid extracellular domain that includes two consensus sites for N-linked glycosylation. The electrophoretic mobility of the type VIII-B variant is not altered by treatment with N-glycosidase F, contrary to that of types VIII-A or VIII-C (29,37). This is consistent with the existence of nonglycosylated forms of brain AC that have been identified by protein purification (12,13). One hypothesis is that type VIII-B might be differentially localized within a cell because a protein must proceed down a distinct intracellular pathway to be glycosylated, and could be targeted to a distinct membranous compartment as a result of this processing. The type VIII-C message is missing an exon encoding 66 amino acids in C_{1b}. This region is of interest because it is just on the carboxy-terminal side of the type VIII sequence that aligns with the CaM-binding domain of type I (52–54). The sequence of the latter domain is not conserved in type VIII, and type VIII-C is approximately four times more sensitive to stimulation by Ca^{2+}/CaM than either type VIII-A or -B (discussed later). Evidence consistent with the expression of the three variants in brain has been obtained at both the message and protein levels. The sequences of the variant messages have been verified by analyzing RT-PCR products amplified from rat brain RNA (37). The VIII-B variant has also been cloned from a bovine brain cDNA library; conservation across species suggests that these variants have distinct functional roles. Three species are detected by immunoblotting rat brain membranes with an antibody directed against a carboxy-terminal epitope common to the variants, consistent with the endogenous expression of all three proteins (37). Treatment with N-glycosidase F reduces all of them to a "single" species of approximately 125 kDa, which is exactly what is observed for the variants as expressed in HEK-293 cells (37).

LOCALIZATION OF ADENYLYL CYCLASES

Expression of the ACs has most frequently been examined at the message level because of difficulties in developing isoform-specific antibodies capable of detecting the low native protein concentrations in mammalian tissues. The messages for some forms (types IV, V, VI and IX) are broadly distributed, and others (types I, II, VII, and VIII) have a more limited distribution with types I and VIII expressed abundantly only in brain and type III being most abundant specifically in olfactory neuroepithelium (for a summary of the tissue distribution of AC mRNAs see ref. 7). However, there is not necessarily a simple correlation between endogenous concentrations of a protein and its message. A good example is illustrated by AC9, where the distribution has been examined at both the protein and mRNA levels. Whereas little or no type IX mRNA is detected in spleen or testes, significant immunoreactivity is present in both tissues. Conversely, AC9 mRNA is expressed most abundantly in skeletal muscle, but the protein is barely detected in muscle membranes relative to other tissues (33). Likewise, for type VIII, message is readily detectable by *in situ* hybridization analyses of rat brain in hippocampus and cerebral cortex, but only after prolonged exposure of autoradiographic emulsions in cerebellum (29). Immunoblots of membranes from hippocampus, cerebral cortex and cerebellum, however, show that these three regions express roughly equivalent amounts of type VIII protein (J.C. and J.K., unpublished observation). Variations in the tissue distribution and relative expression levels of the AC proteins need to be characterized in more detail to understand type-specific contributions to the physiologic regulation of cAMP concentrations.

Although ACs are typically assumed to be intrinsic membrane proteins, they need not be free to randomly diffuse over the entire cell surface. Immunofluorescence microscopy and cAMP imaging suggest that activation of AC activity by distinct agonists leads to different, highly localized changes in the distribution of cAMP within a cell (56,57). Immunohistochemical analysis of olfactory sensory neurons demonstrates that type III is confined to the cilia or sensory apparatus, where it is strategically positioned to respond to stimulatory olfactants (21). Also, a highly selective concentration of AC(s) at synapses of the CA1 and CA2 hippocampal fields in rat brain has been observed using an antibody raised against a domain common to all known mammalian ACs (58). This positions the ACs to respond rapidly to neurotransmitters and Ca^{2+} entry that can modulate enzymatic activity. Localization of ACs in the dendritic spines of postsynaptic neurons is also expected to facilitate the subsequent activation of cAMP-dependent protein kinase, because the latter enzyme is tethered to cytoskeletal elements by A kinase anchor proteins (AKAPs) in the postsynaptic densities (59). Co-localization of the signaling components is expected to enhance the efficiency of signal transduction.

Differential subcellular localization of ACs may also contribute to the specificity of receptor/effector coupling that has been observed in cardiac myocytes. A compartmentalization of cAMP within cardiac myocytes has been inferred from the differential ability of certain hormones to activate cAMP-dependent protein kinase. Brunton and co-workers have found that both isoproterenol and prostaglandin E_1 can stimulate cAMP accumulation in cardiac myocytes, but only the former agonist causes activation of phosphorylase kinase, glycogen phosphorylase, and inhibition of glycogen synthase (60). Activation of the particulate fraction of cAMP-dependent protein kinase is required to stimulate phosphorylase (61). The distinctions between different receptors have been further refined by the demonstration that although both β_1 and β_2 receptors are coupled to increases in cAMP in rat cardiac myocytes, only cAMP synthesized in response to activation of the former receptor is coupled to changes in contractility (62). Again this is correlated with an increase in the activation of the particulate fraction of cAMP-dependent protein kinase. Differences between particulate and soluble fractions of cAMP-dependent protein kinase can be explained by association of the enzyme with an AKAP that can tether the kinase to specific intracellular structures (59). However, if the ACs were free to diffuse in the membrane, both the soluble and particulate fractions of all intracellular compartments would be accessible to cAMP. One hypothesis is that different receptors couple to distinct populations of G proteins and cardiac ACs that are co-localized to discrete regions in the membrane where they interact with specific intracellular compartments. It will be of interest to define the mechanisms and the structural domains within the ACs that mediate subcellular targeting in various physiologic settings.

REGULATORY PROPERTIES

The sequence diversity of the AC family originally suggested that there would be significant regulatory diversity manifested through type-specific cross-talk with other signaling pathways, as will be documented later. These other pathways are also frequently controlled by G protein–coupled receptors, but the component that directly regulates the AC is not necessarily a subunit of a heterotrimeric G protein. An important point is that these findings have significantly changed our understanding of the hormonal control of physiologic processes because the regulation of AC activity by signal cross-talk can be at least as potent and efficacious as direct regulation by G protein subunits (4,5).

All the mammalian AC isoforms are stimulated by the diterpene forskolin and by the activated α subunit of the stimulatory G protein, G_s. Conserved sequences in C_1 and C_2 can account for these common properties (47,48,51). Marked differences are observed, however, in the effects of other regulators of AC activity. These include the G proteins G_{i1-3}, G_o, G_z, G protein $\beta\gamma$ subunits, Ca^{2+}, various protein kinases, and the protein phosphatase calcineurin. The characteristic response of a given cell type to stimuli that alter cAMP synthesis through one or more of these regulators will vary depending on what AC(s) is present and the unique manner in which that isoform integrates the signaling inputs.

Direct Regulation by G Protein α Subunits

The discovery of heterotrimeric G proteins came in the course of investigations into the mechanisms of hormone-regulated cAMP synthesis (1). The heterotrimeric G proteins consist of α, β, and γ subunits and couple receptor activation to effector regulation. Originally, the regulation of AC by G proteins was discussed in terms of stimulation by G_s (*s* for *stimulatory*) and inhibition by G_i (*i* for *inhibitory*), but now it is known that there are more than 20 G protein α subunits (including splice variants), at least 6 β subunits, and 12 γ subunits, implying that there are more than 1000 potential G protein heterotrimers. Many of these functional heterotrimers are capable of regulating cAMP concentrations, although they may exert their effects through indirect mechanisms. All G proteins undergo a common catalytic cycle. In the basal state, guanosine diphosphate (GDP) remains tightly bound to the α subunit, which is in turn associated with a $\beta\gamma$ heterodimer. The GDP-bound form of the heterotrimer can also associate with a receptor. Upon receptor activation the associated G protein α subunit exchanges GDP for GTP and dissociates from $\beta\gamma$. Both α-GTP and $\beta\gamma$ go on to regulate various effectors. Stimulation of the effector by the α subunit is terminated by hydrolysis of GTP caused by an intrinsic GTPase activity, and reassociation with $\beta\gamma$ follows. Although all AC isoforms tested to date are stimulated by the GTP-bound α subunit of G_s, considerable variations in response to $\beta\gamma$ and other GTP-bound α subunits have now been observed. More detailed discussions of this topic can be found in Chapter 4.

Inhibition of AC activity by $G_{i\alpha}$ subunits, $G_{o\alpha}$, and $G_{z\alpha}$ has been observed in intact cells and with membrane preparations (63–67). Type I is inhibited by $G_{i\alpha1}$, $G_{i\alpha2}$, $G_{i\alpha3}$, $G_{z\alpha}$, and $G_{o\alpha}$, but the potency of the α subunit varies and the extent of inhibition is highly dependent on the activator (50% for Ca^{2+}/CaM, 20% for forskolin, and a negligible inhibition of $G_{s\alpha}$-stimulated activity) (66). $G_{o\alpha}$ inhibits Ca^{2+}/CaM and forskolin-stimulated type I activity to an extent similar to the G_is, albeit with a 10-fold higher IC_{50}. In spite of this reduced potency, its effect is of potential physiologic significance, given that G_o constitutes approximately 1–2% of brain membrane protein (68), where AC1 is also expressed. $G_{s\alpha}$-stimulated type II activity is insensitive to $G_{o\alpha}$, and only a modest inhibitory effect of $G_{i\alpha1}$ is observed in membrane preparations (66). Co-expression of type II with a constitutively active mutant of $G_{i\alpha2}$ does result in inhibition of hormone-stimulated cAMP accumulation in intact cells (64). The reason for these differences remains to be elucidated. Forskolin-stimulated type II membrane activity is actually enhanced at high concentrations of $G_{i\alpha1}$, possibly because of its interaction at a regulatory site for G_s. Forskolin- and $G_{s\alpha}$-

stimulated type V and VI activities are most sensitive to inhibition by $G_{i\alpha1}$, $G_{i\alpha2}$, and $G_{i\alpha3}$ (66), although subsequent experiments have demonstrated that $G_{z\alpha}$ is actually a more potent inhibitor of type V than $G_{i\alpha1}$ (67). $G_{o\alpha}$ has no effect on types V and VI.

Inhibition of $G_{s\alpha}$, forskolin, and CaM-stimulated type I activity by G protein $\beta\gamma$ subunits is more pronounced than its inhibition by a $G_{i\alpha}$ or $G_{o\alpha}$ (66). Purified preparations of type I activated by $GTP\gamma S \cdot G_{s\alpha}$ are inhibited by $\beta\gamma$ subunits in detergent solution (69). This implies that inhibition probably involves a direct interaction between $\beta\gamma$ subunits and the AC, because $\beta\gamma$ subunits do not interact appreciably with $GTP\gamma S \cdot G_{s\alpha}$. This conclusion is supported by the observations that $\beta\gamma$ subunits also inhibit Ca^{2+}/CaM- or forskolin-stimulated type I in membrane preparations (42), and $\beta\gamma$-dependent inhibition of type I is not altered even if the enzyme is activated by a truncated $G_{s\alpha}$ that has reduced affinity for $\beta\gamma$ (66). The concentration dependence for $\beta\gamma$-dependent inhibition implies that the stimulatory effect of $GTP \cdot G_{s\alpha}$ should not be counteracted by the inhibitory effect of its released $\beta\gamma$ subunits because stimulation occurs at much lower concentrations, at least in detergent solutions. In the brain, the source of $\beta\gamma$ for inhibition of $G_{s\alpha}$-stimulated type I activity is expected to be a more abundant G protein such as G_o. In intact cells, β-adrenergic stimulation of type I is essentially undetectable under conditions where efficacious activation of types V and VI can be observed (23,27,64). It is possible that the presence of tightly bound detergent may be increasing the apparent IC_{50} for $\beta\gamma$-dependent inhibition observed in membrane preparations, and the effect may actually occur at relevant concentrations in an intact cell. Detergents can alter $\beta\gamma$-dependent effects on type I based on the fact that it was originally impossible to detect inhibition with AC1 that was purified in Lubrol (42), although the effect is observed in dodecyl maltoside (69).

G protein $\beta\gamma$ subunits stimulate the activity of types II and IV in contrast to their inhibition of type I (20,70,71). The effect is most pronounced on $G_{s\alpha}$-stimulated activity, although a modest stimulation of basal activity can be detected with purified protein preparations in detergent solutions (69). In a setting in which type II is endogenously expressed (72), it has been shown that activation of G_i-linked receptors potentiates β-adrenergic effects on the afterhyperpolarization of hippocampal pyramidal cells (73). This effect may reflect the conditional activation of AC2 by the combination of activated $G_{s\alpha}$ and $\beta\gamma$ subunits released from the G_i-linked receptors. This has been proposed as a mechanism to sensitize postsynaptic neurons to strong or repeated stimuli (73,74). The structural motif necessary for $\beta\gamma$-dependent interactions with AC2 includes amino acids 956 to 982, based on the observation that a peptide with this sequence inhibits several $\beta\gamma$-regulated processes (75). This same region has been shown to interact directly with β, but not with γ subunits when analyzed using the yeast two-hybrid system (76). The corresponding regions of types IV and VII are similar to that in AC2; the $\beta\gamma$-binding domain includes the one insert region in C_{2a} that is not conserved in all AC sequences, consistent with this subfamily-specific regulatory mechanism (see Fig. 1) (75). Although AC7 has not yet been shown to be activated by $\beta\gamma$ subunits in membrane preparations, data consistent with this mechanism have been obtained in intact cells (77). Given the broad tissue distribution of the type VII message, this suggests that conditional activation by $\beta\gamma$ subunits may be a widely utilized mechanism to modulate cAMP concentrations in response to coincident stimuli.

Stimulation by Ca^{2+}/Calmodulin

Type I and the three splice variants of type VIII are stimulated by Ca^{2+}/CaM *in vivo* and *in vitro* (37,42,78). Types I and VIII-C are similar in their sensitivity to stimulation (EC_{50} = 20 and 30 nM CaM, respectively) but differ from types VIII-A and VIII-B (EC_{50} = 140 and 116 nM CaM, respectively). Types I and VIII are synergistically activated by Ca^{2+}/CaM and $G_{s\alpha}$, making them targets for positive crosstalk between Ca^{2+} and G_s-signaling pathways (29,42). The CaM-binding domain of AC1 has been localized to the C_{1b} region (52–54). There is no significant sequence homology between the CaM-binding domain of type I and the corresponding sequence in type VIII-A, nor is this region in type VIII-A predicted to have the amphipathic structure typical of CaM-binding domains. If the aligned sequences of types I and VIII-A are compared (Fig. 1), one can see that the type I CaM-binding domain (amino acids 495–522) (52) terminates 12 amino acids before the start of the 66-amino-acid region that is missing from the C_{1b} region of type VIII-C (amino acids 635–700 in type VIII-A). Deletion of this region actually enhances the CaM sensitivity of VIII-C about fourfold relative to that of types VIII-A and VIII-B both in membranes and intact cells (37). This could be a conformational effect or the additional 66 amino acids in VIII-A, and VIII-B could play an inhibitory role (37). The different CaM sensitivities might reflect a need to allow the type VIII splice variants to differentially respond to localized changes in Ca^{2+} concentrations. Potential physiologic roles for the Ca^{2+}/CaM-stimulated ACs will be discussed in more detail later.

Ca^{2+}-Dependent Inhibition

Type III AC can be stimulated *in vitro* by Ca^{2+}/CaM; however, concomitant activation by a nonhydrolyzable GTP analog or forskolin is required (79). The physiologic relevance of this property must be re-evaluated in light of the demonstration that increasing intracellular Ca^{2+} concentrations inhibits glucagon- and isoproterenol-stimulated type III activity in intact cells (80). This effect is antagonized by inhibition of CaM kinase II and mimicked by co-expression of a constitutively active form of the kinase. The latter observation suggests that in olfactory sensory neurons where type III is abundantly expressed, the dominant effect of Ca^{2+} on this isoform will be inhibitory. This type of regulation may provide a mechanism for feedback inhibition of olfactant-stimulated type III activity. The G protein G_{olf} is activated when odorants bind to their receptors on the surface of olfactory sensory neurons (81). GTP-bound $G_{olf\alpha}$ then stimulates AC3 and the increase in intracellular cAMP concentrations leads to the opening of a cAMP-gated cation channel (82). Ca^{2+} entering through this channel would be expected to inhibit AC3 through CaM kinase II and contribute to adaptation and termination of the cAMP signal.

The same concentrations of Ca^{2+} that stimulate AC1 and AC8 (low micromolar range) inhibit types V and VI in a CaM-independent fashion (23,26,27,83) but have no effect on types II, IV, and VII in the presence or absence of CaM. This effect is distinct from the inhibition that occurs for all the ACs at submillimolar Ca^{2+}, which is thought to occur because Ca^{2+} competes with Mg^{2+} and binds to ATP, but the $Ca^{2+}\cdot ATP$ complex is a poor substrate for the catalytic site. There is no obvious Ca^{2+}-binding site in types V or VI, such as an E-F hand, and evidence for direct binding of Ca^{2+} in the low micromolar range is lacking. It remains possible that an unidentified cellular factor mediates the inhibitory effect of Ca^{2+}. Regardless of the precise molecular mechanism, inhibition of types V and VI by Ca^{2+} has been proposed to play an important physiologic role in the regulation of cardiac contractility. This will be discussed in more detail later.

Calcineurin-dependent dephosphorylation has been proposed as an additional mechanism by which Ca^{2+}-dependent inhibition of cAMP signaling may occur (34). When AC9 is expressed in COS7 or HEK-293 cells, cAMP synthesis is enhanced by FK506 and cyclosporin A, inhibitors of the Ca^{2+}/CaM-stimulated protein phosphatase 2B (PP2B, calcineurin) (34). It follows that agents that increase intracellular Ca^{2+} should inhibit cAMP synthesis through activation of PP2B in cells expressing the type IX isoform. This has not yet been demonstrated directly in cells transfected with AC9, but this mechanism does occur in the AtT20 cell line from which the type IX cDNA was cloned (84). Amino acids 504–611 in the C_1 region of the type IX sequence (see Fig. 1) are similar to the PP2B-binding protein, FKBP12 (34). It has been suggested that this might be a PP2B docking site, with the implication that the AC9/PP2B complex would actively dephosphorylate an as yet unidentified phosphoprotein that, in its phosphorylated state, must be positively coupled to cAMP synthesis (84). The possibility that the phosphoprotein is AC9 itself has not been excluded, and the identity of the kinase(s) responsible for the initial activation that is reversed by calcineurin has not been reported. Given the wide tissue distribution of the type IX protein (33), this implies that the antagonism between calcineurin and the cAMP signaling pathways should be a general phenomenon (84). This regulatory mechanism is discussed in more detail in Chapter 8.

Regulation by Protein Kinase C

Early reports on the effects of phorbol esters (direct activators of protein kinase C [PKC]) on AC activity indicated that they could either desensitize agonist-stimulated AC activity (85) or increase agonist-stimulated and basal activity (86,87), depending on the cell type examined. It is reasonable to expect that the effects of PKC will vary among the AC isoforms and that this could play a role in the differential responsiveness among cell types. Each of the cloned AC forms contains multiple consensus PKC phosphorylation sites; however, their locations are not conserved. PKC-dependent phosphorylation of an AC in frog erythrocytes and human platelets has been demonstrated (88,89), although these effects could be an indirect result of a phosphorylation cascade involving other kinases. Cyclic adenosine monophosphate synthesis can be stimulated indirectly by PKC-dependent mechanisms as well. Notably, direct phosphorylation of G_i by PKC results in its inactivation and can, by relief of AC inhibition, facilitate net stimulation (64,65).

Not surprisingly, PKC-dependent effects on ACs are type specific. Forskolin-stimulated activity of types I and III is enhanced by phorbol esters when the ACs are expressed in HEK-293 cells (90,91). Because the effects have not been duplicated in a reconstituted system with purified components and direct phosphorylation of AC1 or AC3 by PKC has not been demonstrated, the mechanism of this stimulation remains unclear. Effects of phorbol esters on types IV, VI, and VIII have not been observed (37,90). Phorbol ester–dependent stimulation is most pronounced in cells expressing type II, V, or VII, and these effects are described in more detail.

If cells overexpressing type II are treated with a phorbol ester, there is a marked increase in cAMP accumulation. Basal, G_s, and forskolin-stimulated activities are all enhanced by phorbol ester treatment, and the effect is antagonized by staurosporine, a potent PKC inhibitor (92–94). Phorbol ester–dependent incorporation of phosphate into recombinant AC2 parallels the increase in activity, consistent with a direct stimulatory role for PKC (95). A four-amino-acid region in the C-terminal portion of C_{2a} is required for phorbol ester–dependent stimulation of type II; however, these residues do not include a consensus PKC phosphorylation site (54). It has been suggested that the interaction between this region and the as yet unidentified PKC phosphorylation site(s) may be required to increase catalytic activity (54).

Type VII is activated by a phorbol ester–dependent pathway when expressed in HEK-293 cells like its subfamily member type II (30,31). However, the time course of activation differs considerably, depending on which AC is expressed (30). Phorbol ester treatment results in a transient increase in intracellular cAMP concentrations in AC7/HEK-293 cells, with a peak occurring at 4 minutes and a return to basal cAMP concentrations by 10 minutes (30). Cyclic adenosine monophosphate content increases gradually in response to PKC activation in AC2/HEK-293 cells. Maximal cAMP concentrations are achieved at 20 minutes and remain elevated for at least 1 hour (30,93). These results imply that cross-talk between the PKC and cAMP signaling pathways will regulate downstream events on distinct time scales depending on whether AC2 or AC7 has been activated. Synergistic stimulation of type VII is achieved by the combination of a phorbol ester and the β-adrenergic agonist, isoproterenol, presumably acting through $G_{s\alpha}$. A similar synergism between isoproterenol and agents that increase intracellular Ca^{2+} is observed; however, neither agent applied alone increases cAMP synthesis relative to control cells (30). This suggests that a Ca^{2+}-responsive PKC acts most effectively to stimulate type VII in consort with $G_{s\alpha}$. Although these results are consistent with PKC acting on type VII, direct phosphorylation has not yet been demonstrated.

Stimulation of type V by PKC has been demonstrated *in vitro* using purified PKC and AC isoforms (96). PKC α, a conventional, phorbol ester–activated and Ca^{2+}-activated isoform, and PKC ζ, an atypical, Ca^{2+}- and phorbol ester–insensitive form, can both phosphorylate and activate AC5. PKC ζ is a more potent stimulator of type V than α, the effects of the two are additive, and both act synergistically with forskolin or $G_{s\alpha}$. The additive nature of the stimulation by the two PKCs suggests that they phosphorylate distinct sites, and this is consistent with the phosphopeptide mapping, although the peptides were not sequenced (96). The *in vitro* findings are consistent with results *in vivo*, where phorbol ester treatment causes a staurosporine-inhibitable stimulation of cAMP accumulation in cells overexpressing AC5 (97). Co-expression of PKC ζ with type V causes an additional stimulation, indicating that type V can respond to both phorbol ester–sensitive and phorbol ester–insensitive PKCs in intact cells.

Phorbol ester–dependent stimulation of type V in intact cells emphasizes the differences between the rat and canine variants of the enzyme (35,97). When each is expressed in HEK-293 cells, there are significant quantitative differences in the responses to phorbol esters: cAMP content increases up to ninefold for rat type V, whereas canine type V is activated less than twofold (35,97). An incomplete rat AC5 cDNA has also been expressed in HEK-293 cells, and stimulation by phorbol esters is similar to what is observed with the canine enzyme (92). The species variants only differ significantly in the N domain, and when the first 244 amino acids of rat AC5 are deleted, forskolin-stimulated cAMP synthesis is not altered, but phorbol ester–dependent cAMP increases are reduced from ninefold to less than twofold (35). Based on the fact that the canine variant is potently activated by PKCs *in vitro*, it is unlikely that the N domain includes the PKC phosphorylation site(s). Instead, this region may be required for localization or recognition *in vivo*, although the precise mechanism is secondary to the fact that PKC efficaciously activates rat AC5 in intact cells.

In the heart, PKC-dependent activation of AC may be important in myocardial ischemia and preconditioning (98–100). Following 15 minutes of global ischemia there is a covalent modification, resulting in activation of an AC based on the fact that the cyclase can be purified away from G proteins while remaining in an activated state (101). Activation is mimicked by phorbol esters but occurs by a pathway that is independent of both α- and β-adrenergic agents, suggesting that a specific PKC isoform may mediate this response (101,102). Concomitant activation of both cAMP and PKC signaling pathways would be expected to have complex long-term effects on the regulation of gene expression, such as those required for the response to ischemia and reperfusion. The precise role of specific AC isoforms in ischemia remains to be determined.

Regulation by Protein Kinase A

It is logical to consider that AC(s) might be acted upon by protein kinase A (PKA), the major downstream target for cAMP. Either stimulation or inhibition of AC activity could theoretically result from phosphorylation by PKA, but only the latter has been documented thus far. Direct phosphorylation of purified AC5 by purified PKA has been demonstrated and shown to attenuate basal, forskolin-, and $G_{s\alpha}$-stimulated activity, with the effect being most pronounced on the latter two activities

(103). Type V activity is therefore subject to dual regulation by phosphorylation (also see Chapter 5). Phosphopeptide mapping shows that the residues in type V that are phosphorylated by PKA differ from those modified by PKC. Type VI is also inhibited by PKA, suggesting that the one consensus PKA phosphorylation site shared by AC5 and AC6 is modified by the kinase (103). The heterologous desensitization of glucagon-stimulated AC activity in hepatocytes may be an example of PKA-dependent inhibition in cells in which types V and VI are endogenously expressed (104). Consensus PKA phosphorylation sites are found in other ACs, but there are no data to indicate that they are utilized. It will be of interest to determine how common this type of regulation is and whether any of the isoforms are activated by PKA-dependent phosphorylation in a potential feed-forward mechanism.

PHYSIOLOGIC ROLES FOR ADENYLYL CYCLASES

As novel ACs are characterized, a knowledge of their regulatory properties can be used to gain an understanding of the role specific isoforms play in distinct physiological settings. This is illustrated by a discussion of the function of Ca^{2+}-inhibitable ACs in heart, and the role of Ca^{2+}/CaM-sensitive ACs in brain. The interested reader is also referred to Chapter 2, which provides a more extensive discussion of the cross-talk between the Ca^{2+} and cAMP signaling pathways.

Ca^{2+}-Inhibitable Adenylyl Cyclases in the Heart

Ca^{2+}-dependent inhibition of types V and VI is expected to have important ramifications for the regulation of cardiac contractility (105). Sympathetic innervation of the heart leads to the activation of β-adrenergic receptors by norepinephrine, and results in a well-documented increase in intracellular cAMP (106). Since L-type Ca^{2+} channels are a principal target for cAMP-dependent protein kinase in heart, the activation of either type V or VI by a β-agonist will lead to an increase in intracellular Ca^{2+} that will feed back and inhibit the activity of these cyclases. In fact, Ca^{2+}-dependent inhibition of AC activity has been measured in a cardiac myocyte model in which both types V and VI are expressed (107,108). Feedback inhibition mediated by PKA-dependent phosphorylation is also expected to attenuate the activity of types V and VI (103). As cAMP decreases and PKA is deactivated, the stimulus for Ca^{2+} entry is removed, and the sequestration of Ca^{2+} into intracellular stores returns the system to its initial state. Types V and VI AC could be reactivated as long as a β-agonist is still present, and the cycle would resume. Thus, an oscillatory pattern in intracellular cAMP would contribute to oscillations in Ca^{2+} concentrations that would affect the contractile apparatus and alter heart rate (105). Co-localization of the proteins involved is expected to facilitate the various interactions predicted by this regulatory scheme. The first observation of oscillations in cAMP concentrations was reported more than 20 years ago, when a rapid freezing method was used to fix preparations of frog ventricle at different stages in their relatively slow contraction cycle (109). The peak in cAMP concentration is achieved early in systole as twitch tension begins to rise. A mathematical model of this system has been described (5). The development of methods to simultaneously image both Ca^{2+} and cAMP in single cells will make it possible to re-examine this issue in detail (110).

Ca^{2+}/Calmodulin-Stimulated Adenylyl Cyclases in Long-Term Potentiation

The Ca^{2+}/CaM-stimulated ACs, types I and VIII, have been suggested to play a role in mammalian learning and memory based on their enzymatic properties, their abundant expression in the hippocampus, and analogies to *Aplysia* and *Drosophila* models. Characterization of the *Drosophila* mutant, *rutabaga*, provides the strongest evidence implicating a Ca^{2+}/CaM-stimulated AC in learning (111–113). A single-point mutation in the gene encoding a Ca^{2+}/CaM-stimulated AC renders the *rutabaga* fly incapable of mastering a simple olfactory learning task (113). Interestingly, both mammalian AC1 and AC8 are more closely related to the product of the *Drosophila rutabaga* locus than they are to any mammalian AC (29,113).

Long-term potentiation (LTP), a use-dependent enhancement of postsynaptic excitatory potentials, has been studied as a possible synaptic model of processes underlying memory formation in the mammalian hippocampus (114). One type of LTP is dependent on the class of glutamate receptors that can also be activated by N-methyl-D-aspartate (NMDA), because an antagonist specific for this receptor subclass blocks induction of the phenomenon (115). Membrane depolarization is required before this channel will allow Ca^{2+} to enter the dendritic spine of the postsynaptic neuron (116). The channel's selectivity for Ca^{2+} provides a link to CaM-sensitive ACs because activation of the receptor by either glutamate or NMDA is accompanied by an increase in cAMP in neurons within hippocampal field CA1 (117). Recent studies have demonstrated that cAMP has distinct roles in both the early and late stages of LTP. cAMP plays a "gating" role in the development of LTP in the CA1 field (118). This role is dependent upon cAMP that is synthesized postsynaptically in response to NMDA receptor activation and formation of a Ca^{2+}/CaM complex. Inhibitors of cAMP-dependent protein kinase prevent the development of early LTP, but the addition of cAMP analogs does *not* induce the expression of early LTP per se, implying a requisite regulatory role for the active state of the kinase (118). In a mouse model in which the AC1 gene has been disrupted by homologous recombination, both the rate at which LTP develops and the amount of Ca^{2+}/CaM-stimulated AC activity in hippocampal membranes are approximately half those in a normal mouse (119). These results may reflect the ability of type I to synthesize cAMP that plays a gating role in early LTP and the fact that other Ca^{2+}/CaM-stimulated ACs, such as type VIII, are expressed in the hippocampus (29,72,119,120). In the late stage of LTP, the increase in cAMP is expected to lead to the regulation of gene expression through activation of PKA and subsequent phosphorylation of the cAMP-response element binding (CREB) protein (121). Development of the late stage of LTP, which requires new protein synthesis, is blocked by inhibitors of PKA, and activators of this enzyme are sufficient to induce late LTP in the absence of electrical stimuli (122). G_s-dependent activation of AC(s) is also important for the development of late LTP, since a D_1 dopamine receptor antagonist inhibits this process (122). This suggests that any AC expressed in the hippocampus (29,72,119,120) may be capable of contributing to the development and expression of late LTP.

Other studies indicate that activation of a CaM-sensitive AC is important in the presynaptic regulation of nonassociative LTP that develops at the mossy fiber synapse (123). In this case, it is the depolarization-dependent activation of Ca^{2+} channels in the presynaptic axon terminal that leads to a Ca^{2+} influx, activation of CaM-sensitive cyclases, and subsequent stimulation of neurotransmitter release through the activation of cAMP-dependent protein kinase. It is worth noting that the product of the *rutabaga* locus has been localized to presynaptic axon terminals by immunohistochemistry (124), and a role for AC1 and/or AC8 in nonassociative LTP has been suggested (123).

Differential localization of specific ACs will be important in the regulation of these two types of LTP by cAMP. Long-term potentiation in the CA1 field involves a Ca^{2+}/CaM-sensitive AC located in the dendritic spine of the postsynaptic neuron, whereas LTP at the mossy fiber synapse requires localization of the AC in the presynaptic axon terminal. The first study demonstrating that ACs are located in both of these sites has appeared (58). However, the antiserum used for the immunohistochemical localization recognizes all ACs, thus preventing the identification of the specific isoform(s) involved. Both the localization of the CaM-stimulated AC1 and AC8, and their functional properties imply that they are suited to contribute to the development and expression of LTP in the mammalian hippocampus, but other ACs may also play a role in specific aspects of this phenomenon.

Clearly, these examples do not define the only physiologic functions for types I, V, VI, or VIII. These enzymes are also expressed in different cell types where their unique properties are presumably used to achieve distinct goals. Instead the examples simply illustrate how knowledge of the type-specific regulatory mechanisms that affect AC activity can be used to suggest models for the role of a particular isoform that is expressed in a given physiologic setting. It was this approach that led to the hypothesis that a previously unknown isoform would be expressed in the AtT20 cell line, subsequently resulting in the identification of type IX (84). A detailed understanding of the signal cross-talk that underlies the hormone-dependent regulation of cellular events will ultimately provide a physiologic rationale for the molecular diversity of this enzyme family.

ACKNOWLEDGMENTS

Work in J.K.'s laboratory has been supported by grants from the Southwestern Pennsylvania Affiliate of the American Heart Association, the National Affiliate of the American Heart Association (96007140), and the National Institutes of Health (GM46395).

REFERENCES

1. Gilman AG: G proteins and regulation of adenylate cyclase (Nobel lecture). *Angew Chem Int Ed Engl* 1995;34:1406–1419.
2. Yau K-W: Cyclic nucleotide-gated channels: an expanding new family of ion channels. *Proc Natl Acad Sci U S A* 1994;91:3481–3483.
3. Walsh DA, Van Patten SM: Multiple pathway signal transduction by the cAMP-dependent protein kinase. *FASEB J* 1994;8:1227–1236.
4. Taussig R, Gilman AG: Mammalian membrane-bound adenylyl cyclases. *J Biol Chem* 1995;270:1–4.
5. Cooper DMF, Mons N, Karpen JW: Adenylyl cyclase and the interaction between calcium and cAMP signalling. *Nature* 1995;374:421–424.
6. Krupinski J: The adenylyl cyclase family. *Mol Cell Biochem* 1991;104:73–79.
7. Iyengar R: Molecular and functional diversity of mammalian G_s-stimulated adenylyl cyclases. *FASEB J* 1993;7:768–775.
8. May DC, Ross EM, Gilman AG, Smigel MD: Reconstitution of catecholamine-stimulated adenylate cyclase activity using three purified proteins. *J Biol Chem* 1985;260:15829–15833.
9. Feder D, Im M-J, Klein HW et al: Reconstitution of β_1-adrenocepter-dependent adenylate cyclase from purified components. *EMBO J* 1986;5:1509–1514.
10. Yeager RE, Heideman W, Rosenberg GR, Storm DR: Purification of the calmodulin-sensitive adenylate cyclase from bovine cerebral cortex. *Biochemistry* 1985;24:3776–3783.
11. Mollner S, Pfeuffer T: Two different adenylyl cyclases in brain distinguished by monoclonal antibodies. *Eur J Biochem* 1988;171:265–271.
12. Smigel MD: Purification of the catalyst of adenylate cyclase. *J Biol Chem* 1986;261:1976–1982.
13. Pfeuffer E, Mollner S, Pfeuffer T: Adenylate cyclase from bovine brain cortex: purification and characterization of the catalytic unit. *EMBO J* 1985;4:3675–3679.
14. Gross MK, Toscano DG, Toscano WA Jr: Calmodulin-mediated adenylate cyclase from mammalian sperm. *J Biol Chem* 1987;262:8672–8676.
15. Braun T: Purification of soluble form of adenylyl cyclases from testes. *Methods Enzymol* 1991;195:130–136.
16. Pfeuffer T, Metzger H: 7-O-hemisuccinyl-deacetyl forskolin-Sepharose: a novel affinity support for purification of adenylate cyclase. *FEBS Lett* 1982;146:369–375.
17. Pfeuffer E, Drehev R-M, Metzger H, Pfeuffer T: Catalytic unit of adenylate cyclase: purification and identification by affinity crosslinking. *Proc Natl Acad Sci U S A* 1985;82:3086–3090.
18. Krupinski J, Coussen F, Bakalyar HA et al: Adenylyl cyclase amino acid sequence: possible channel- or transporter-like structure. *Science* 1989;244:1558–1564.
19. Feinstein PG, Schrader KA, Bakalyar HA et al: Molecular cloning and characterization of a Ca^{2+}/calmodulin-insensitive adenylyl cyclase from rat brain. *Proc Natl Acad Sci U S A* 1991;88:10173–10177.
20. Gao B, Gilman AG: Cloning and expression of a widely distributed (type IV) adenylyl cyclase. *Proc Natl Acad Sci U S A* 1991;88:10178–10182.
21. Bakalyar HA, Reed RR: Identification of a specialized adenylyl cyclase that may mediate odorant detection. *Science* 1990;250:1403–1406.
22. Ishikawa Y, Katsushika S, Chen L et al: Isolation and characterization of a novel cardiac adenylyl-cyclase cDNA. *J Biol Chem* 1992;267:13553–13557.
23. Yoshimura M, Cooper DMF: Cloning and expression of a Ca^{2+}-inhibitable adenylyl cyclase from NCB-20 cells. *Proc Natl Acad Sci U S A* 1992;89:6716–6720.
24. Katsushika S, Chen L, Kawabe JI et al: Cloning and characterization of a 6th adenylyl cyclase isoform: type-V and type-VI constitute a subgroup within the mammalian adenylyl cyclase family. *Proc Natl Acad Sci U S A* 1992;89:8774–8778.
25. Premont RT, Chen J, Ma H-W et al: Two members of a new subfamily of hormone-stimulated adenylyl cyclases. *Proc Natl Acad Sci U S A* 1992;89:9809–9813.
26. Wallach J, Droste M, Kluxen FW et al: Molecular cloning of a novel type V adenylyl cyclase from rabbit myocardium. *FEBS Lett* 1994;338:257–263.
27. Krupinski J, Lehman TC, Frankenfield CD et al: Molecular diversity in the adenylylcyclase family: evidence for 8 forms of the enzyme and cloning of type VI. *J Biol Chem* 1992;267:24858–24862.
28. Glatt CE, Snyder SH: Cloning and expression of an adenylyl cyclase localized to the corpus striatum. *Nature* 1993;361:536–538.
29. Cali JJ, Zwaagstra JC, Mons N et al: Type VIII adenylyl cyclase: a Ca^{2+}/calmodulin-stimulated enzyme expressed in discrete regions of rat brain. *J Biol Chem* 1994;269:12190–12195.

30. Watson PA, Krupinski J, Kempinski AM, Frankenfield CD: Molecular cloning and characterization of the type VII isoform of mammalian adenylyl cyclase expressed widely in mouse tissues and in S49 mouse lymphoma cells. *J Biol Chem* 1994;269:28893–28898.
31. Hellevuo K, Yoshimura M, Mons N et al: The characterization of a novel human adenylyl cyclase which is present in brain and other tissues. *J Biol Chem* 1995;270:11581–11589.
32. Defer N, Marinx O, Stengel D et al: Molecular cloning of the human type VIII adenylyl cyclase. *FEBS Lett* 1994;351:109–113.
33. Premont RT, Matsuoka I, Mattei M-G et al: Identification and characterization of a widely expressed form of adenylyl cyclase. *J Biol Chem* 1996;271:13900–13907.
34. Paterson JM, Smith SM, Harmar AJ, Antoni FA: Control of a novel adenylyl cyclase by calcineurin. *Biochem Biophys Res Commun* 1995;214:1000–1008.
35. Krupinski J, Asundi JV, Rothblum KN: Rat type V adenylyl cyclase: the full-length enzyme is activated by a phorbol ester-dependent pathway in intact cells. 1997. Submitted.
36. Villacres EC, Xia Z, Bookbinder LH et al: Cloning, chromosomal mapping, and expression of human fetal brain type I adenylyl cyclase. *Genomics* 1993;16:473–478.
37. Cali JJ, Parekh RS, Krupinski J: Splice variants of type VIII adenylyl cyclase: differences in glycosylation and regulation by Ca^{2+}/calmodulin. *J Biol Chem* 1996;271:1089–1095.
38. Stengel D, Parma J, Gannage MH et al: Different chromosomal localization of 2 adenylyl cyclase genes expressed in human brain. *Hum Genet* 1992;90:126–130.
39. Haber N, Stengel D, Defer N et al: Chromosomal mapping of human adenylyl cyclase genes type III, type V, and type VI. *Hum Genet* 1994;94:69–73.
40. Gaudin C, Homcy CJ, Ishikawa Y: Mammalian adenylyl cyclase family members are randomly located on different chromosomes. *Hum Genet* 1994;94:527–529.
41. Hellevuo K, Berry R, Sikela JM, Tabakoff B: Localization of the gene for a novel human adenylyl cyclase (ADCY7) to chromosome 16. *Hum Genet* 1995;95:197–200.
42. Tang W-J, Krupinski J, Gilman AG: Expression and characterization of calmodulin-activated (type I) adenylyl cyclase. *J Biol Chem* 1991;266:8595–8603.
43. Chinkers M, Garbers DL: The protein kinase domain of the ANP receptor is required for signaling. *Science* 1989;245:1392–1394.
44. Schultz JE, Klumpp S, Benz R et al: Regulation of adenylyl cyclase from *Paramecium* by an intrinsic potassium conductance. *Science* 1992;255:600–603.
45. Reddy R, Smith D, Wayman G et al: Voltage-sensitive adenylyl cyclase activity in cultured neurons: a calcium-independent phenomenon. *J Biol Chem* 1995;270:14340–14346.
46. Iwami G, Akanuma M, Kawabe J et al: Multiplicity in type V adenylyl cyclase: type V-a and type V-b. *Mol Cell Endocrinol* 1995;110:43–47.
47. Tang W-J, Gilman AG: Construction of a soluble adenylyl cyclase activated by $G_s\alpha$ and forskolin. *Science* 1995;268:1769–1772.
48. Yan SZ, Hahn D, Huang ZH, Tang W-J: Two cytoplasmic domains of mammalian adenylyl cyclase form a $G_{s\alpha}$- and forskolin-activated enzyme *in vitro*. *J Biol Chem* 1996;271:10941–10945.
49. Whisnant RE, Gilman AG, Dessauer CW: Interaction of the two cytosolic domains of mammalian adenylyl cyclase. *Proc Natl Acad Sci U S A* 1996;93:6621–6625.
50. Dessauer CW, Gilman AG: Purification and characterization of a soluble form of mammalian adenylyl cyclase. *J Biol Chem* 1996;271:16967–16974.
51. Tang W-J, Stanzel M, Gilman AG: Truncation and alanine-scanning mutants of type I adenylyl cyclase. *Biochemistry* 1995;34:14563–14572.
52. Vorherr T, Knopfel L, Hofmann F et al: The calmodulin binding domain of nitric oxide synthase and adenylyl cyclase. *Biochemistry* 1993;32:6081–6088.
53. Wu Z, Wong ST, Storm DR: Modification of the calcium and calmodulin sensitivity of the type I adenylyl cyclase by mutagenesis of its calmodulin binding domain. *J Biol Chem* 1993;268:23766–23768.
54. Levin LR, Reed RR: Identification of functional domains of adenylyl cyclase using *in vivo* chimeras. *J Biol Chem* 1995;270:7573–7579.
55. Katsushika S, Kawabe J, Homcy CJ, Ishikawa Y: *In vivo* generation of an adenylyl cyclase isoform with a half-molecule motif. *J Biol Chem* 1993;268:2273–2276.
56. Barsony J, Marx SJ: Immunocytology on microwave-fixed cells reveals rapid and agonist-specific changes in subcellular accumulation patterns for cAMP or cGMP. *Proc Natl Acad Sci U S A* 1990; 87:1188–1192.
57. Bacskai BJ, Hochner B, Mahaut-Smith M et al: Spatially resolved dynamics of cAMP and protein kinase A subunits in *Aplysia* sensory neurons. *Science* 1993;260:222–226.

58. Mons N, Harry A, Dubourg P et al: Immunohistochemical localization of adenylyl cyclase in rat brain indicates a highly selective concentration at synapses. *Proc Natl Acad Sci U S A* 1995;92: 8473–8477.

59. Rubin CS: A kinase anchor proteins and the intracellular targeting of signals carried by cyclic AMP. *Biochim Biophys Acta* 1994;1224:467–479.

60. Bode DC, Brunton LL: Post-receptor modulation of the effects of cyclic AMP in isolated cardiac myocytes. *Mol Cell Biochem* 1988;82:13–18.

61. Buxton ILO, Brunton LL: Compartments of cyclic AMP and protein kinase in mammalian cardiomyocytes. *J Biol Chem* 1983;258:10233–10239.

62. Xiao R-P, Hohl C, Altschuld R et al: β_2-adrenergic receptor-stimulated increase in cAMP in rat heart cells is not coupled to changes in Ca^{2+} dynamics, contractility, or phospholamban phosphorylation. *J Biol Chem* 1996;269:19151–19156.

63. Wong YH, Conklin BR, Bourne HR: G_ζ-mediated hormonal inhibition of cyclic AMP accumulation. *Science* 1992;255:339–342.

64. Chen J, Iyengar R: Inhibition of cloned adenylyl cyclases by mutant-activated G_i-α and specific suppression of type 2 adenylyl cyclase inhibition by phorbol ester treatment. *J Biol Chem* 1993;268:12253–12256.

65. Taussig R, Iniguez-Lluhi JA, Gilman AG: Inhibition of adenylyl cyclase by $G_{i\alpha}$. *Science* 1993; 261:218–221.

66. Taussig R, Tang W-J, Hepler JR, Gilman AG: Distinct patterns of bidirectional regulation of mammalian adenylyl cyclases. *J Biol Chem* 1994;269:6093–6100.

67. Kozasa T, Gilman AG: Purification of recombinant G proteins from Sf9 cells by hexahistidine tagging of associated subunits. Characterization of α_{12} and inhibition of adenylyl cyclase by α_z. *J Biol Chem* 1995;270:1734–1741.

68. Sternweis PC, Robishaw JD: Isolation of two proteins with high affinity for guanine nucleotides from membranes of bovine brain. *J Biol Chem* 1984;259:13806–13813.

69. Taussig R, Quarmby LM, Gilman AG: Regulation of purified type I and type II adenylylcyclases by G protein bg subunits. *J Biol Chem* 1993;268:9–12.

70. Tang W-J, Gilman AG: Type-specific regulation of adenylyl cyclase by G protein bg subunits. *Science* 1991;254:1500–1503.

71. Federman AD, Conklin BR, Schrader KA et al: Hormonal stimulation of adenylyl cyclase through G_i-protein $\beta\gamma$ subunits. *Nature* 1992;356:159–161.

72. Mons N, Yoshimura M, Cooper DMF: Discrete expression of Ca^{2+}/calmodulin-sensitive and Ca^{2+}-insensitive adenylyl cyclases in the rat brain. *Synapse* 1993;14:51–59.

73. Andrade R: Enhancement of β-adrenergic responses by G_i-linked receptors in rat hippocampus. *Neuron* 1993;10:83–88.

74. Bourne HR, Nicoll R: Molecular machines integrate coincident synaptic signals. *Cell (Suppl)* 1993;72:65–75.

75. Chen J, DeVivo M, Dingus J et al: A region of adenylyl cyclase 2 critical for regulation by G protein $\beta\gamma$ subunits. *Science* 1995;268:1166–1169.

76. Yan K, Gautam N: A domain on the G protein β subunit interacts with both adenylyl cyclase 2 and the muscarinic atrial potassium channel. *J Biol Chem* 1996;271:17597–17600.

77. Yoshimura M, Ikeda H, Tabakoff B: μ-Opioid receptors inhibit dopamine-stimulated activity of type V adenylyl cyclase but enhance dopamine-stimulated activity of type VII adenylyl cyclase. *Mol Pharmacol* 1996;50:43–51.

78. Choi E-J, Wong ST, Hinds TR, Storm DR: Calcium and muscarinic agonist stimulation of type I adenylyl cyclase in whole cells. *J Biol Chem* 1992;267:12440–12442.

79. Choi E-J, Xia ZG, Storm DR: Stimulation of the type III olfactory adenylyl cyclase by calcium and calmodulin. *Biochemistry* 1992;31:6492–6498.

80. Wayman GA, Impey S, Storm DR: Ca^{2+} inhibition of type III adenylyl cyclase *in vivo*. *J Biol Chem* 1995;270:21480–21486.

81. Jones DT, Reed RR: G_{olf}: An olfactory neuron specific G protein involved in odorant signal transduction. *Science* 1989;244:790–795.

82. Dhallan RS, Yau K-W, Schrader KA, Reed RR: Primary structure and functional expression of a cyclic nucleotide-activated channel from olfactory neurons. *Nature* 1990;347:184–187.

83. Mons N, Cooper DMF: Selective expression of one Ca^{2+}-inhibitable adenylyl cyclase in dopaminergically innervated rat brain regions. *Mol Brain Res* 1994;22:236–244.

84. Antoni FA, Barnard RJO, Shipston MJ et al: Calcineurin feedback inhibition of agonist-evoked cAMP formation. *J Biol Chem* 1995;270:28055–28061.

85. Kelleher DJ, Pessin JE, Ruoho AE, Johnson GL: Phorbol ester induces desensitization of adenylate cyclase and phosphorylation of the β-adrenergic receptor in turkey erythrocytes. *Proc Natl Acad Sci U S A* 1984;81:4316–4320.

86. Bell JC, Buxton IL, Brunton LL: Enhancement of adenylate cyclase activity in S49 lymphoma cells by phorbol esters. *J Biol Chem* 1985;260:2625–2628.

87. Sibley DR, Jeffs RA, Daniel K et al: Phorbol diester treatment promotes enhanced adenylate cyclase activity in frog erythrocytes. *Arch Biochem Biophys* 1986;244:373–381.

88. Yoshimasa T, Sibley DR, Bouvier M et al: Cross-talk between cellular signalling pathways suggested by phorbol-ester-induced adenylate cyclase phosphorylation. *Nature* 1987;327:67–70.

89. Simmoteit R, Schulzki H-D, Palm D et al: Chemical and functional analysis of components of adenylyl cyclase from human platelets treated with phorbolesters. *FEBS Lett* 1991;285:99–103.

90. Jacobowitz O, Chen JQ, Premont RT, Iyengar R: Stimulation of specific types of G_s-stimulated adenylyl cyclases by phorbol ester treatment. *J Biol Chem* 1993;268:3829–3832.

91. Choi E-J, Wong ST, Dittman AH, Storm DR: Phorbol ester stimulation of the type I and type III adenylyl cyclases in whole cells. *Biochemistry* 1993;32:1891–1894.

92. Jacobowitz O, Chen J, Premont RT, Iyengar R: Stimulation of specific types of G_s-stimulated adenylyl cyclases by phorbol ester treatment. *J Biol Chem* 1993;268:3829–3832.

93. Yoshimura M, Cooper DMF: Type-specific stimulation of adenylyl cyclase by protein kinase C. *J Biol Chem* 1993;268:4604–4607.

94. Lustig KD, Conklin BR, Herzmark P et al: Type II adenylyl cyclase integrates coincident signals from G_s, G_i, and G_q. *J Biol Chem* 1993;268:13900–13905.

95. Jacobowitz O, Iyengar R: Phorbol ester–induced stimulation and phosphorylation of adenylyl cyclase 2. *Proc Natl Acad Sci U S A* 1994;91:10630–10634.

96. Kawabe J-I, Iwami G, Ebina T et al: Differential activation of adenylyl cyclase by protein kinase C isoenzymes. *J Biol Chem* 1994;269:16554–16558.

97. Kawabe J, Ebina T, Toya Y et al: Regulation of type V adenylyl cyclase by PMA-sensitive and -insensitive protein kinase C isoenzymes in intact cells. *FEBS Lett* 1996;384:273–276.

98. Strasser RH, Marquetant R: Sensitization of the β-adrenergic system in acute myocardial ischaemia by a protein kinase C–dependent mechanism. *Eur Heart J* 1991;12(supp. F):48–53.

99. Kubler W, Strasser RH: Signal transduction in myocardial ischaemia. *Eur Heart J* 1994;15:437–445.

100. Strasser RH, Marquetant R, Kubler W: Adrenergic receptors and sensitization of adenylyl cyclase in acute myocardial ischemia. *Circulation* 1990;82:II-23–II-28.

101. Strasser RH, Krimmer J, Braun-Dullaeus R et al: Dual sensitization of the adrenergic system in early myocardial ischemia: independent regulation of the β-adrenergic receptors and the adenylyl cyclase. *J Mol Cell Cardiol* 1990;22:1405–1423.

102. Strasser RH, Braun-Dullaeus R, Walendzik H, Marquetant R: α_1-Receptor-independent activation of protein kinase C in acute myocardial ischemia. Mechanisms for sensitization of the adenylyl cyclase system. *Circ Res* 1992;70:1304–1312.

103. Iwami G, Kawabe J-I, Ebina T et al: Regulation of adenylyl cyclase by protein kinase A. *J Biol Chem* 1995;270:12481–12484.

104. Premont RT, Jacobowitz O, Iyengar R: Lowered responsiveness of the catalyst of adenylyl cyclase to stimulation by G_s in heterologous desensitization: a role for adenosine 3′,5′-monophosphate-dependent phosphorylation. *Endocrinology* 1992;131:2774–2784.

105. Cooper DMF, Brooker G: Ca^{2+}-inhibited adenylyl cyclase in cardiac tissue. *Trends Pharmacol Sci* 1993;14:34–36.

106. Homcy CJ, Vatner SF, Vatner DE: β-Adrenergic receptor regulation in the heart in pathophysiologic states: abnormal adrenergic responsiveness in cardiac disease. *Annu Rev Physiol* 1991;53:137–159.

107. Yu HJ, Ma H, Green RD: Calcium entry via L-type calcium channels acts as a negative regulator of adenylyl cyclase activity and cyclic AMP levels in cardiac myocytes. *Mol Pharmacol* 1993;44:689–693.

108. Yu HJ, Unnerstall JR, Green RD: Determination and cellular localization of adenylyl cyclase isozymes expressed in embryonic chick heart. *FEBS Lett* 1995;374:89–94.

109. Brooker G: Oscillation of cyclic adenosine monophosphate concentration during the cardiac cycle. *Science* 1973;182:933–934.

110. DeBernardi MA, Brooker G: Single cell Ca^{2+}/cAMP cross-talk monitored by simultaneous Ca^{2+}/cAMP fluorescence ratio imaging. *Proc Natl Acad Sci U S A* 1996;93:4577–4582.

111. Livingstone MS, Sziber PP, Quinn WG: Loss of calcium/calmodulin responsiveness in adenylate cyclase of *rutabaga*: a *Drosophila* learning mutant. *Cell* 1984;37:205–215.

112. Livingstone MS: Genetic dissection of *Drosophila* adenylate cyclase. *Proc Natl Acad Sci U S A* 1985;82:5992–5996.
113. Levin LR, Han P-L, Hwang PM, Feinstein PG et al: The *Drosophila* learning and memory gene *rutabaga* encodes a Ca^{2+}/calmodulin-responsive adenylyl cyclase. *Cell* 1992;68:479–489.
114. Bliss TVP, Collingridge GL: A synaptic model of memory: long-term potentiation in the hippocampus. *Nature* 1993;361:31–39.
115. Morris RGM, Anderson E, Lynch GS, Baudry M: Selective impairment of learning and blockade of long-term potentiation by an N-methyl-D-aspartate receptor antagonist, AP5. *Nature* 1986;319:774–776.
116. Nowak L, Bregestovski P, Ascher P et al: Magnesium gates glutamate-activated channels in mouse central neurones. *Nature* 1984;307:462–465.
117. Chetkovich DM, Gray R, Johnston D, Sweatt JD: N-methyl-D-aspartate receptor activation increases cAMP levels and voltage-gated Ca^{2+} channel activity in area CA1 of hippocampus. *Proc Natl Acad Sci U S A* 1991;88:6467–6471.
118. Blitzer RD, Wong T, Nouranifar R et al: Postsynaptic cAMP pathway gates early LTP in hippocampal CA1 region. *Neuron* 1995;15:1403–1414.
119. Wu Z-L, Thomas SA, Villacres EC et al: Altered behavior and long-term potentiation in type I adenylyl cyclase mutant mice. *Proc Natl Acad Sci U S A* 1995;92:220–224.
120. Xia Z, Refsdal CD, Merchant KM et al: Distribution of mRNA for the calmodulin-sensitive adenylate cyclase in rat brain: expression in areas associated with learning and memory. *Neuron* 1991;6:431–443.
121. Sassone-Corsi P: Transcription factors responsive to cAMP. *Annu Rev Cell Dev Biol* 1995;11:355–377.
122. Frey U, Huang Y-Y, Kandel ER: Effects of cAMP simulate a late stage of LTP in hippocampal CA1 Neurons. *Science* 1993;260:1661–1664.
123. Weisskopf MG, Castillo PE, Zalutsky RA, Nicoll RA: Mediation of hippocampal mossy fiber long-term potentiation by cyclic AMP. *Science* 1994;265:1878–1882.
124. Han P-L, Levin LR, Reed RR, Davis RL: Preferential expression of the *Drosophila rutabaga* gene in mushroom bodies, neural centers for learning in insects. *Neuron* 1992;9:619–627.

*Advances in Second Messenger and
Phosphoprotein Research*, Vol. 32,
edited by Dermot M. F. Cooper
Lippincott–Raven Publishers, Philadelphia © 1998

4

Type-Specific Regulation of Mammalian Adenylyl Cyclases by G Protein Pathways

Ronald Taussig and Gregor Zimmermann

*Department of Biological Chemistry, The University of Michigan Medical School,
Ann Arbor, Michigan 48109-0636*

Modulation of intracellular cyclic adenosine monophosphate (cAMP) levels has been shown to impact on a number of cellular processes underlying changes in protein phosphorylation state, regulation of ion channel conductance, and gene expression. These cAMP-regulated cellular processes play an important role in the control of a number of metabolic processes, including the regulation of blood glucose homeostasis (1), learning and memory (2), and cell growth (3). The observation that hormones regulating intracellular cAMP levels have profound effects on these metabolic processes clearly highlights the important role of adenylyl cyclase (AC).

Regulation of intracellular cAMP concentrations is principally controlled at the level of its synthesis, through the hormonal regulation of AC, the enzyme responsible for the conversion of adenosine triphosphate (ATP) into cAMP. The AC system is comprised of three components: heptahelical, G protein–coupled receptors for a variety of hormones, neurotransmitters, and autocoids; heterotrimeric G proteins; and the catalytic entity itself. Currently, nine isoforms of membrane-bound ACs have been identified by molecular genetic approaches. These isoforms are subjected to hormonal regulation that are mediated by G proteins of the G_s, G_i/G_o, G_z, and G_q subclasses. Most important, AC isoforms have the ability to integrate the multitude of hormonal inputs relayed by these pathways to ultimately control intracellular cAMP levels. This review will provide an overview of the structure and expression of these enzymes and focus primarily on their regulation by these G protein pathways.

STRUCTURE OF MEMBRANE-BOUND MAMMALIAN ADENYLYL CYCLASES

Our current knowledge of the structure and function of the mammalian AC family has reached a new understanding following the application of molecular approaches.

Using an affinity matrix, synthesized from forskolin, a plant diterpine activator of AC (4), Krupinski and co-workers (5) were able to purify enough of the AC protein from bovine brain extracts to obtain partial amino acid sequence. This led to their isolation of cDNA clones encoding the first full-length sequence of a mammalian AC termed type I. Additional members of the AC family (types II–IX; AC2–9) were subsequently isolated from various mammalian species using low-stringency hybridization and polymerase chain reaction (PCR) technology (6–17). Further diversity of ACs is provided by alternatively spliced messenger ribonucleic acids (mRNAs) derived from transcripts of some the family members (13,18,19). Currently, nine distinct genes encoding membrane-bound ACs have been identified (Table 1). These genes are broadly distributed on different chromosomes (or arms of chromosome) throughout the human genome. This distribution is very different from that of the G protein α subunit family that arose by tandem duplication prior to their expansion and are found in the genome as paired genes (20).

Determination of the precise expression patterns of the nine AC genes has been difficult because of the low levels of expression and the unavailability of high-affinity, isoform-specific antibodies. Nevertheless, analysis of the patterns of mRNA expression (by northern blot or *in situ* hybridization analysis) reveals that the nine genes are differentially expressed in various tissues in mammalian species. For example, type IX appears to be rather widely distributed, whereas type III is largely restricted to olfactory neuroepithelium. All isoforms of AC appear to be expressed in the brain; however, expression of each isoform is limited to specific regions of the central nervous system.

All the cloned ACs have molecular masses of roughly 120–150 kDa and contain between 1064 and 1353 amino acid residues. The larger observed mass and broad

TABLE 1. *Localization and expression of the adenylyl cyclase gene family*

Type	Length (A.A.)	Location (chromosome)	Expression (tissue)
I	1135	7p12–13	Brain
II	1090	5p15.3	Brain (lung)
III	1144	2p22–24	Olfactory epithelium (brain, heart, lung)
IV	1064	14q11.2	Widely distributed
V	1184	3q13.2–21	Heart (brain, lung, kidney, liver, testes)
VI	1165	12q12–13	Heart (brain, lung, kidney, liver, testes)
VII	1099	16q12–13	Widely distributed
VIII	1248	8q24.2	Brain
IX	1353	16p13.3–13.2	Skeletal muscle, brain (widely distributed)

Please refer to the references (in parentheses) for additional information.

Chromosomal location: type I (78); types II and VIII (79); types III, V, and VI (80); type IV (81); type VII (82); and type IX (17).

Tissue expression: type I (83); type II (6); type III (84); type IV (8); type V (10,11); type VI (9,12); type VII, (15,17,85); type VIII (9,12,14,86); and type IX (17).

migration on SDS polyacrylamide gels is due to the glycosylation of these proteins. At least one form of AC has been shown to be modified by the posttranslational addition of palmitate to a cysteine residue on the protein (21); the identification of this AC isoform is not known. Additionally, the generality of this modification, as well as the role of this post-translational modification still remains to be clarified.

The deduced amino acid sequence of all identified AC isoforms predicts a complex topology within the membrane (Fig. 1). The short amino terminus is followed by six transmembrane spans (designated M_1) and a large (roughly 40-kDa) cytoplasmic domain (C_1). This pattern is repeated in the second half of the molecule; a second set of six transmembrane spans (M_2) is followed by a second large cytoplasmic domain (C_2). The cytoplasmic domains are now known to catalyze the conversion of ATP into cAMP. This predicted structure is reminiscent of the ATP-dependent transporters typified by the P glycoprotein and cystic fibrosis transmembrane conductance regulator; however, the ACs do not contain the "Walker" motif characteristic of these transporters. Nonetheless, this relationship prompted the speculation that the AC proteins may also function as transporters or ion channels, but there is no evidence to support this function. Of particular note is the observation that the AC activity from *Paramecium* co-purifies with a potassium channel (22) and may indeed be one in the same molecule; however, the structural similarity with the mam-

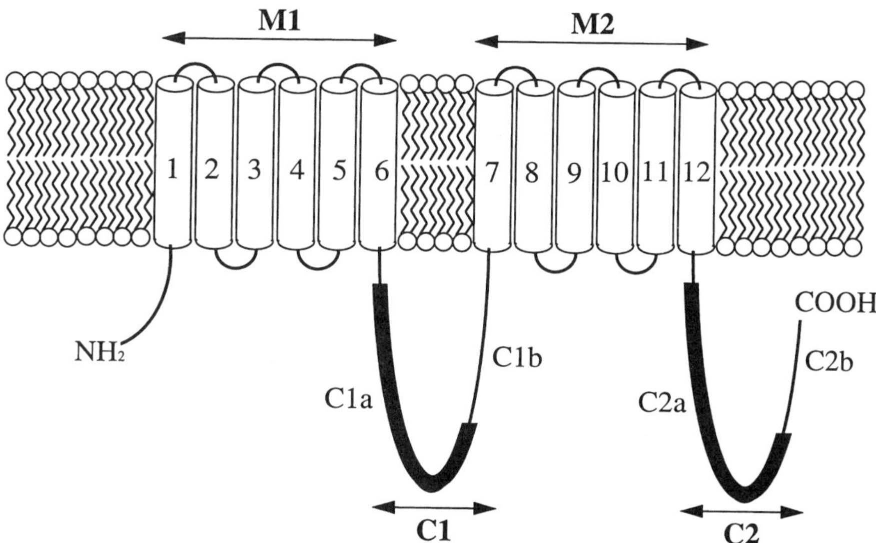

FIG. 1. Structure of mammalian ACs. Predicted topology of membrane-bound AC. Numbered cylinders represent membrane spanning regions; bold lines indicate regions of high amino acid conservation among all the AC family members. Nomenclature, from amino terminus (NH_2) to carboxy terminus (COOH) is the following: M_1, first set of membrane spanning regions; C_1, the first large cytoplasmic domain comprised of conserved C_{1a} and variable C_{1b} subdomains; M_2, the second set of transmembrane spanning regions; and C_2, the second large cytoplasmic domain comprised of conserved C_{2a} and variable C_{2b} subdomains.

malian ACs is still an open question. Related to this, Reddy and co-workers have proposed that the membrane spans of mammalian ACs may play a role as sensors of membrane potential (like voltage-dependent channels), providing an additional level of regulating AC activity through agents that depolarize the cell (23).

The overall amino acid sequence similarity among the different isoforms of AC is roughly 50% and mostly limited to the cytoplasmic domains. Within each cytoplasmic domain, two subregions highlighted in Fig. 1 (designated C_{1a} and C_{2a}) are more highly conserved (up to 93% sequence identity); within a particular cyclase isoform, these regions also display sequence similarity to each other. This relationship can also be seen in the cytoplasmic domains of topographically similar ACs from *Drosophila* (24) and *Dictyostelium* (25). Likewise, the C_{1a} and C_{2a} domains are also highly homologous to the catalytic domains of membrane-bound guanylyl cyclases (26) and domains that are found in each of the subunits of cytosolic heterodimeric guanylyl cyclases (27).

The homologies among the cytoplasmic domains of all these cyclases suggests that one or both of these domains of mammalian ACs constitute the site for catalysis of cAMP synthesis. In support of this, Tang and co-workers demonstrated that a "shortened" soluble AC construct consisting of the C_{1a} domain from type I attached to the C_{2a} domain of type II was catalytically active (28). When expressed in bacteria, neither C_{1a} nor C_{2a} alone was catalytically competent; however, when coexpressed (as separate molecules) or mixed *in vitro*, the two soluble domains could form an active AC that was responsive to forskolin, $G_{s\alpha}$, and G protein $\beta\gamma$ subunits (29,30). Despite these observations and the finding that targeted mutagenesis of conserved residues in both C_{1a} and C_{2a} affect ATP binding (31), location of the ATP-binding and catalytic site(s) on the molecule awaits further structural studies.

Based on the similarities of their amino acid sequences and patterns of regulation by G protein pathways (see later), the ACs can be grouped into subfamilies (Fig. 2). The first group, comprised of types I and VIII, is characterized by its stimulation by calmodulin (CaM) (5,14); the type III isoform is also CaM stimulatable *in vitro* (32), but its sequence is more distantly related and the enzyme displays regulatory properties distinct from members of this subfamily (33). Types II, IV, and VII ACs form the second group. Characteristic of these isoforms is their stimulation by G protein $\beta\gamma$ subunits (8,34–36). Types V and VI, which are the two most related isoforms, constitute the last group and are characterized by their inhibition by calcium (9,12,13). Type IX is the largest and most diverse of the AC isoforms and, like the type III enzyme, does not belong to any of the three groups outlined earlier.

REGULATION OF ADENYLYL CYCLASE ACTIVITY BY
G PROTEIN–MEDIATED PATHWAYS

With the cloning of the membrane-bound ACs, investigators have begun to examine the complex regulatory repertoire of each family member (Tables 2 and 3). Two lines of experimental approach have been followed to examine this issue, both utilizing the expression on the cDNAs encoding the AC isoforms in appropriate cell

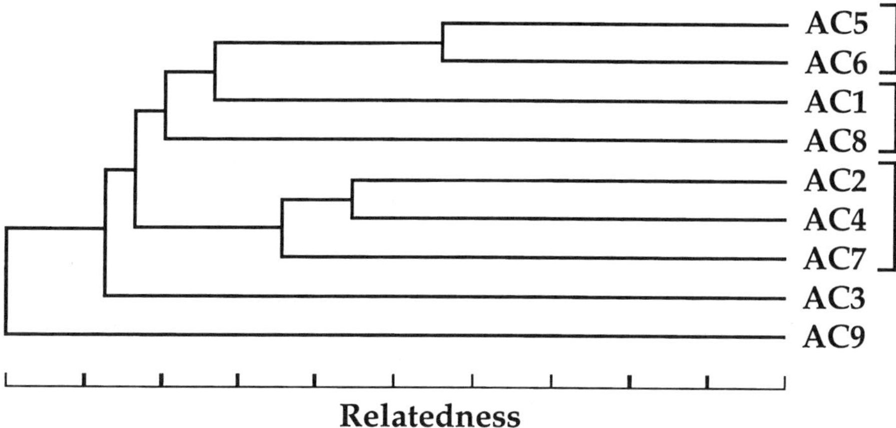

FIG. 2. Degree of sequence similarities among the AC family. Comparison of AC sequences of type I (bovine), types II–IV (rat), types V and VI (dog), and types VII–IX were performed, and corresponding dendogram generated, using the Clustal algorithm of the Megalign program from Lasergene (DNASTAR).

lines. Biochemical characterization of AC isoforms has allowed investigators to directly reconstitute the ACs with purified regulatory molecules (i.e., G protein subunits, CaM, or protein kinases) in the test tube. Homogeneous AC preparations are typically obtained from the viral expression systems (Sf9 insect cells as host) (37) or as soluble domains expressed in bacteria (28). Transfection (or co-transfection) of mammalian cell lines with constructs encoding AC isoforms and appropriate regulators (38) has been the second approach. The former has the advantage of allowing control of cyclase and regulator concentrations, and the examination of the effects in isolation of "undefined" inputs, but it lacks the constraints imposed on the system by cellular compartmentalization. The latter allows for the analysis of the regulation in the context of the whole cell but is limited by uncertainty of the levels of the exogenously expressed protein. Clearly, both approaches are necessary, and more often than not yield qualitatively similar results.

TABLE 2. *Regulation of adenylyl cyclase isoforms by forskolin and G protein subunits*

Regulator	Cyclase	Effect	Comments
$G_{s\alpha}$	All	Stimulatory	For I and VIII, *in vivo* effects require a second activator
$G_{i\alpha}$	I, V, VI	Inhibitory	Inhibition of type I is activator dependent: CaM > FSK >> $G_{s\alpha}$
$G_{o\alpha}$	I	Inhibitory	See $G_{i\alpha}$
$G_{z\alpha}$	I, V	Inhibitory	
$G_{\beta\gamma}$	II, IV, VII	Stimulatory	Requires stimulation by $G_{s\alpha}$
	I	Inhibitory	
Forskolin	All	Stimulatory	Small effect on type IX

TABLE 3. *Regulation of adenylyl cyclase isoforms by G_q-coupled pathways*

Regulator	Cyclase	Effect	Comments
Ca/CaM	I, III, VIII	Stimulatory	Stimulatory
Ca	V, VI	Inhibitory	
PKC	I, III	Stimulatory	
	II, VII	Stimulatory	Blocks $G_{\beta\gamma}$ superstimulation of type II
	V	Stimulatory	
	IV	Inhibitory	No effect on basal or forskolin stimulation
CaM kinase	III	Inhibitory	

FORSKOLIN AND $G_{s\alpha}$: ACTIVATORS OF ALL MEMBRANE-BOUND ADENYLYL CYCLASES

Two common features of all mammalian membrane-bound ACs are their stimulatory responses to forskolin and the α subunit of G_s ($G_{s\alpha}$). Forskolin is an effective activator of all isoforms of AC at low micromolar concentrations. Because of the hydrophobic character of forskolin, it was predicted that the binding site would likely be found in the membrane-spanning regions of the protein. Therefore, it was quite surprising to find that the soluble (C_{1a}–C_{2a}) cyclases construct was responsive to stimulation by forskolin.

Studies examining the direct binding of forskolin to ACs revealed that forskolin binds with a K_d of less than 1 μM. The conclusion of these studies is that there are two forskolin binding sites per AC molecule: a high-affinity (submicromolar) site detected by binding and a lower-affinity site (micromolar) responsible for the stimulation of the enzyme activity (39). In light of the predicted repeated structure of the ACs, it is reasonable to envision a forskolin-binding site located in each of the cytoplasmic domains, C_1 and C_2. An alternative model suggested by the observation that forskolin stimulates AC by increasing the affinity of the C_1 and C_2 domains for each other (29,30) is that each cytoplasmic domain contains two "half" sites for forskolin; binding of forskolin to these half sites would therefore promote association of the two cytoplasmic domains.

Forskolin, a naturally occurring plant diterpene, is not synthesized in mammalian cells, yet the ability of all AC isoforms to bind this molecule and be stimulated by it has been retained through evolution (40). These observations lead one to the inescapable speculation that there is some as yet unidentified analog of forskolin endogenously synthesized by mammalian cells. The chemical properties of forskolin would lead to the prediction that this molecule would be likewise hydrophobic in character, possibly a lipid second messenger.

All AC isoforms are activated by $G_{s\alpha}$ *in vitro*; this activation is due to the direct interaction of the G protein α subunit with the cyclase (35). The regulation of ACs by $G_{s\alpha}$ *in vivo* is less clear-cut. In response to hormonal activation of G_s by β-adrenergic agonists, cells exogenously expressing AC1 do not demonstrate enhanced intracellular cAMP levels (7,12,41,42). However, $G_{s\alpha}$ stimulation of this isoform is evident if this signal is concomitant with hormonal activation of CaM (43). These observations highlight an important feature of AC regulation: the ability to integrate multiple input signals derived from different G protein pathways in a synergistic, rather than simply an additive, fashion. This same pattern of regulation has been observed for the other member of the CaM-stimulated AC subfamily, the type VIII isoform (14); *in vivo*, $G_{s\alpha}$ stimulation of this isoform depends on the simultaneous hormonal activation of CaM, despite the fact that membranes prepared from these same cells display $G_{s\alpha}$ stimulation in the absence of the CaM input. This apparent paradox may be due to compartmentalization of the cyclases in the intact cells that would be disrupted en route to the biochemical characterization of the enzymes; clearly, analysis of the subcellular localization of these isoforms will be required to resolve these discrepancies.

A related $G_{s\alpha}$-like G protein α subunit, G_{olf} is expressed in olfactory epithelium (44), the principal site of type III AC expression. Expression of G_{olf} in S49 cyc⁻ cells (void of stimulatory $G_{s\alpha}$ protein) demonstrated that this α subunit could indeed activate the endogenous AC in these cells. In light of its restricted expression in olfactory neuroepithelium and localization to the sensory apparatus of the receptor neurons in particular, it has been proposed that G_{olf} is the physiologic activator of the AC3-mediating olfactory detection in these cells (7).

Some insight into the mechanism by which $G_{s\alpha}$ activates ACs has been revealed by studies with the soluble domain constructs expressed in bacteria. Like the case for forskolin, $G_{s\alpha}$ increases the affinity of the C_1 domain for the C_2 domain (29,30), and presumably is at least partly responsible for the activation of the cyclase. One hypothesis is that $G_{s\alpha}$ binds to determinants present on each of the cytoplasmic domains of the cyclase, thereby increasing their affinity for each other; supporting this notion is the identification of specific residues of C_{1a} and C_{2a} that, when mutated, disrupt $G_{s\alpha}$ but not forskolin activation (Taussig R, unpublished observations).

The ability of the ACs to integrate multiple stimulatory inputs by $G_{s\alpha}$ and forskolin is isoform specific. For the type I isoform, $G_{s\alpha}$ and forskolin act as independent activators of this isoform; addition of $G_{s\alpha}$ to the cyclase maximally stimulated with forskolin results in little further stimulation of the enzyme (41). Binding of forskolin to this cyclase is also unaffected by the addition of $G_{s\alpha}$ to the assay (40). This behavior is in marked contrast to those of types II, IV, V, and VI. Coincidental stimulation of these isoforms by $G_{s\alpha}$ and forskolin results in synergistic superstimulation of enzymatic activity (6,8,45). The effect of forskolin is to increase the responsiveness of these isoforms to $G_{s\alpha}$ (observed as a leftward shift in the dose–response curve for $G_{s\alpha}$), and the affinity of the enzyme for the G protein. Conversely, $G_{s\alpha}$ increases the cyclases' affinity for and responsiveness to forskolin (40). Increases in the V_{max} of the enzymes are also observed; this is most striking for the C_1/C_2 soluble cyclase construct (46).

REGULATION OF MEMBRANE-BOUND ADENYLYL CYCLASES BY PERTUSSIS TOXIN–SENSITIVE PATHWAYS

Hormonal inhibition of AC activity has been shown to be blocked by treatment with pertussis toxin, implicating one of the G_i proteins (G_{i1}, G_{i2}, or G_{i3}) or G_o in coupling inhibitory receptors (e.g., an α_2-adrenergic receptor) to AC (47,48). Hormonal activation of these pertussis toxin–sensitive G proteins leads to nucleotide exchange and the concomitant release of $\beta\gamma$ subunits from the activated guanosine triphosphate (GTP)-bound α subunit, thus releasing two signaling molecules capable of interacting with the AC effector molecules. Indeed, both the α and the $\beta\gamma$ subunits have been shown to regulate the activity of some but not all AC isoforms.

Inhibition of AC activity by $G_{i\alpha}$ subunits was conclusively demonstrated in *in vivo* transfection studies wherein constitutively activated mutants of $G_{i\alpha2}$ subunits were shown to inhibit the AC endogenously expressed in HEK-293 cells endogenous. Using similar transfection approaches (this time co-transfecting the $G_{i\alpha}$ mutants with AC isoforms), Chen and Iyengar demonstrated the inhibition of types II and VI by $G_{i\alpha2}$ (42).

Approaches utilizing *in vitro* reconstitution of purified recombinant G protein α subunits with AC isoforms have complemented these transfection studies and have provided insight into the mechanism of cyclase inhibition by pertussis toxin substrates (45,49). Both type V and type VI isoforms can be inhibited by $G_{i\alpha s}$; all $G_{i\alpha}$ subunits (G_{i1}, G_{i2}, and G_{i3}) are equally potent and efficacious. The inhibition is noncompetitive with $G_{s\alpha}$, arguing that $G_{s\alpha}$ and $G_{i\alpha}$ bind to separate nonoverlapping sites on the AC protein. Consistent with the mechanism of G protein activation, the GTP-bound form of the $G_{i\alpha}$ is at least two orders of magnitude more potent at inhibiting cyclase activity than the guanosine diphosphate (GDP)-bound form. Myristoylation of the α subunits is required for these inhibitory effects. Type I AC is likewise inhibited by $G_{i\alpha s}$ but is unique in that $G_{o\alpha}$ can also inhibit this isoform, although higher concentrations of this subunit are required. For types V and VI, $G_{i\alpha}$ inhibition is observed under stimulatory conditions regardless of whether $G_{s\alpha}$ or forskolin is used as the activator; however, although inhibition of type I is evident when forskolin or CaM is used as the activator, none is observed in the presence of $G_{s\alpha}$.

Type II AC, when stimulated by $G_{s\alpha}$, is not inhibited by activated (GTPγ_s-bound) $G_{o\alpha}$ and only weakly inhibited by activated $G_{i\alpha}$; no inhibition is observed under stimulation by forskolin. However, unlike the case for types I, V, and VI, the GDP-bound forms of these α subunits inhibit this enzyme with a higher potency. Because AC2 activity is synergistically stimulated by G protein $\beta\gamma$ subunits in the presence of $G_{s\alpha}$ (see later), and the GDP-bound form of G protein α subunits have high affinity for the $\beta\gamma$ subunits, it seems unlikely that $G_{i\alpha}$ has a directly inhibitory action on this cyclase per se, but rather functions to deactivate the free $\beta\gamma$ subunits present in the preparation. Similar arguments can be presented to explain the observed inhibition of this isoform when assayed in the transfection approaches outlined earlier.

$\beta\gamma$ subunits released from hormonal activation of G_i or G_o heterotrimers have profound effects on some isoforms of AC. Type I is inhibited *in vivo* by G protein $\beta\gamma$ subunits under $G_{s\alpha}$-, forskolin-, or CaM-stimulated conditions (34,41), and this is due to the direct binding of $\beta\gamma$ to the cyclase (35). Inhibition of type I *in vivo*, however, appears to be due mainly to activated $G_{i\alpha}$ and not to the released $\beta\gamma$ subunits (50). Interestingly, the related CaM-activated cyclase, type VIII, is not inhibited *in vivo* by hormonal activation of G_i. Neither type V nor type VI is sensitive to $\beta\gamma$ regulation.

Types II, IV, and presumably VII are conditionally activated by $\beta\gamma$ subunits in the presence of $G_{s\alpha}$ (8,34,51). The synergistic activation of these (35) isoforms by $G_{s\alpha}$ and $\beta\gamma$ is due to direct interactions of these subunits with the cyclase and does not require association of these two regulators with each other, as amino terminal truncated $G_{s\alpha}$ that does not bind $\beta\gamma$ subunits functions equally as well (45). In the absence of G_s stimulation, few stimulatory effects of $\beta\gamma$ on these cyclases are observed; binding of $G_{s\alpha}$ to the cyclase presumably increases the affinity of the cyclase for $\beta\gamma$ or exposes a second high-affinity $\beta\gamma$ binding site (52).

A region of AC2 involved in the binding of $\beta\gamma$ has been determined by the use of synthetic peptides (53). A 27-amino-acid peptide, based on the sequence 956–982 from the C_2 domain blocked $\beta\gamma$ stimulation of this cyclase isoform. In addition, this peptide blocked $\beta\gamma$ inhibition of AC1, $\beta\gamma$ regulation of G protein–regulated potassium channels, and phospholipase C–β, and $\beta\gamma$ interactions with βARK but has no effect on Gα-$\beta\gamma$ interactions. The QXXER motif contained in this peptide is crucial for its ability to interact with $\beta\gamma$ and is conserved in the other $\beta\gamma$-stimulated ACs (types IV and VII). This motif is also present in calcium channel α (1A) subunits, which are modulated by G proteins, but absent from G protein–insensitive α (1C) subunits; removal of this site renders the channel G protein insensitive (54). This site has been proposed to interact with the β subunit of the G protein, within the amino terminal 100 residues (55,56).

REGULATION OF MEMBRANE-BOUND ADENYLYL CYCLASES BY G_z

Hormonal inhibition of AC via a pertussis toxin–insensitive mechanism may now be explained by G_z-mediated pathways. *In vivo* studies demonstrated that constitutively active $G_{z\alpha}$ mutants exogenously expressed could inhibit the endogenous AC present in HEK-293 cells (57). *In vitro*, the specificity for the inhibition of AC isoforms seems to parallel those of the $G_{i\alpha}$ family; $G_{z\alpha}$ can inhibit types I and V, and presumably VI, but not type II (58). $G_{z\alpha}$ inhibited these cyclases to an extent similar to that observed for $G_{i\alpha}$ and with similar (or greater) potencies.

REGULATION OF MEMBRANE-BOUND ADENYLYL CYCLASES BY G_q

Hormonal activation of receptors coupled to phospholipase C by G_q gives rise to diacylglycerol and inositol trisphosphate. These second messengers, through the release of intracellular calcium and subsequent activation of calcium-dependent signaling molecules, such as CaM, CaM kinase, and protein kinase C (PKC), can have profound effects on the activities of various AC isoforms.

Inhibition of AC by calcium had been observed prior to the cloning of ACs. All ACs are inhibited by high (100–1000 μM) concentrations of calcium as a result of competition for magnesium, which is required for catalysis. At more physiologic calcium concentrations (low micromolar), only types V and VI are inhibited by calcium (9,10,12), and this appears to be due to a direct effect of calcium on the AC, independent of CaM or protein phosphorylation. *In vivo*, cAMP accumulation has been shown to be inhibited following treatments that elevate intracellular calcium concentrations (59–61), and in some cases, the expression of either type V or type VI has been demonstrated in these cell types (9,10,12,62).

Stimulation of AC activity by calcium CaM has been demonstrated for the types I and VIII (14,41). *In vitro*, type III is conditionally activated by CaM, requiring $G_{s\alpha}$ stimulation for the effect (32). Sequences located within the C_{1a} region of AC1 are important for CaM activation. A synthetic peptide containing the amino acid sequences from residues 495–522 inhibited CaM activation of brain AC (63). A mutation within this region abolished CaM activation of the type I AC (64). This sequence has a hydrophobic/basic composition and an aromatic amino acid in its N-terminal portion, typical of most CaM-binding domains, and is therefore likely to function as the CaM-binding domain in the cyclase.

Despite the observation that type III is activated by CaM *in vitro*, activation of G_q-coupled pathways *in vivo* leads to inhibition of type III activity (65). This inhibition is accompanied by an increase in phosphorylation of AC3 that is antagonized by a CaM kinase inhibitor. Coexpression of constitutively active mutants of CaM kinase II with type III cyclase mimics the hormonal inhibitory effect. These observations point to a role of calcium inhibition mediated through a CaM/CaM kinase–dependent pathway.

Much attention has focused on the role of phosphorylation by PKC in regulating AC activity since the initial report that AC purified from brain can be directly phosphorylated by this kinase (66). Recently, *in vivo* transfection experiments have demonstrated that the activities of many ACs can be stimulated following phorbol ester treatment, suggesting that PKC can regulate ACs in an isoform-specific manner. Mixed results have been reported for the effect of phorbol ester treatment on type I activity; however, the consensus is that both the forskolin- and CaM-stimulated activities can be enhanced by activation of PKC (67–69). Phorbol esters can likewise enhance the forskolin-stimulated type III activity (67). Calcium-sensitive PKC activation has not been observed for the type VI isoform, and none have been reported for type VIII or IX.

Both *in vivo* and *in vitro* approaches have demonstrated that type V AC activity can be augmented by calcium-sensitive PKC (70,71). *In vitro*, type V can be directly phosphorylated by PKC α. An additional point of these studies demonstrated that protein kinase ζ, a calcium-insensitive isoform regulated by receptor tyrosine kinases, also phosphorylated and augmented type V activity, suggesting cross-talk between tyrosine kinase signaling and the AC system.

The most extensively studied effects of regulation by PKC have focused on the types II, IV, and VII subfamilies. *In vivo*, types II and VII $G_{s\alpha}$-stimulated activities are augmented by phorbol esters (15,68,69,72,73). Differences in the rates of activation and subsequent inactivation of types II and VII in response to phorbol ester treatment have been noted (15) and may reflect differences in sequences surrounding the phosphorylation sites or specificity of these cyclases for different PKC isoforms. Type II AC phosphorylation by PKC α has been demonstrated both *in vitro* and *in vivo* (52,74). *In vivo* approaches using type I/type II chimeric constructs have identified a region located within the C_2 domain of type II important for phorbol ester responsiveness (75); however, this region lacks any serine or threonine residues, and therefore is unlikely to be the site of phosphorylation by PKC.

In vitro analysis of the effects of PKC on types II and IV has revealed an intricate modulatory role of AC phosphorylation on G protein subunit responsiveness (52). PKC enhances type II stimulation by forskolin and responsiveness to $G_{s\alpha}$. By contrast, activation of type IV by forskolin is unaffected by PKC; however, $G_{s\alpha}$ is dramatically inhibited, principally because of decreases in $G_{s\alpha}$ responsiveness and probably V_{max}. Type IV is the only cloned isoform tested and shown to be inhibited by PKC and may serve to explain the inhibitory effects of phorbol esters on the $G_{s\alpha}$-stimulated AC activity observed in some mammalian cell lines (76,77). Despite increasing $G_{s\alpha}$ responsiveness of type II, PKC blocks $\beta\gamma$ superstimulation of this isoform; this block of $\beta\gamma$ superstimulation may be due to an inability of $G_{s\alpha}$ to expose the high-affinity $\beta\gamma$ site on type II cyclase following phosphorylation. Type IV is similarly regulated by PKC with respect to the block of $\beta\gamma$ superstimulation; type VII has not been examined.

PATTERNS OF ADENYLYL CYCLASE REGULATION

Adenylyl cyclase activity, and ultimately intracellular cAMP concentrations, can be regulated by a number of hormonal inputs coupled to heterotrimeric G protein–mediated pathways. An important feature of the membrane-bound AC isoforms is their ability to integrate multiple signals derived from distinct G protein–mediated pathways. Hormone receptors coupled to the G_s and G_i subclass of G proteins can directly activate or inhibit AC activity through the interactions of the released α and $\beta\gamma$ subunits with the cyclase. Another level of regulation is provided by the crosstalk of distinct G protein–mediated signaling pathways with AC isoforms. Hormonal activation of a G_q-coupled receptor leading to the activation of phospholipase C, production of diacylglycerol and IP_3, and release of intracellular calcium can have effects on the activities of different AC isoforms.

The simplest pattern of regulation, diagrammed in Fig. 3, is evident for the type V/VI subfamily. Hormonal stimulation of G_s pathways leads to activation of these cyclases that is mediated by the direct binding of the released GTP-bound α subunit. Inhibition by G_i- or G_z-coupled pathways proceeds via an analogous mechanism. Hormonal stimulation of G_q leads to intracellular calcium increases and PKC activation that seemingly give rise to two opposing regulatory signals (at least for the type V isoform that is stimulated by PKC); *in vivo*, the inhibitory calcium input is predominantly observed.

Regulatory patterns for types I and VIII provide additional features and are shown in Fig. 4. Hormonal stimulation of G_s proceeds as outlined earlier, leading to $G_{s\alpha}$-mediated activation. Additional inhibitory inputs are provided by $G_{o\alpha}$ and $\beta\gamma$ subunits released upon hormonal activation of G_o or G_i pathways. Hormonal activation of G_q leads to the elevation of intracellular calcium, resulting in a stimulatory signal that is mediated by CaM, although effects of PKC have been demonstrated.

As diagrammed in Fig. 5, types II, IV, and VII exhibit almost identical regulatory

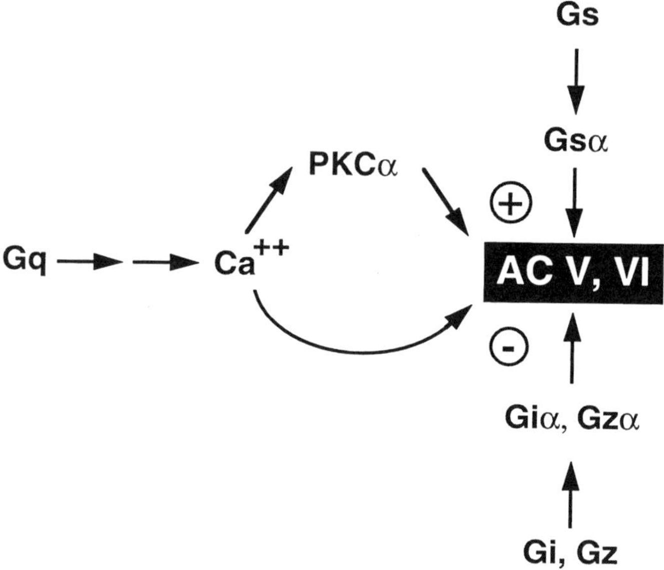

FIG. 3. Regulatory patterns of the type V/VI AC subfamily.

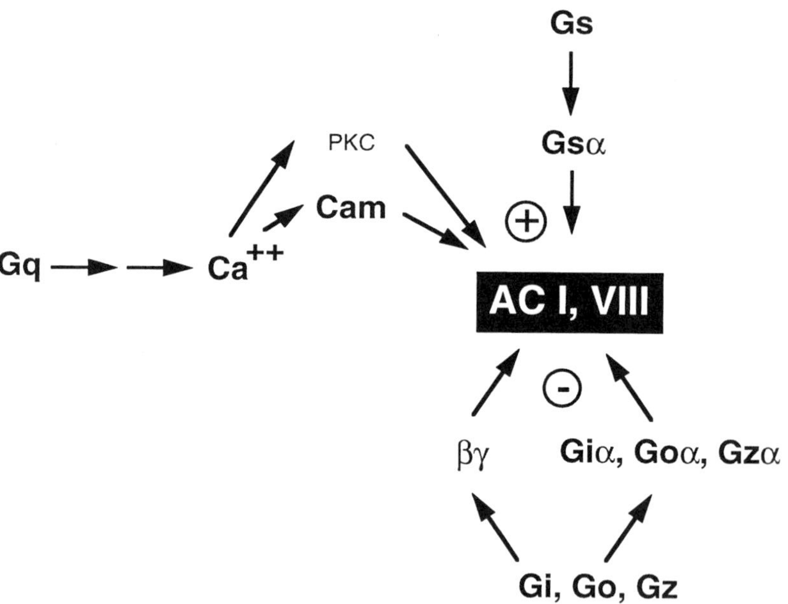

FIG. 4. Regulatory patterns of the type I/VIII AC subfamily.

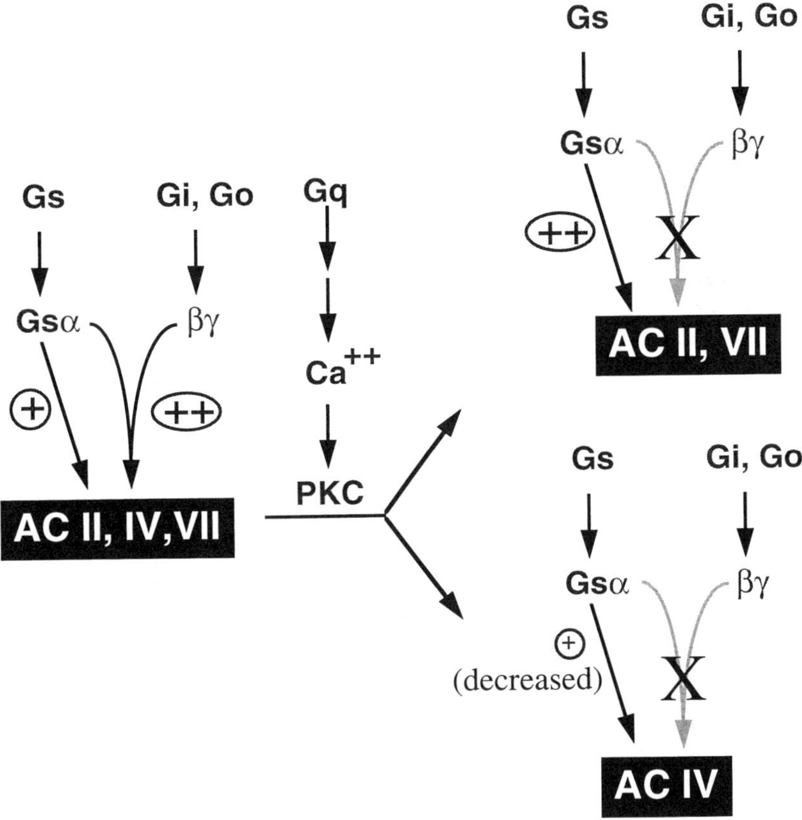

FIG. 5. Regulatory patterns of the type II/IV/VII AC subfamily.

patterns with respect to their direct responses to G_s and G_i subunits. All these ACs can further integrate hormonal inputs from G_q-coupled receptors that are mediated by PKC. All are activated by hormonal stimulation mediated by $G_{s\alpha}$ and are super-stimulated by $G_{s\alpha}$ and $\beta\gamma$ under conditions of coincidental G_s and G_i activation by hormones; G_i and G_o inputs are therefore (conditionally) stimulatory. These isoforms are, however, affected quite differently by PKC. Hormonal activation by $G_{s\alpha}$ is sensitized for types II and VIII following G_q activation; however, the superstimulation by activation of G_i pathways is blocked for type II (and presumably type VII). By contrast, G_q activation of type IV diminishes the responses to both hormonal pathways mediated by G_s and G_i.

Types III and IX do not fall into any of these regulatory classes, but their regulatory patterns are displayed in Fig. 6 for completeness. Very little is known of the regulation of type IX other than that it is stimulated by forskolin and $G_{s\alpha}$. Type III is activated by G_s-coupled receptor stimulation (through the action of $G_{s\alpha}$) and is

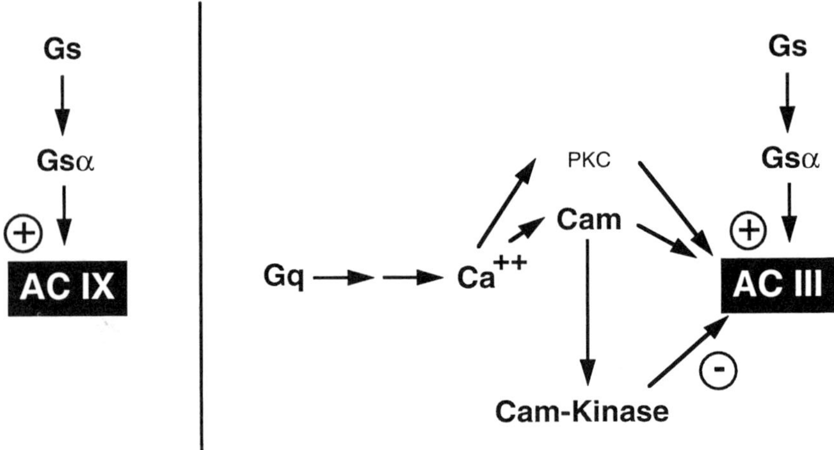

FIG. 6. Regulatory patterns of the type IX/III AC subfamily.

seemingly unresponsive to G_i or G_o regulation. Hormonal input to G_q and calcium appears to proceed via three regulatory molecules—PKC, CaM, and CaM kinase—with the observed *in vivo* effect being the inhibitory signal.

Our understanding of the hormonal control of intracellular cAMP has come a long way since the discovery of AC and even further since the application of molecular genetic approaches. Identification of a family of genes encoding nine AC isoforms has increased our appreciation for the intricate complexity of the AC signaling system. Clearly, it is impossible to label a hormonal pathway stimulatory or inhibitory to AC activity per se because of the different ways these cyclase isoforms integrate this information. A major avenue of research will be to continue to define the regulatory repertoires of AC isoforms and to couple this with anatomic information concerning tissue and subcellular localization, and genetic "knockout" approaches, in an effort to understand the precise physiological role of the AC family members.

ACKNOWLEDGMENTS

Research in R. T.'s laboratory was supported by the United States Public Health Service grant GM 53645, University of Michigan Comprehensive Cancer Center, and the Burroughs Wellcome Foundation.

REFERENCES

1. Pilkis SJ, Claus TH, el Maghrabi MR: The role of cyclic AMP in rapid and long-term regulation of gluconeogenesis and glycolysis. *Adv Second Messenger Phosphoprotein Res* 1988;22:175–91.

2. Kandel E, Abel T: Neuropeptides, adenylyl cyclase, and memory storage. *Science* 1995;268:825–826.
3. Roger PP, Reuse S, Maenhaut C, Dumont JE: Multiple facets of the modulation of growth by cAMP. *Vitam Horm* 1995;51:59–191.
4. Pfeuffer T, Metzger H: 7-O-Hemisuccinyl-deacetyl forskolin-Sepharose: a novel affinity support for purification of adenylate cyclase. *Febs Lett* 1982;146:369–375.
5. Krupinski J, Coussen F, Bakalyar HA et al: Adenylyl cyclase amino acid sequence: possible channel- or transporter-like structure. *Science* 1989;244:1558–1564.
6. Feinstein PG, Schrader KA, Bakalyar HA et al: Molecular cloning and characterization of a Ca^{2a}/calmodulin insensitive adenylyl cyclase from rat brain. *Proc Natl Acad Sci U S A* 1991;88: 10173–10177.
7. Bakalyar HA, Reed RR: A specialized adenylyl cyclase may mediate odorant detection. *Science* 1990;250:1403–1406.
8. Gao B, Gilman AG: Cloning and expression of a widely distributed (type IV) adenylyl cyclase. *Proc Natl Acad Sci U S A* 1991;88:10178–10182.
9. Katsushika S, Chen L, Kawabe JI et al: Cloning and characterization of a 6th adenylyl cyclase isoform: type-V and Type-VI constitute a subgroup within the mammalian adenylyl cyclase family. *Proc Natl Acad Sci U S A* 1992;89:8774–8778.
10. Ishikawa Y, Katsushika S, Chen L et al: Isolation and characterization of a novel cardiac adenylyl cyclase cDNA. *J Biol Chem* 1992;267:13553–13557.
11. Premont RT, Chen JQ, Ma HW et al: Two members of a widely expressed subfamily of hormone-stimulated adenylyl cyclase. *Proc Natl Acad Sci U S A* 1992;89:9809–9813.
12. Yoshimura M, Cooper D: Cloning and expression of a Ca^{2+}-inhibitable adenylyl cyclase from NCB-20 cells. *Proc Natl Acad Sci U S A* 1992;89:6716–6720.
13. Wallach J, Droste M, Kluxen FW et al: Molecular cloning and expression of a novel type V adenylyl cyclase from rabbit myocardium. *Febs Lett* 1994;338:264–266.
14. Cali JJ, Zwaagstra JC, Mons N et al: Type VIII adenylyl cyclase: a Ca^{2+}/calmodulin-stimulated enzyme expressed in discrete regions of rat brain. *J Biol Chem* 1994;269:12190–12195.
15. Watson PA, Krupinski J, Kempinski AM, Frankenfield CD: Molecular cloning and characterization of the type VII isoform of mammalian adenylyl cyclase expressed widely in mouse tissues and in S49 mouse lymphoma cells. *J Biol Chem* 1994;269:28893–28898.
16. Völkel H, Beitz E, Klumpp S, Schultz JE: Cloning and expression of a bovine adenylyl cyclase type VII specific to the retinal pigment epithelium. *FEBS Lett* 1996;378:245–249.
17. Premont RT, Matsuoka I, Mattei MG et al: Identification and characterization of a widely expressed form of adenylyl cyclase. *J Biol Chem* 1996;271:13900–13907.
18. Cali JJ, Parekh RS, Krupinski J: Splice variants of type VIII adenylyl cyclase: differences in glycosylation and regulation by Ca^{2+}/calmodulin. *J Biol Chem* 1996;271:1089–1095.
19. Iwami G, Akanuma M, Kawabe J et al: Multiplicity in type V adenylylcyclase: type V-a and type V-b. *Mol Cell Endocrinol* 1995;110:43–47.
20. Wilkie TM, Gilbert DJ, Olsen AS et al: Evolution of the mammalian G protein α subunit multigene family. *Nat Gen* 1992;1:85–91.
21. Mollner S, Beck K, Pfeuffer T: Acylation of adenylyl cyclase catalyst is important for enzymic activity. *FEBS Lett* 1995;371:241–244.
22. Schultz JE, Klumpp S, Benz R et al: Regulation of adenylyl cyclase from *Paramecium* by an intrinsic potassium conductance. *Science* 1992;255:600–603.
23. Reddy R, Smith D, Wayman G et al: Voltage-sensitive adenylyl cyclase activity in cultured neurons: a calcium-independent phenomenon. *J Biol Chem* 1995;270:14340–14346.
24. Levin LR, Han PL, Hwang PM et al: The *Drosophila* learning and memory gene *rutabaga* encodes a Ca^{2+}/calmodulin-responsive adenylyl cyclase. *Cell* 1992;68:479–489.
25. Pitt GS, Milona N, Borleis J et al: Structurally distinct and stage-specific adenylyl cyclase genes play different roles in dictyostelium development. *Cell* 1992;69:305–315.
26. Chinkers M, Garbers DL: Signal transduction by guanylyl cyclases. *Annu Rev Biochem* 1991;60:553–575.
27. Koesling D, Herz J, Gausepohl H et al: The primary structure of the 70 kDa subunit of bovine soluble guanylate cyclase. *Febs Lett* 1988;239:29–34.
28. Tang WJ, Gilman AG: Construction of a soluble Gsα-and forskolin-activated adenylyl cyclase. *Science* 1995; 268: 1769–1772.
29. Whisnant RE, Gilman AG, Dessauer CW: Interaction of the two cytosolic domains of mammalian adenylyl cyclase. *Proc Natl Acad Sci U S A* 1996;93:6621–6625.

30. Yan SZ, Hahn D, Huang ZH, Tang WJ: Two cytoplasmic domains of mammalian adenylyl cyclase form a Gsα- and forskolin-activated enzyme *in vitro*. *J Biol Chem* 1996;271:10941–10945.
31. Tang WJ, Stanzel M, Gilman AG: Truncation and alanine-scanning mutants of type I adenylyl cyclase. *Biochemistry* 1995;34:14563–14572.
32. Choi EJ, Xia ZG, Storm DR: Stimulation of the type-III olfactory adenylyl cyclase by calcium and calmodulin. *Biochemistry* 1992;31:6492–6498.
33. Wayman GA, Impey S, Storm DR: Ca^{2+} inhibition of type III adenylyl cyclase in vivo. *J Biol Chem* 1995;270:21480–21486.
34. Tang WJ, Gilman AG: Type-specific regulation of adenylyl cyclase by G protein βγ subunits. *Science* 1991;254:1500–1503.
35. Taussig R, Quarmby LM, Gilman AG: Regulation of purified type-I and type-II adenylyl cyclases by G protein αυ subunits. *J Biol Chem* 1993;268:9–12.
36. Yoshimura M, Ikeda H, Tabakoff B: μ-Opioid receptors inhibit dopamine-stimulated activity of type V adenylyl cyclase but enhance dopamine-stimulated activity of type VII adenylyl cyclase. *Mol Pharmacol* 1996;50:43–51.
37. Taussig R, Tang WJ, Gilman AG: Expression and purification of recombinant adenylyl cyclases in Sf9 cells. *Meth Enzymol* 1994;238:95–108.
38. Jacobowitz O, Chen J, Iyengar R: Transient expression assays for mammalian adenylyl cyclases. *Meth Enzymol* 1994;238:108–16.
39. Robbins JD, Boring DL, Tang WJ et al: Forskolin carbamates: Binding and activation studies with type I adenylyl cyclase. *J Med Chem* 1996;39:2745–2752.
40. Sutkowski EM, Tang WJ, Broome CW et al: Regulation of forskolin interaction with type I, II, V, and VI adenylyl cyclases by Gsα. *Biochemistry* 1994;33:12852–12859.
41. Tang WJ, Krupinski J, Gilman AG: Expression and characterization of calmodulin-activated (type-I) adenylyl cyclase. *J Biol Chem* 1991;266:8595–8603.
42. Chen JQ, Iyengar R: Inhibition of cloned adenylyl cyclases by mutant-activated Giα and specific suppression of type-2 adenylyl cyclase inhibition by phorbol ester treatment. *J Biol Chem* 1993; 268:12253–12256.
43. Wayman GA, Impey S, Wu Z et al: Synergistic activation of the type I adenylyl cyclase by Ca^{2+} and Gs-coupled receptors in vivo. *J Biol Chem* 1994;269:25400–25405.
44. Jones DT, Reed RR: Golf, an olfactory neuron specific G protein involved in odorant signal transduction. *Science* 1989;244:790–795.
45. Taussig R, Tang WJ, Hepler JR, Gilman AG: Distinct patterns of bidirectional regulation of mammalian adenylyl cyclases. *J Biol Chem* 1994;269:6093–6100.
46. Dessauer CW, Gilman AG: Purification and characterization of a soluble form of mammalian adenylyl cyclase. *J Biol Chem* 1996;271:16967–16974.
47. Hazeki O, Ui M: Modification by islet-activating protein of receptor-mediated regulation of cyclic AMP accumulation in isolated rat heart cells. *J Biol Chem* 1981;256:2856–2862.
48. Katada T, Amano T, Ui M: Modulation by islet-activating protein of adenylate cyclase activity in C6 glioma cells. *J Biol Chem* 1982;257:3739–3746.
49. Taussig R, Iniguez-Lluhi J, Gilman AG: Inhibition of adenylyl cyclase by Giα. *Science* 1993; 261:218–221.
50. Nielsen MD, Chan GCK, Poser SW, Storm DR: Differential regulation of type I and type VIII Ca^{2+}-stimulated adenylyl cyclases by G_i-coupled receptors *in vivo*. *J Biol Chem* 1996;271:33308–33316.
51. Federman AD, Conklin BR, Schrader KA et al: Hormonal stimulation of adenylyl cyclase through G_i-protein βγ subunits. *Nature* 1992;356:159–161.
52. Zimmermann G, Taussig R: Protein kinase C alters the responsiveness of adenylyl cyclases to G protein α and βγ subunits. *J Biol Chem* 1996;271:27161–27166.
53. Chen J, DeVivo M, Dingus J et al: A region of adenylyl cyclase 2 critical for regulation by G protein betagamma subunits. *Science* 1995;268:1166–1169.
54. Herlitze S, Hockerman GH, Scheuer T, Catterall WA: Molecular determinants of inactivation and G protein modulation in the intracellular loop connecting domains I and II of the calcium channel alpha 1A subunit. *Proc Natl Acad Sci U S A* 1997;94:1512–1516.
55. Weng GZ, Li JR, Dingus J et al: β subunit interacts with a peptide encoding region 956–982 of adenylyl cyclase 2: cross-linking of the peptide to free βγ but not the heterotrimer. *J Biol Chem* 1996;271:26445–26448.

56. Yan K, Gautam N: Structural determinants for interaction with three different effectors on the G protein beta subunit. *J Biol Chem* 1997;272:2056–2059.
57. Wong YH, Conklin BR, Bourne HR: G_z-mediated hormonal inhibition of cyclic AMP accumulation. *Science* 1992;255:339–341.
58. Kozasa T, Gilman AG: Purification of recombinant G proteins from Sf9 cells by hexahistidine tagging of associated subunits. Characterization of alpha 12 and inhibition of adenylyl cyclase by alphaz. *J Biol Chem* 1995;270:1734–1741.
59. Narayanan N, Lussier B, French M et al: Growth hormone-releasing factor–sensitive adenylate cyclase system of purified somatotrophs: effects of guanine nucleotides, somatostatin, calcium, and magnesium. *Endocrinology* 1989;124:484–495.
60. Debernardi MA, Seki T, Brooker G: Inhibition of cAMP accumulation by intracellular calcium mobilization in C6–2B cells stably transfected with substance K receptor cDNA. *Proc Natl Acad Sci U S A* 1991;88:9257–9261.
61. Boyajian CL, Garritsen A, Cooper D: Bradykinin stimulates Ca^{2+} mobilization in NCB-20 cells leading to direct inhibition of adenylyl cyclase. *J Biol Chem* 1991;266:4995–5003.
62. Debernardi MA, Munshi R, Yoshimura M et al: Predominant expression of type-VI adenylate cyclase in C6–2B rat glioma cells may account for inhibition of cyclic AMP accumulation by calcium. *Biochem J* 1993;293:325–328.
63. Vorherr T, Knopfel L, Hofmann F et al: The calmodulin binding domain of nitric oxide synthase and adenylyl cyclase. *Biochemistry* 1993;32:6081–6088.
64. Wu Z, Wong ST, Storm DR: Modification of the calcium and calmodulin sensitivity of the type I adenylyl cyclase by mutagenesis of its calmodulin binding domain. *J Biol Chem* 1993;168:23766–23768.
65. Wei J, Wayman G, Storm DR: Phosphorylation and inhibition of type III adenylyl cyclase by calmodulin-dependent protein kinase II in vivo. *J Biol Chem* 1996;271:24231–24315.
66. Yoshimasa T, Sibley DR, Bouvier M et al: Cross-talk between cellular signalling pathways suggested by phorbol-ester-induced adenylate cyclase phosphorylation. *Nature* 1987;327:67–70.
67. Choi EJ, Wong ST, Dittman AH, Storm DR: Phorbol ester stimulation of the type-I and type-III adenylyl cyclases in whole cells. *Biochemistry* 1993;32:1891–1894.
68. Jacobowitz O, Chen J, Premont RT, Iyengar R: Stimulation of specific types of G_s-stimulated adenylyl cyclases by phorbol ester treatment. *J Biol Chem* 1993;268:3829–3832.
69. Yoshimura M, Cooper D: Type-specific stimulation of adenylyl cyclase by protein kinase C. *J Biol Chem* 1993;268:4604–4607.
70. Kawabe J, Iwami G, Ebina T et al: Differential activation of adenylyl cyclase by protein kinase C isoenzymes. *J Biol Chem* 1994;269:16554–16558.
71. Kawabe J, Ebina T, Toya Y et al: Regulation of type V adenylyl cyclase by PMA-sensitive and -insensitive protein kinase C isoenzymes in intact cells. *FEBS Lett* 1996;384:273–276.
72. Lustig KD, Conklin BR, Herzmark P et al: Type-II adenylylcyclase integrates coincident signals from G_s, G_i, and G_q. *J Biol Chem* 1993;268:13900–13905.
73. Tsu RC, Wong YH: G_i-mediated stimulation of type II adenylyl cyclase is augmented by G_q-coupled receptor activation and phorbol ester treatment. *J Neurosci* 1996;16:1317–1323.
74. Jacobowitz O, Iyengar R: Phorbol ester-induced stimulation and phosphorylation of adenylyl cyclase 2. *Proc Natl Acad Sci U S A* 1994;91:10630–10634.
75. Levin LR, Reed RR: Identification of functional domains of adenylyl cyclase using *in vivo* chimeras. *J Biol Chem* 1995;270:7573–7579.
76. Gorman RR, Benjamin CW, Tarpley WG: Inhibition of adenylate cyclase and phospholipase A2/C in NIH-3T3 cells expressing the EJ-ras oncogene. *Adv Prostaglandin Thromboxane Leukot Res* 1987;17B:963–8 1987.
77. Gusovsky F, Gutkind JS: Selective effects of activation of protein kinase C isozymes on cyclic AMP accumulation. *Mol Pharmacol* 1991;39:124–129.
78. Villacres EC, Xia Z, Bookbinder LH et al: Cloning, chromosomal mapping, and expression of human fetal brain type I adenylyl cyclase. *Genomics* 1993;16:473–478.
79. Stengel D, Parma J, Gannage MH et al: Different chromosomal localization of two adenylyl cyclase genes expressed in human brain. *Hum Genet* 1992;90:126–30.
80. Haber N, Stengel D, Defer N et al: Chromosomal mapping of human adenylyl cyclase genes type III, type V and type VI. *Hum Genet* 1994;94:69–73.
81. Edelhoff S, Villacres EC, Storm DR, Disteche CM: Mapping of adenylyl cyclase genes type I, II, III, IV, V, and VI in mouse. *Mamm Genome* 1995;6:111–113.

82. Hellevuo K, Berry R, Sikela JM, Tabakoff B: Localization of the gene for a novel human adenylyl cyclase (ADCY7) to chromosome 16. *Hum Genet* 1995;95:197–200.

83. Xia ZG, Choi EJ, Fan W et al: Type-I calmodulin-sensitive adenylyl cyclase is neural specific. *J Neurochem* 1993;60:305–311.

84. Xia ZG, Choi EJ, Wang F, Storm DR: The type-III calcium/calmodulin-sensitive adenylyl cyclase is not specific to olfactory sensory neurons. *Neurosci Lett* 1992;144:169–173.

85. Hellevuo K, Yoshimura M, Mons N et al: The characterization of a novel human adenylyl cyclase which is present in brain and other tissues. *J Biol Chem* 1995;270:11581–11589.

86. Krupinski J, Lehman TC, Frankenfield CD et al: Molecular diversity in the adenylylcyclase family: evidence for 8 forms of the enzyme and cloning of type VI. *J Biol Chem* 1992;267:24858–24862.

Advances in Second Messenger and Phosphoprotein Research, Vol. 32,
edited by Dermot M. F. Cooper
Lippincott–Raven Publishers, Philadelphia © 1998

5

Regulation of cAMP Signaling by Phosphorylation

Yoshihiro Ishikawa

Cardiovascular and Pulmonary Research Institute, Allegheny University of the Health Sciences, Pittsburgh, Pennsylvania 15212

The forces generated by proteins determine the directed biologic movements within cells. Binding of a low-molecular-weight ligand, for example, can change a protein molecule into an active form, as seen in receptors. Such a ligand may bind loosely to the protein molecule and readily dissociate, thus allowing reversible conformational changes of the protein. This is an easy way to regulate the function of proteins. An input of chemical energy, instead of a ligand, is also used to convert the conformation of a protein. A typical example is the transfer of a phosphate group from adenosine triphosphate (ATP) to serine, threonine, or tyrosine residues in the protein and the formation of a covalent linkage. A large amount of energy released from ATP hydrolysis is stored in maintaining the protein conformation. Indeed, it is possible to make ATP *in vitro* by adding adenosine diphosphate (ADP) to such a phosphorylated protein. In eukaryotic cells, protein phosphorylation is a common regulatory mechanism to activate or inhibit the function of specific proteins.

Kinases are the phosphorylating enzymes that catalyze such transfer of the phosphoryl group from ATP or another nucleoside triphosphate to alcohol or amino group acceptors of protein. Numerous pathways involve kinase activation in eukaryotic cells, including the well-studied β-adrenoceptor/G protein/adenylyl cyclase (AC) signaling pathway. This pathway activates cyclic adenosine monophosphate (cAMP)–dependent kinase or protein kinase A (PKA) and is simultaneously regulated by PKA and other kinases. Because the components within the β-adrenoceptor/G protein/ AC signaling pathway are functionally interdependent, the individual components are usually difficult to examine separately. Therefore, in this review, I will describe the general aspect of kinase-mediated regulation of the cAMP signaling pathway, with emphasis on that of the catalyst AC.

THE β-ADRENOCEPTOR SIGNALING PATHWAY

β-Adrenoceptor signaling is initiated by the release of catecholamine from the synaptic terminal. It is a major sympathetic signaling pathway to regulate the function of many organs, including the heart. Occupation of cell surface catecholamine receptors by norepinephrine evokes signaling through the stimulatory GTP regulatory protein $G_{s\alpha}$, which in turn activates AC. AC is a membrane-bound enzyme that catalyzes the conversion of ATP to cAMP. cAMP, an intracellular second messenger, activates PKA through dissociating the catalytic subunit from the regulatory subunit. The catalytic subunit of PKA now initiates an enzymatic cascade of phosphorylation reactions within the cell—for example, various enzymes involved in myocyte contraction in the heart, such as troponin (1), phospholamban (2), and those involved in glucose metabolism in the liver, such as glycogen phosphorylase (3). By phosphorylating uniquely differentiated proteins in each cell type, the catecholamine signaling and thus PKA regulate the unique function of each organ. cAMP is eventually degraded to AMP by phosphodiesterase (for review see ref. 4). Protein kinase A is inactivated by reassociation of the catalytic subunit with the regulatory subunit (for review see ref. 5). Phosphorylated proteins are dephosphorylated by phosphatases, thereby pulling back the protein conformation into an inactive form. It is well known that this whole process is regulated by various kinases, including PKA.

β-ADRENOCEPTOR

The regulation of the β-adrenoceptor via phosphorylation is a prime example of kinase-mediated regulation of the cAMP signaling pathway (for review see ref. 6). Although this paradigm has been demonstrated mostly with the β_2-subtype, the β_1-subtype seems to follow this paradigm as well (7). The phosphorylation of the receptor has been examined as a principal mechanism of desensitization. When agonist stimulation of the β-adrenoceptor is prolonged, intracellular cAMP levels generally plateau or even return to near basal levels within minutes. This decay of the stimulated response despite the presence of continuous agonist stimulation has been termed *desensitization*. This phenomenon is not unique to the cAMP signaling system. Similar attenuation of the signaling observed in other hormone receptor systems in many species is variously called tachyphylaxis, quenching, or adaptation. It may also be a general mechanism to protect cells from the overload of signal inputs.

At least four groups of kinases are involved in the phosphorylation of the β-adrenoceptor: β-adrenergic receptor kinase (βARK), PKA, protein kinase C (PKC), and tyrosine kinase (8). However, only the initial two have been extensively studied as a mechanism of desensitization.

Protein Kinase A

Protein kinase A is the major effector kinase in the cAMP signaling pathway; thus, PKA-mediated phosphorylation of the β-adrenoceptor may be a good example of desensitization as a result of negative feedback. Protein kinase A–mediated phosphorylation also occurs to other receptors because the stimulation of these receptors, such as prostaglandin E1 receptor, can also activate PKA; activation of PKA potentially leads to phosphorylation of multiple receptors, including the β-adrenoceptor. As a result of phosphorylation of heterologous receptors, cells become refractory to multiple classes of agonists; this process is termed *heterologous desensitization.*

Within the β$_2$-adrenoceptor there are two sites with a consensus sequence for PKA phosphorylation. They are located in the third intracellular loop and the carboxyl terminal domain (9). Since these phosphorylation sites are proximate to the site for G protein interaction, it is reasonable to find that phosphorylated β-adrenoceptor is uncoupled from the G protein (10). The functional significance of PKA-mediated phosphorylation has been confirmed in studies using reconstitution into lipid vesicles, site-directed mutagenesis, and specific kinase inhibitors (11,12). These studies demonstrate, for example, that the two PKA phosphorylation sites had different functions. The one located in the third cytoplasmic loop was involved more in the rapid desensitization in response to low concentration of agonist (13). Deletion of the consensus site for PKA did not alter the receptor-stimulated AC activity, but impaired the ability of the receptor to undergo phosphorylation by PKA and desensitization (11). Similarly, desensitization was attenuated by specific inhibitors of PKA in permeabilized cells (14). These data confirmed the involvement of this kinase in desensitization at the level of the receptor.

β-Adrenergic Receptor Kinase

In the past decade, βARK has been examined most extensively as a mechanism of desensitization. βARK was originally identified in a crude cytosolic fraction of S49 lymphoma cells as an activity that specifically phosphorylated the agonist-occupied form of the β$_2$-adrenoceptor (15). In the case of βARK, the functional phosphorylation sites lie in the distal portion of the carboxyl terminal segment of the receptor. βARK is an analog of rhodopsin kinase in the retina, which phosphorylates the active form of rhodopsin. Therefore, many, if not all, regulations of the β-adrenoceptor by βARK follow the paradigm of rhodopsin. Indeed, augmentation of phosphorylation by activation of the receptor is quite dramatic in both rhodopsin and βARK. Therefore, βARK is a major mechanism of inducing homologous desensitization.

It is now known that βARK, as well as rhodopsin kinase, is a member of a large family of enzymes that phosphorylate numerous G protein–coupled receptors and thus is termed GRK (*G* protein–coupled *r*eceptor *k*inase) (16). So far, six full-length complementary DNAs (cDNAs) encoding members of the GRK subfamilies have been published. Although the amino acid sequence varies significantly among subtypes, a putative catalytic domain shows relatively conserved amino acid sequence among subtypes and even among other kinases. In contrast to the amino terminus catalytic domain, the carboxyl terminus domain is proposed to be important for membrane targeting. It is interesting that this domain shows the highest degree of variation among subtypes. For example, GRK-1 (or rhodopsin kinase) has a consensus sequence of CAAX motif, through which GRK-1 is farnesylated and thus attached to the membrane. In contrast, GRK-2 and -3 (or βARK-1 and -2) have a unique domain though which GRK binds to G protein βγ subunits.

Phosphorylation of receptors by βARK is not a simple negative feedback mechanism. It also phosphorylates receptors that inhibit cAMP signaling. βARK phosphorylates the α_2-adrenoceptor, which inhibits AC through $G_{i\alpha}$. Phosphorylation of the reconstituted α_2-adrenoceptor was dependent on agonist occupancy and was completely blocked by co-incubation with α_2-antagonists. The time course of phosphorylation of the α_2-adrenoceptor receptor was virtually identical to that observed with the β-adrenoceptor with maximum stoichiometry of 7–8 mol of phosphate/mol of receptor in each case (17). βARK also phosphorylated the muscarinic cholinergic receptor (18,19). These findings suggest that βARK has broad substrate specificity of the substrate receptor as long as the receptor is in an active form.

Then what determines the target specificity of βARK? It is generally accepted that βARK translocates to the membrane using G protein βγ subunits as its anchor. Addition of βγ subunits to βARK-1 augmented both its activity and its association to the membrane *in vitro* (20). Unlike rhodopsin kinase, βARK does not have a consensus sequence for acylation for the attachment to the membrane and thus exists as a cytosolic kinase. Agonist-induced release of the βγ subunit, followed by translocation of βARK therefore occurs by a mechanism distinct from that of other kinases. The presence of the βγ binding domain in βARK seems to add another important role of βARK. βARK by itself may regulate signaling through the βγ subunits. Overexpression of the carboxyl terminus of βARK, for example, attenuated the phospholipase C–mediated phosphoinositol hydrolysis signaling through βγ subunits in cells (21). It is recognized that βγ subunits by themselves transmit multiple signals, such as phospholipase C, ion channel activity, and MAP kinase signaling pathways (22). The carboxyl terminus of βARK indeed has a sequence similar to that of the pleckstrin homology (PH) domain, which is a common domain conserved in Ras-GAP, GRF, and rac kinase molecules (23,24). These findings suggest that βARK, together with the βγ subunits, provides a site of cross-talk between the cAMP signaling and the PH signaling pathways.

The role of βARK *in vivo* has been examined. Overexpression of βARK augmented desensitization of β_2-adrenoceptor receptors in cells (25). Dominant negative overexpression of βARK attenuated desensitization of the β-adrenoceptor but not that of the prostaglandin E_2 receptor (26). Involvement of βARK in modulating catecholamine action at the moment of birth (27) and in pathophysiologic changes was also demonstrated (28). In a recent study, transgenic mice overexpressing βARK showed attenuation of isoproterenol-stimulated AC activity as well as cardiac contractility (29).

Protein Kinase A vs β-Adrenergic Receptor Kinase

What is then the functional difference between PKA- and βARK-mediated phosphorylation of β-adrenoceptor? There are several findings to suggest this difference. Phosphorylation of β-adrenoceptor by PKA can occur in the presence of low agonist concentration (~nanomolar). Because of spare receptors this concentration of agonist is enough to fully activate AC catalytic activity within the cell. In contrast, at higher agonist concentration, (~micromolar) βARK was also activated (11). Since such high concentration of agonist is rarely achieved in circulating blood, βARK-mediated desensitization may be a mechanism that takes place only in certain locations, such as postsynaptic membranes; catecholamines are released from the synaptic terminus into a narrow synaptic cleft, achieving a high concentration of agonist (30).

Desensitization initiated by βARK appears to be independent of or parallel to that mediated by PKA (31). Furthermore, phosphorylation by βARK promoted interaction of the receptor with β-arrestin, which binds to the phosphorylated receptor, thereby preventing further activation of $G_{s\alpha}$ (32,33). Phosphorylation of β-adrenoceptor followed by the binding of β-arrestin also initiates the sequestration of the receptor, leading to the down-regulation of the receptor. Overexpression of β-arrestin and βARK rescued the sequestration of a β-adrenoceptor mutant that lacks the ability to sequestrate (34). Overexpression of β-arrestin mutants in return inhibited wild-type β-adrenoceptor sequestration. These findings suggest that β-arrestin acted as an adapter molecule in this process.

Nevertheless, it is important to emphasize that, for both kinases, β-adrenoceptor phosphorylation occurs only when the signaling pathway is activated. It is triggered either by agonist occupation of the receptor or by cAMP formation as a result of receptor activation, thus forming a closed loop of negative feedback in both cases.

Protein Kinase C

The role of PKC-mediated phosphorylation of β-adrenoceptor is less well understood. Protein kinase C is the major effector kinase in the other catecholamine signaling pathway, α-adrenoceptor signaling. Stimulation of the α-adrenoceptor, in particular the α_1-subtype, activates G_q and phospholipase C, which leads to the formation of diacylglycerol, which in turn activates PKC. Therefore, the release of catecholamine from the synaptic terminal results in simultaneous activation of PKA and PKC within the cell. Regulation of the β-adrenoceptor by PKC constitutes cross-talk within the catecholamine signaling. Physiologically, the two pathways, α- and β-adrenoceptor signaling, often lead to opposite regulation *in vivo*. For example, in the airway smooth muscle, stimulation of α_1-adrenoceptor followed by Ca mobilization and activation of PKC leads to increased airway smooth muscle tone (35). In contrast, β-adrenoceptor stimulation results in relaxation of airway smooth muscle (36). Thus, phosphorylation of β-adrenoceptor by PKC may be a feedback mechanism to balance the two signaling pathways *in vivo*.

Desensitization of β-adrenoceptor has been demonstrated after phorbol ester treatment of cells (37–40), although the effect may vary among cells (41). With purified receptor and kinases reconstituted *in vitro*, phosphorylation of the receptor inhibited its coupling to $G_{s\alpha}$ by 60% (PKA), 40% (PKC), and 30% (βARK) (42). This desensitization was most pronounced with low Mg concentration in the reaction mixture; high Mg concentration (>5 mM) obscured the PKC-mediated desensitization (40,42). There are at least two sites that may be phosphorylated by both PKA and PKC within the β_2-adrenoceptor. These two sites are located very close to regions of the third cytoplasmic loop and proximal carboxyl terminal that are involved in G protein coupling (10). It is therefore not surprising that the phosphorylation of these residues alters the $G_{s\alpha}$ coupling. Within the PKA/C consensus phosphorylation site in the third intracellular loop of the β_2-adrenoceptor, there are two serine residues (RRSSK263); serine 262 was the primary site of the PKA-induced desensitization, whereas either serine 261 or serine 262 was sufficient to confer the PKC-mediated desensitization. Coincident stimulation of PKA and PKC caused an additive desensitization that was significantly reduced only by the double substitution mutation (43). In the preceding experiment, the β-adrenoceptor expression level also played a critical role in determining the pattern of β-adrenoceptor desensitization (43).

Tyrosine Kinase

The newest paradigm for the regulation of the β-adrenoceptor by phosphorylation may be that by tyrosine kinase. A series of experiments supports this paradigm. Cells stimulated with insulin showed loss of function and increased phosphotyrosine content of β_2-adrenoceptor (44). The insulin receptor phosphorylated synthetic peptides corresponding to cytoplasmic domains of the β_2-adrenoceptor *in vitro*. Peptide mapping of the β_2-adrenoceptor phosphorylated by insulin in cells revealed that tyrosyl residues 350/354 and 364 in the cytoplasmic C-terminal region were the primary targets (45). With use of the recombinant human insulin receptor and the hamster β_2-adrenoceptor in reconstitution and phosphorylation assays, insulin was shown to stimulate insulin receptor–catalyzed phosphorylation of the β_2-adrenoceptor. Phosphoamino acid analysis established again that the site of phosphorylation of the β_2-adrenoceptor receptor *in vitro* was confined to tyrosine residues (46). Therefore, clearly, tyrosine residues of the β-adrenoceptor were phosphorylated; however, the biologic significance of this phosphorylation and how it differs from that with other kinases need to be further studied.

G PROTEINS

$G_{s\alpha}$ Protein

Several reports suggest the involvement of the stimulatory G protein $G_{s\alpha}$ in the process of desensitization (47–49). After prolonged hormonal stimulation, functional alterations of $G_{s\alpha}$ protein associated with heterologous desensitization were determined by reconstitution assays with cyc⁻ cell membranes. Indeed, addition of purified $G_{s\alpha}$ to desensitized cell membranes partially restored $G_{s\alpha}$-stimulated AC activity (48). Because $G_{s\alpha}$ is more abundant than receptors in the cell (50), it is reasonable to assume that postreceptor mechanisms, including phosphorylation of $G_{s\alpha}$, are involved in heterologous desensitization. However, changes in the function of $G_{s\alpha}$ were not universally observed in other cell systems (51). No direct phosphorylation of $G_{s\alpha}$ by PKA and PKC has been shown so far.

$G_{s\alpha}$ may be phosphorylated and regulated by tyrosine kinases. Overexpression of pp60c-src resulted in hypersensitivity to agonist-induced signaling through β-adrenoceptors (52). In reconstituted lipid vesicles, phosphorylation of $G_{s\alpha}$ by pp60c-src resulted in enhanced rates of receptor-mediated GTPγS binding and GTP hydrolysis (53). Tryptic phosphopeptide analysis of phosphorylated $G_{s\alpha}$ showed that $G_{s\alpha}$ was phosphorylated by pp60c-src on two residues (54) located either near the site of βγ binding in the N-terminus that modulate GDP dissociation and activation by GTP (55), or near the site in the C-terminus that interacts with the receptor (56). Tyrosine phosphorylation of $G_{s\alpha}$ may also be catalyzed by other tyrosine kinases, such as the epidermal growth factor receptor (EGFR) (57). EGFR kinase phosphorylated $G_{s\alpha}$ with a stoichiometry of 2 mol of phosphate incorporated/mol of $G_{s\alpha}$. Phosphorylated $G_{s\alpha}$ showed greater AC activity when reconstituted in S49 cyc⁻ cell membranes than it did in nonphosphorylated $G_{s\alpha}$. Again, like that of the β-adrenoceptor, the significance of tyrosine phosphorylation of $G_{s\alpha}$ *in vivo* needs to be further studied.

$G_{i\alpha}$ Protein

In contrast to $G_{s\alpha}$, phosphorylation of $G_{i\alpha}$, in particular that by PKC, has been well demonstrated. It was originally shown that the $G_{i\alpha}$ served as an excellent substrate for the PKC (58). Treatment of S49 cyc⁻ lymphoma cells with the phorbol ester did not alter stimulation of AC catalytic activity by forskolin, whereas the $G_{i\alpha}$-mediated inhibition of the cyclase by somatostatin was impaired (58). Thus, it was proposed that PKC attenuated the inhibitory cAMP signaling pathway through phosphorylation of $G_{i\alpha}$, and it was further hypothesized that PKC activates cAMP signaling by relieving AC from the $G_{i\alpha}$-mediated tonic inhibition. This concept was indeed supported by studies using multiple cell systems (59–62). Phorbol ester treatment of cells also increased pertussis toxin–catalyzed ADP ribosylation. Since the heterotrimeric form of $G_{i\alpha}$ serves as a better substrate for the toxin than the monomeric form, these data suggested that phorbol ester treatment stabilized $G_{i\alpha}$ in an inactive heterotrimeric form.

However, it is important to note that the phosphorylation and inactivation of $G_{i\alpha}$ may not always explain the stimulatory effect of PKC on cAMP production. In pituitary adenoma cells, for example, phorbol ester enhanced forskolin- and cholera toxin-induced cAMP accumulation. However, this enhancement was unaffected by treatment with pertussis toxin (63). Another study demonstrated that isozymes of PKC are major determinants of the effect of PKC on cAMP signaling (64). The two PKC isozymes, α and γ, had opposite effects on the responses of the cAMP generation in NIH 3T3 cells. This may be an important concept because it is now clear, as described later, that PKC comprises a large enzyme family with multiple isozymes that have divergent tissue distribution and biochemical characteristics (65).

More recent studies have suggested that, even in the absence of hormonal stimulation, the activity of $G_{i\alpha}$ protein is under the control of a dynamic phosphorylation/dephosphorylation system with PKC and protein phosphatases. Immunoprecipitation of $G_{i\alpha}$ from hepatocytes labeled with ^{32}P showed that phosphorylation of $G_{i\alpha}$ under basal (resting) conditions was further enhanced by phorbol ester treatment of the cell (66). In neuroblastoma-glioma cells in culture, activation of PKC by phorbol ester resulted in a markedly attenuated activation of the inhibitory AC response to delta-opiate agonists and epinephrine. It is interesting that okadaic acid mimicked the effects of phorbol ester on the inhibitory response (67). A similar observation was made in intact hepatocytes (68). These findings suggest that the regulation of kinases and phosphatases may provide another important mechanism of cross-talk between PKC and cAMP signaling.

ADENYLYL CYCLASE

Direct regulation of AC by phosphorylation is another paradigm that has been examined by many investigators. Because AC is the effector enzyme within the cAMP signaling pathway, the regulation of AC by kinases allows the direct and the most efficient regulation of cAMP production within the cell. However, as described initially, this paradigm is difficult to examine because multiple other components within the signaling pathway, which are also subject to kinase-mediated regulation, modify the catalytic activity of AC. Nevertheless, there have been numerous reports to suggest that the site of regulation by kinases, in particular by PKC, is the catalyst itself within the cell.

Previous Studies

As mentioned earlier, treatment of different cell types with phorbol esters activated PKC and induced a desensitization of the AC activity stimulated with various receptor agonists (37,38,69,70). Phosphorylation of receptors by PKC may be a mechanism for these changes. However, this finding was not universal among different cell systems. For example, phorbol ester treatment enhanced the receptor-mediated stimulation of AC in some cell systems (71,72). In addition, potentiation of forskolin-, GTP analog-, and fluoride-stimulated AC activity by phorbol esters has been widely reported (73,74). These findings suggest that alterations in postreceptor components, including G proteins and AC, are involved in the action of phorbol esters; however, the degree of involvement of each component may differ among cell systems. Alternatively, different PKC isozymes expressed in each cell system may alter cAMP signaling differently.

Using PKC purified to homogeneity and rat adipocyte membranes purified through sucrose-gradient centrifugation, researchers demonstrated that PKC activated AC activity (75). Addition of the purified PKC was sufficient to stimulate AC activity in a dose-dependent manner in the absence of phorbol esters. This activation was calcium dependent; at a low concentration of calcium, calcium enhanced the effect of PKC, whereas at a high concentration, calcium inhibited the effect.

The preceding finding was thereafter confirmed in other cell and tissue systems (76). Phorbol ester treatment of frog erythrocyte produced phosphorylation and activation of the AC. Purified PKC directly phosphorylated AC purified from bovine brain, which indicates that the catalyst was directly phosphorylated by PKC. However, the study did not demonstrate the direct "activation" of the purified AC by the purified PKC. Therefore the involvement of $G_{i\alpha}$ remained as a possible mechanism of activating AC (77).

Protein kinase C–mediated phosphorylation of AC was also demonstrated with human platelet membranes. Phosphorylation of the catalyst AC was determined in membranes with use of specific antibodies raised against AC. Phorbol ester stimulation increased the incorporation of ^{32}P-phosphate into catalyst; the incorporation was also found in the absence of phorbol ester. AC in membranes from untreated platelets was significantly more inhibited by exogenously added $G_{i\alpha}$ than that from phorbol ester–treated cells. This suggests that phosphorylated AC was less subject to $G_{i\alpha}$-mediated inhibition (78).

The preceding findings suggest that PKC directly phosphorylates AC and alters its function, leading to either a modified response to $G_{i\alpha}$ or an increase in the catalytic activity itself.

Diversity in Protein Kinase C and Adenylyl Cyclase

The recent cloning studies of PKC and AC have made this issue more complicated to address. Protein kinase C, which was classically defined as calcium-activated serine/threonine kinase within the cell, now contains isozymes that do not require calcium for activation. Indeed, at least 10 isozymes of PKC have been isolated, which are subdivided into three classes: classic (α, β, and γ), novel (δ, ϵ, η, and ξ), and atypical isozymes (ζ, λ, and μ) (for review see ref. 65). They all share the common carboxyl terminus kinase domain (termed C_4) with the ATP binding site (C_3) but differ in the amino terminus regulatory domain. They show dramatic differences in the cysteine-rich, putative membrane-binding domain (C_1) as well as in the putative calcium-binding domain (C_2), which exists only in the classic isozymes. Thus the classic characteristics of calcium sensitivity are maintained only in this class of isozymes. The novel and atypical isozymes are insensitive to calcium. Furthermore, the atypical isozymes, which have only one cysteine-rich zinc fingerlike motif in C_1, are unaffected by diacylglycerol and phorbol ester. Rather, this class of isozymes is regulated by specific lipid products generated through the phosphatidylinositol 3-kinase pathway (79).

It is also clear that a single cell type expresses multiple PKC isozymes. Protein kinase C isozyme expression differs among tissues and cells (80) and among developmental stages (81). Differences in subcellular localization of various PKC isozymes have also been reported (82). More important, there is substrate specificity among PKC isozymes, even within the same class (83–85). These new findings indicate that phorbol ester treatment, which has been the major method of activating PKC within the cell in numerous experiments, may not always reflect the exact interaction between PKC and AC. Furthermore, different components within the cAMP signaling pathway—for example, $G_{i\alpha}$ and AC—may be phosphorylated by different PKC isozymes because of different substrate specificity among these isozymes. Alternatively, the same component, AC, for example, may be phosphorylated by multiple PKC isozymes at the same time, leading to different or mixed regulation of its function. All these issues should now be considered.

Similar diversity has been demonstrated in AC. So far, nine distinct full-length isoforms have been isolated from multiple sources (86–99). Detailed discussions of their biochemical diversity and tissue distribution are found in other chapters of this book. However, it should be emphasized that multiple isoforms of AC are also expressed in a single cell type, possibly by different subcellular components. The expression of these isoforms changes under normal development and pathophysiologic conditions (100,101).

Regulation of Adenylyl Cyclases Isoforms

Three independent studies simultaneously demonstrated that the catalytic activity of each AC isoform was enhanced by phorbol ester treatment of the cell (102–104).

Cyclic AMP accumulation in HEK-293 cells stably overexpressing either type I or type III calmodulin-sensitive isoforms was significantly increased in both time-dependent and phorbol ester concentration–dependent manners (104). Because the intracellular concentration of calcium was not elevated, it was reasonable to conclude that these isoforms were activated not through calmodulin signaling, but through PKC-mediated signaling. Thus, calmodulin-sensitive isoforms were also subject to PKC-mediated regulation. Direct phosphorylation of these AC isoforms by PKC was not clear in HEK-293 cells, but this was because of the lower efficiency of AC overexpression in the mammalian cell system. In another study, using HEK-293 cells transiently overexpressing AC isoforms, the effect of phorbol ester stimulation was compared among types I, II, and VI. A striking finding was that cAMP accumulation was dramatically higher in cells overexpressing the type II isoform, which suggests that a calmodulin-insensitive isoform was also regulated through PKC (103). The comparison was further expanded to all six isoforms isolated at that time (102). In HEK-293 cells transiently overexpressing AC isoforms, AC catalytic activity was measured with use of the lysed cell membranes in tissue culture wells. The type II isoform showed the greatest degree of stimulation by phorbol ester treatment. It is interesting that the degree of stimulation of this isoform was influenced by the concentration of Mg. This finding is reminiscent of the desensitization of the β-adrenoceptor by PKC, which was most pronounced at low Mg concentration; high Mg concentration (>5 mM) obscured the PKC-mediated desensitization (40,42).

These findings suggest that many, if not all, AC isoforms are subject to PKC-mediated activation and that the type II isoform is most sensitive to such activation (see Table 1 for summary). It is also important to point out that these studies examined the effect of phorbol ester–sensitive PKC isozymes in HEK-293 cells (i.e., classic and novel isozymes). Because of the lack of specific activator and inhibitor, the phorbol ester–insensitive atypical isozymes ζ and λ are currently difficult to examine in the preceding experimental system. Nevertheless, it was confirmed that AC isoforms are, to variable degrees among isoforms, regulated by the PKC signaling

TABLE 1. *Isoform-dependent regulation of adenylyl cyclase catalytic activity by various kinases*

	PKC	PKA	CaMK	PP2B	EGFR	References
I	$\pm \sim + +^a$ $+ +^b$				$\pm^b$	102–104,130
II	$+ + +^{a,b,c}$ $\pm \sim +^d$				$\pm^b$	102,103,111,112, 114,115,130
III	$+ +^a$ $+ + +^b$		$-^b$			102,104,132
IV	$+^a$ $\pm^c$					102,115
V	$+^a$ $+ +^b$ $+ + +^d$	$-^d$			$+ +^b$	102,113,121,124, 130
VI	$\pm \sim +^a$	$-^e$			$\pm^b$	102,103,123,130
VII	$+ + +^b$					95
VIII						
IX				$-^{a,e}$		92

−, inhibition; ±, no effect; +, weak stimulation; + +, good stimulation; + + +, very good stimulation.
[a]In transient overexpression system.
[b]In stable overexpression systems.
[c]In insect overexpression systems.
[d]In purified systems.
[e]In cell lines.

pathway. The effects of phorbol ester stimulation on receptor-stimulated cAMP production were not examined in these studies because AC isoforms overexpressed in cells were poorly coupled to hormonal activation (104).

Interaction With G Proteins

Differences in sensitivity to PKC stimulation among AC isoforms do not necessarily indicate that a particular isoform is more potently and directly phosphorylated and activated by PKC. It is also possible that an AC isoform may be more sensitive to G protein–mediated regulation than other isoforms. For example, $G_{i\alpha}$ protein may inhibit a certain isoform more potently; phosphorylation of $G_{i\alpha}$ protein by PKC, and thus release of "$G_{i\alpha}$-mediated tonic inhibition" from this isoform, may result in higher AC catalytic activity. In fact, a study demonstrated that each AC isoform had different sensitivity to G protein–mediated regulation (105). The type II isoform was more potently stimulated by $G_{s\alpha}$ protein than the type I, V, and VI isoforms when the degree of stimulation was standardized by that of forskolin.

With use of a GTPase-deficient mutant of $G_{i\alpha}$ protein (Q205L) and luteinizing hormone receptor transiently overexpressed in COS cells, many AC isoforms (types II, III, and VI) were shown to be sensitive to $G_{i\alpha}$-mediated inhibition, although not to the same degree (106). This result seems reasonable because $G_{i\alpha}$-mediated inhibition of AC catalytic activity varies among tissues or cell types. The most striking finding in this study was that phorbol ester treatment abolished the $G_{i\alpha}$-mediated inhibition of the type II isoform but not that of other isoforms, which suggests that the stimulation of the type II isoform by phorbol ester was, at least partially, due to the release of the tonic inhibition of $G_{i\alpha}$ protein (106).

Conditional Regulation

Does the stimulatory effect of PKC on AC then result solely from the G protein–mediated mechanisms? The answer is probably no. The regulation of AC isoforms, including type II, appears to be more conditional than originally speculated. For example, overexpressed AC isoforms in cells did not seem to be well coupled to hormonal stimulation in the absence of concomitant stimulation from other pathways (107). In contrast, when isoforms were overexpressed in insect cells or purified from insect cells, addition of the purified $G_{s\alpha}$ did stimulate AC (108). Many AC isoforms were subject to $G_{i\alpha}$-mediated inhibition in cells (106), whereas only one or two isoforms were inhibited when the purified $G_{i\alpha}$ and membranes overexpressing AC isoforms were examined (109). These conflicting findings suggest that the pattern of regulation of AC alters with the coexistence of other regulatory signals and in different sample preparations.

At least two independent studies showed that the regulation of AC by PKC was indeed conditional. In HT4 cells, basal cAMP production by types I and VI was unaffected by phorbol esters, nor did phorbol esters have any effect on forskolin-induced cAMP production (110). However, phorbol esters synergistically increased cAMP production when adrenoceptor was simultaneously activated. The synergistic activation of type II AC between PKC and G protein–mediated signaling was also demonstrated in HEK-293 cells (111). Phorbol ester treatment slightly elevated cAMP in cells overexpressing type II alone; this elevation was greatly augmented when the cells were co-transfected with $G_{s\alpha}$-Q227L, a constitutively active mutant of $G_{s\alpha}$. These findings suggest that PKC acts well on an activated catalytic component but poorly on the resting state of the catalyst.

Similarly, two sets of experiments using particular AC isoforms overexpressed in insect cells supported this concept. Experiments conducted with insect cells have an advantage over those conducted in COS cells or other mammalian cells in that they can achieve a significantly greater efficiency of overexpression. Therefore, it is easy to distinguish the effect of protein kinases on the specific AC isoform from that on the endogenous isoforms.

The type II isoform overexpressed in insect cells was stimulated after phorbol ester treatment of the insect cells (112). Kinetic analysis revealed that this stimulation was due to an increase in V_{max} without any changes in the K_m value to ATP. There was a significant synergism between phorbol ester–mediated activation and forskolin- or $G_{s\alpha}$-mediated activation. This study also demonstrated the phosphorylation of the type II isoform after phorbol ester treatment. The effect of phorbol ester was most prominent at lower concentrations of Mg: an approximately 3-fold increase in AC catalytic activity at a low concentration versus a 0.5-fold increase at a high concentration (>20 mM). This study demonstrated that besides the synergistic stimulation of AC by PKC and $G_{s\alpha}$, phorbol ester–sensitive insect PKC could also stimulate and phosphorylate a mammalian AC isoform.

Direct stimulation of AC by a specific mammalian PKC isozyme was also demonstrated (113). Both the type V isoform and the PKC-α isozyme were overexpressed and purified independently from insect cells. Protein kinase C-α directly phosphorylated and activated the type V isoform dose-dependently and saturably. The effect of PKC-α was, again, significantly augmented in the presence of forskolin and $G_{s\alpha}$ stimulation. Most important, this study clearly demonstrated that, even in the absence of $G_{i\alpha}$ protein, PKC-mediated phosphorylation could increase AC catalytic activity (113).

In contrast to the clear phosphorylation and activation of purified type V isoform by PKC-α, those of type II were difficult to demonstrate using purified enzymes (111). However, a recent study showed that the activation of type II AC by PKC was indeed conformation dependent; the effect of PKC, but not that of $G_{s\alpha}$ and forskolin, was decreased or totally lost when this isoform was purified or solubilized (114,115). Similar conformation-dependent regulation of AC was also seen in type V with kinases (113) and type I with calmodulin and βγ subunits (115,116).

The question now arises: Does the $G_{s\alpha}$-mediated activation of AC always augment the PKC effect on AC? There is a study to argue against this generalization. Although the stimulatory effect of $G_{s\alpha}$ on the type II isoform was augmented by phorbol ester treatment (112), that on the type IV isoform overexpressed in insect cells was decreased by PKC (115). Furthermore, the stimulation of $G_{s\alpha}$-activated type II activity by $\beta\gamma$ was eliminated by PKC. This study clarified two important issues. First, the PKC-mediated regulation of AC catalytic activity was conditional on the presence of different G_s subunit (α versus $\beta\gamma$). Second, the effect of the same PKC isozyme (α-isozyme) differed between the AC isoforms within the same subgroup (types II and IV). It is thus easy to speculate that such differences are even greater among subgroups.

Regulation by Atypical Protein Kinase C Isozymes

The studies cited so far have focused on the regulation of AC isoforms by classic isozymes of PKC with phorbol ester treatment. As described earlier, recent cloning studies have elucidated the presence of PKC isozymes that are insensitive to diacyl glycerol and phorbol ester (atypical isozymes). These isozymes are difficult to study because of the lack of specific activators and inhibitors (79,117). More recent studies have shown that atypical isozymes were activated by growth-mediating extracellular stimuli such as insulin and growth factors (79,118–120). Therefore, the biologic significance of these PKC isozymes may be even greater under certain circumstances than those of classic isozymes. Regulation of AC isoforms by these isozymes may also be biologically important if one considers the importance of cAMP signaling in cell growth and differentiation.

Two studies demonstrated the involvement of atypical PKC isozymes and the tyrosine kinase signaling pathway with the AC signaling. Type V AC was directly phosphorylated and activated by PKC-ζ, an atypical isozyme, with the purified protein (113). This isozyme showed a similar or even greater degree of activation of the type V isoform than that by PKC-α. It is interesting that activation by PKC-α was abolished in the absence of calcium, whereas activation by the ζ-isozyme was maintained, an indication that activation of AC occurred in a calcium-dependent manner for the α-isozyme, but in a calcium-independent manner for the ζ-isozyme. The most striking finding was that the activation of AC by PKC-α and PKC-ζ was additive. A phosphopeptide mapping study demonstrated that these two isozymes phosphorylated distinct residues (113). These findings suggest that a calmodulin-insensitive AC isoform can be regulated by calcium dependently through PKC; this regulation also occurs independently among isozymes.

Another study demonstrated the regulation of the type V isoform stably overexpressed in HEK-293 cells (121). Both phorbol ester and insulin enhanced cAMP accumulation in these cells. Co-transfection with PKC-ζ, but not with PKC-α, further potentiated the effects of insulin. In contrast, neither co-transfection with PKC-α nor that with PKC-ζ potentiated the effects of phorbol ester, which suggests the functional saturation of HEK-293 cells with phorbol ester–sensitive PKC isozymes. This study demonstrated that insulin signaling potentiated AC activity through activation of PKC-ζ. AC activity may be regulated *in vivo* by multiple PKC isozymes whose activation is transmitted from different extracellular stimuli.

Other Kinases

Although previous studies have mostly focused on the regulation of AC by PKC, several studies indicate that other kinases are also involved in the regulation of AC.

An interesting paradigm is that AC is regulated by PKA. Numerous reports show that prolonged stimulation of cells attenuate cAMP production within the cells (see earlier). The receptor-stimulated cAMP production was most promptly and dramatically attenuated. However, many studies also showed the attenuation of G protein– and forskolin-stimulated cAMP production. It is also difficult to explain the heterologous desensitization solely by phosphorylation of multiple receptors by PKA.

In fact, a previous study showed that purified AC from the striatum in the brain was directly phosphorylated and inactivated by PKA purified from the heart (122). In another study, treatment of chick hepatocytes with glucagon or 8-bromo-cAMP resulted in desensitization of the receptor-stimulated AC (123). The addition of exogenous $G_{s\alpha}$ to desensitized hepatocyte membranes, however, could not fully restore $G_{s\alpha}$-stimulated AC activity, suggesting that hormone-induced desensitization occurred at the level of AC (123). Because these cells expressed type VI, it was suggested that the type VI isoform undergoes phosphorylation and inactivation by PKA. A subsequent study demonstrated that PKA directly phosphorylated and inactivated the purified type V isoform, which is very close to type VI (124). Inactivation of type V resulted from a decreased catalytic rate, not from a decreased affinity with the substrate ATP. The sites of phosphorylation by this kinase, as assessed by phosphopeptide mapping analysis, were different from those phosphorylated by PKC. Protein kinase A–mediated phosphorylation and inactivation of AC may thus represent an alternative mechanism for heterologous desensitization of the G protein–coupled receptor pathways that lead to cAMP production. In addition, the preceding studies suggested that AC was subject to dual regulation by phosphorylation: activation by PKC and inhibition by PKA. These were mediated through phosphorylation at unique residues, as demonstrated by phosphopeptide mapping (113,124). A similar dual regulation by these two kinases has been shown in other components, including potassium channels (125,126).

Is AC directly regulated by tyrosine kinase? Although several studies support this paradigm, the exact mechanism of this cross-talk is not yet known. Furthermore, cross-talk between the tyrosine kinase signaling and specific mammalian AC isoforms remains unknown. However, it is worth noting a series of studies that indicated that epidermal growth factor receptor stimulation leads to increased cAMP production with the cells (57,127–129). This stimulation was augmented when the cell overexpressed the type V isoforms but not other isoforms (130). However, this potentiation may not be a direct effect on AC, but rather phosphorylation and activation of $G_{s\alpha}$ protein (57). In our preliminary data, there was no direct phosphorylation by tyrosine kinase of AC, at least when assessed with purified insulin receptor kinase (unpublished observation).

Other kinases, such as calmodulin kinase, may also be involved in the regulation of certain isoforms. The type III isoform was stimulated when examined with purified calmodulin in membrane preparations (131). However, in intact cells, this stimulation was lost and the effect of calcium became inhibitory (132). Because calmodulin kinase inhibitor abolished this calcium-mediated inhibition, it was proposed that this kinase phosphorylates and inhibits the type III isoform *in vivo* (132). Whether this regulation is a common property among other calmodulin-sensitive AC isoforms needs to be examined.

Finally, the type IX isoform has been shown to be inhibited by calcineurin, a calcium-sensitive phosphatase. This finding supports previous observations that the cross-talk between kinase and phosphatase may provide another important regulatory mechanism of AC (66–68). Detailed discussion of this finding is found in Chapter 8.

CONCLUSION

It is now clear that cAMP signaling is regulated by various kinases. The diversity of this regulation arises from different kinases, different subtypes of the same kinase, different target components within the cAMP signaling, and their subtypes (G protein subtypes, AC isoforms). Thus, it is difficult to generalize the effect of certain kinases on cAMP signaling. These isoforms/isozymes are altered at different developmental stages, among different cell types within the same tissue, under different pathophysiologic conditions. Most recent studies indicate that, even within the same cell membrane, components are localized in subcellular fractions, such as caveolae fraction (133). Indeed, in these fractions, receptor, G protein, kinase, and AC are accumulated. The studies so far have tried to expand the diversity among different subtypes of protein and kinase within the cell. Now movements toward the integration of this divergent regulation of cAMP signaling within the cell may also be necessary.

ACKNOWLEDGMENTS

This work was supported by the United States Public Health Service HL54895 and HL59139 and the American Heart Association #13–533–945.

REFERENCES

1. Holroyde MJ, Robertson SP, Johnson JD et al: The calcium and magnesium binding sites on cardiac troponin and their role in the regulation of myofibrillar adenosine triphosphatase. *J Biol Chem* 1980; 255:11668–11693.
2. Kim HW, Steenaart N AE, Ferguson DG, Kranias EG: Functional reconstitution of the cardiac sarcoplasmic reticulum Ca-ATPase with phospholamban in phospholipid vesicles. *J Biol Chem* 1990; 265:1702–1709.
3. Krebs EG: The phosphorylation of proteins: a major mechanism for biological regulation. *Biochem Soc Trans* 1985;13:813–820.
4. Beavo JA, Houslay MD: *Cyclic nucleotide phosphodiesterases: structure, regulation.* Chichester: Wiley Press, 1990.
5. Scott JD: Cyclic nucleotide-dependent protein kinases: *Pharmacol Ther* 1991;50:123–145.

6. Sibley DR, Benovic JL, Caron MG, Lefkowitz RJ: Regulation of transmembrane signaling by receptor phosphorylation. *Cell* 1987;48:913–922.

7. Freedman NJ, Liggett SB, Drachman DE et al: Phosphorylation and desensitization of the human beta 1-adrenergic receptor. Involvement of G protein–coupled receptor kinases and cAMP-dependent protein kinase. *J Biol Chem* 1995;270:17953–17961.

8. Lefkowitz RJ, Hausdorff WP, Caron MG: Role of phosphorylation in desensitization of the beta-adrenoceptor. *Trends Pharmacol Sci* 1990;11:190–194.

9. Kennelly PJ, Krebs EG: Consensus sequences as substrate specificity determinants for protein kinases and protein phosphatases. *J Biol Chem* 1991;266:1555–15558.

10. O'Dowd BF, Hnatowich M, Regan JW et al: Site-directed mutagenesis of the cytoplasmic domains of the human beta 2-adrenergic receptor. Localization of regions involved in G protein–receptor coupling. *J Biol Chem* 1988; 263:15985–15992.

11. Hausdorff WP, Bouvier M, O'Dowd BF et al: Phosphorylation sites on two domains of the β2-adrenergic receptor are involved in distinct pathways of receptor desensitization. *J Biol Chem* 1989; 264:12657–12665.

12. Liggett SB, Bouvier M, Hausdorff WP et al: Altered patterns of agonist-stimulated cAMP accumulation in cells expressing mutant beta 2-adrenergic receptors lacking phosphorylation sites. *Mol Pharmacol* 1989;36:641–646.

13. Clark RB, Friedman J, Dixon RAF, Strader CD: Identification of a specific site required for rapid heterologous desensitization of the β-adrenergic receptor by cAMP-dependent protein kinase. *Mol Pharmacol* 1989;36:343–348.

14. Lohse MJ, Benovic JL, Caron MG, Lefkowitz RJ: Multiple pathways of rapid beta 2-adrenergic receptor desensitization: delineation with specific inhibitors. *J Biol Chem* 1990;265:3202–3211.

15. Benovic JL, Strasser RH, Caron MG, Lefkowitz RJ: β-adrenergic receptor kinase: identification of a novel protein kinase that phosphorylates the agonist occupied form of the receptor. *Proc Natl Acad Sci U S A* 1986;83:2797–2801.

16. Inglese J, Freedman NJ, Koch WJ, Lefkowitz RJ: Structure and Mechanism of the G protein–coupled receptor kinases. *J Biol Chem* 1993;268(32):23735–23738.

17. Benovic JL, Regan JW, Matsui H et al: Agonist-dependent phosphorylation of the alpha 2-adrenergic receptor by the beta-adrenergic receptor kinase. *J Biol Chem* 1987;262:17251–17253.

18. Haga K, Kameyama K, Haga T et al: Phosphorylation of human m1 muscarinic acetylcholine receptors by G protein–coupled receptor kinase 2 and protein kinase C. *J Biol Chem* 1996;271:2776–2782.

19. Haga T, Haga K, Kameyama K, Nakata H: Phosphorylation of muscarinic receptors: regulation by G proteins. *Life Sci* 1993;52:421–428.

20. Pitcher JA, Inglese J, Higgins JB et al: Role of beta gamma subunits of G proteins in targeting the beta-adrenergic receptor kinase to membrane-bound receptors. *Science* 1992;257:1264–1267.

21. Koch WJ, Hawes BE, Inglese J et al: Cellular expression of the carboxyl terminus of a G protein coupled receptor kinase attenuates Gβγ-mediated signaling. *J Biol Chem* 1994;269:6193–6197.

22. Neer EJ: Heterotrimeric G proteins: organizers of transmembrane signals. *Cell* 1995;80:249–257.

23. Mayer BJ, Ren R, Clark KL, Baltimore D: A putative modular domain present in diverse signaling proteins. *Cell* 1993;73:629–630.

24. Musacchio A, Gibson T, Rice P et al: The PH domain: a common piece in the structural patchwork of signaling proteins. *TIBS* 1993;18:343–349.

25. Piping S, Andexinger S, Daniel K et al: Overexpression of β-arrestin and β-adrenergic receptor kinase augment desensitization of β2-adrenergic receptors. *J Biol Chem* 1993;268:3201–3208.

26. Kong G, Penn R, Benovic JL: A β-adrenergic receptor kinase dominant negative mutant attenuates desensitization of the β2-adrenergic receptor. *J Biol Chem* 1994;269:13084–13087.

27. Garcia-Higuera P, Mayor FJ: Rapid desensitization of neonatal rat liver beta-adrenergic receptors: a role for beta-adrenergic receptor kinase. *J Clin Invest* 1994;93:937–943.

28. Ping P, Gelzer-Bell R, Roth DA et al: Reduced β-adrenergic receptor activation decreases G protein expression and β-adrenergic receptor kinase activity in porcine heart. *J Clin Invest* 1995;95:1271–1280.

29. Koch WJ, Rockman HA, Samama P et al: Cardiac function in mice overexpressing the β-adrenergic receptor kinase or a βARK inhibitor. *Science* 1995;268:1350–1353.

30. Kuno M: *The synapse: function, plasticity, and neutrophism.* Oxford: Oxford University Press, 1995.

31. Hausdorff WP, Caron MG, Lefkowitz RJ: Turning of the signal: desensitization of β-adrenergic receptor function. *FASEB J* 1990;4:2881–2889.

32. Benovic JL, Kuhn H, Weyand I et al: Functional desensitization of the isolated beta-adrenergic re-

ceptor by the beta-adrenergic receptor kinase: potential role of an analog of the retinal protein arrestin (48-kDa protein). *Proc Natl Acad Sci U S A* 1987;84:8879–8882.

33. Lohse MJ, Benovic JL, Codina J et al: beta-Arrestin: a protein that regulates beta-adrenergic receptor function. *Science* 1990;248:1547–1550.

34. Ferguson SSG, Downey WE III, Colapietro A et al: Role of β-arrestin in mediating agonist-promoted G protein–coupled receptor internalization. *Science* 1996;271:363–366.

35. Chilvers ER, Lynch BJ, Challiss RAJ: Phosphoinositide metabolism in air way smooth muscle. *Pharmacol Ther* 1994;62:221–245.

36. Schramm CM, Grunstein MM: Assessment of signal transduction mechanisms regulating airway smooth muscle contractility. *Am J Physiol* 1992;262:L119–L139.

37. Garte S, Belman S: Tumor promoter uncouples β-adrenergic receptor from adenylyl cyclase in mouse epidermis. *Nature* 1980;284:171–173.

38. Kelleher DJ, Pessin JE, Ruoho AE, Johnson GL: Phorbol ester induces desensitization of adenylyl cyclase and phosphorylation of the β-adrenergic receptor in turkey erythrocyte. *Proc Natl Acad Sci U S A* 1984;81:4316–4320.

39. Johnson JA, Clark RB, Friedman J et al: Identification of a specific domain in the beta-adrenergic receptor required for phorbol ester–induced inhibition of catecholamine-stimulated adenylyl cyclase. *Mol Pharmacol* 1990;38:289–293.

40. Johnson JA, Goka TJ, Clark RB: Phorbol ester–induced augmentation and inhibition of epinephrine-stimulated adenylate cyclase in S49 lymphoma cells. *J Cyclic Nucl Protein Phosphor Res* 1986;11:199–215.

41. Hernandez-Sotomayor SM, Macias-Silva M, Plebanski M, Garcia-Sainz JA: Homologous and heterologous beta-adrenergic desensitization in hepatocytes: additivity and effect of pertussis toxin. *Biochim Biophys Acta* 1988;972:311–319.

42. Pitcher J, Lohse MJ, Codina J et al: Desensitization of the isolated beta 2-adrenergic receptor by beta-adrenergic receptor kinase, cAMP-dependent protein kinase, and protein kinase C occurs via distinct molecular mechanisms. *Biochemistry* 1992;31:3193–3197.

43. Yuan N, Friedman J, Whaley BS, Clark RB: cAMP-dependent protein kinase and protein kinase C consensus site mutations of the beta-adrenergic receptor: effect on desensitization and stimulation of adenylylcyclase. *J Biol Chem* 1994;269:23032–23038.

44. Hadcock JR, Port JD, Gelman MS, Malbon CC: Cross-talk between tyrosine kinase and G protein linked receptors. *J Biol Chem* 1992;267:26017–26022.

45. Karoor V, Baltensperger K, Paul H et al: Phosphorylation of tyrosyl residues 350/354 of the beta-adrenergic receptor is obligatory for counterregulatory effects of insulin. *J Biol Chem* 1995;270: 25305–25308.

46. Baltensperger K, Karoor V, Paul H et al: The beta-adrenergic receptor is a substrate for the insulin receptor tyrosine kinase. *J Biol Chem* 1996;271:1061–1064.

47. Kassis S, Fishman PH: Different mechanisms of desensitization of adenylate cyclase by isoproterenol and prostaglandin E1 in human fibroblasts: role of regulatory components in desensitization. *J Biol Chem* 1982;257:5312–5318.

48. Premont RT, Iyengar R: Heterologous desensitization of the liver adenylyl cyclase: analysis of the role of G-proteins. *Endocrinology* 1989;125:1151–1160.

49. Kirchick HJ, Iyengar R, Birnbaumer L: Human chorionic gonadotropin-induced heterologous desensitization of adenylyl cyclase from highly luteinized rat ovaries: attenuation of regulatory N component activity. *Endocrinology* 1983;113:1638–1646.

50. Alousi AA, Jasper JR, Insel PA, Motulsky HJ: Stoichiometry of receptor-G_s-adenylate cyclase interactions. *FASEB J* 1991;5:2300–2303.

51. Rebois RV, Fishman PH: Gonadotropin-mediated desensitization in a murine Leydig tumor cell line does not alter the regulatory and catalytic components of adenylyl cyclase. *Endocrinology* 1986; 118:2340–2348.

52. Bushman WA, Wilson LK, Luttrell DK et al: Overexpression of c-src enhances beta-adrenergic-induced cAMP accumulation. *Proc Natl Acad Sci U S A* 1990;87:7462–7466.

53. Hausdorff WP, Pitcher JA, Luttrell DK et al: Tyrosine phosphorylation of G protein α subunit by pp60c-src. *Proc Natl Acad Sci U S A* 1992;89:5720–5724.

54. Moyers JS, Linder ME, Shannon JD, Parsons SJ: Identification of the *in vitro* phosphorylation sites on G_s alpha mediated by pp60c-src. *Biochem J* 1995;305:411–417.

55. Johnson GL, Dhanasekaran N, Gupta SK et al: Genetic and structural analysis of G protein alpha subunit regulatory domains. *J Cell Biochem* 1991;47:136–146.

56. Sullivan KA, Miller RT, Masters SB et al: Identification of receptor contact site involved in receptor-G protein coupling. *Nature* 1987;330:758–760.
57. Poppleton H, Sun H, Fulgham D et al: Activation of G by the epidermal growth factor receptor involves phosphorylation. *J Biol Chem* 1996;271:6947–6951.
58. Katada T, Gilman AG, Watanabe Y et al: Protein kinase C phosphorylates the inhibitory guanine-nucleotide-binding regulatory component and apparently suppresses its function in hormonal inhibition of adenylate cyclase. *Eur J Biochem* 1985;151:431–437.
59. Murphy GJ, Gawler DJ, Milligan G et al: Glucagon desensitization of adenylate cyclase and stimulation of inositol phospholipid metabolism does not involve the inhibitory guanine nucleotide regulatory protein G_i, which is inactivated upon challenge of hepatocytes with glucagon. *Biochem J* 1989;259:191–197.
60. Gordeladze JO, Bjoro T, Torjesen PA et al: Protein kinase C stimulates adenylyl cyclase activity in prolactin-secreting rat adenoma (GH4C1) pituicytes by inactivating the inhibitory GTP-binding protein Gi. *Eur J Biochem* 1989;183:397–406.
61. Wheeler MB, Veldhuis JD: Facilitative actions of the protein kinase C effector system on hormonally stimulated adenosine 3′ 5′-monophosphate production by swine luteal cells. *Endocrinology* 1989;125:2414–2420.
62. Choi EJ, Toscano WAJ: Modulation of adenylate cyclase in human keratinocytes by protein kinase C. *J Biol Chem* 1988;263:17167–17172.
63. Quilliam LA, Dobson PRM, Brown BL: Regulation of GH3 pituitary tumor cell adenylyl cyclase activity by activators of protein kinase C. *Biochem J* 1989;262:829–834.
64. Gusovsky F, Gutkind JS: Selective effects of activation of protein kinase C isozymes on cyclic AMP accumulation. *Mol Pharmacol* 1991;39:124–129.
65. Nishizuka Y: Intracellular signaling by hydrolysis of phospholipids and activation of protein kinase C. *Science* 1992;258:607–614.
66. Pyne NJ, Murphy GJ, Milligan G, Houslay MD: Treatment of intact hepatocytes with either the phorbol ester TPA or glucagon elicits the phosphorylation and functional inactivation of the inhibitory guanine nucleotide regulatory protein G_i. *FEBS Lett* 1989;243:77–82.
67. Strassheim D, Malbon CC: Phosphorylation of G_i alpha 2 attenuates inhibitory adenylyl cyclase in neuroblastoma/glioma hybrid (NG-108–15) cells. *J Biol Chem* 1994;269:14307–14313.
68. Bushfield M, Pyne NJ, Houslay MD: Okadaic acid identifies a phosphorylation/dephosphorylation cycle controlling the inhibitory guanine-nucleotide-binding regulatory protein G_{i2}. *Eur J Biochem* 1991;192:537–542.
69. Heyworth CM, Whetton AP, Kinsella AR, Houslay MD: The phorbol ester, TPA inhibits glucagon-stimulated adenylyl cyclase activity. *FEBS Lett* 1984;170:38–42.
70. Rebois RV, Patel J: Phorbol ester causes desensitization of gonadotropin-responsive adenylate cyclase in a murine Leydig tumor cell line. *J Biol Chem* 1985;260:8026–8031.
71. Sugden D, Vanacek J, Klein DC et al: Activation of protein kinase C potentiates isoprenaline-induced cyclic AMP accumulation in rat pinealocytes. *Nature* 1985;314:359–361.
72. Bell JD, Buxton ILO, Brunton LL: Enhancement of adenylate cyclase activity in S49 lymphoma cells by phorbol esters. *J Biol Chem* 1985;260:2625–2628.
73. Fleming N, Mellow L, Bhullar D: Regulation of the cAMP signal transduction pathway by protein kinase C in rat submandibular cells. *Eur J Physiol* 1992;421:82–89.
74. Hollingsworth EB, Sears EB, Daly JW: An activator of protein kinase C (phobol-12-myristrate-13-acetate) augments 2-chloroadenosine elicited accumulation of cAMP in guinea cerebral cortical particulate preparations. *FEBS Lett* 1895;184:339–342.
75. Naghshineh S, Noguchi M, Huang KP, Londos C: Activation of adipocyte adenylate cyclase by protein kinase C. *J Biol Chem* 1986;261:14534–14538.
76. Yoshimasa T, Sibley DR, Bouvier M et al: Cross-talk between cellular signaling pathways suggested by phorbol-ester-induced adenylate cyclase phosphorylation. *Nature* 1987;327:67–70.
77. Levitzki A: Cross-talk between PKC and cyclic AMP pathways. *Nature* 1987;330:319–320.
78. Simmoteit R, Schulzki HD, Palm D et al: Chemical and functional analysis of components of adenylyl cyclase from human platelets treated with phorbol esters. *FEBS Lett* 1991;285:99–103.
79. Nakanishi H, Brewer KA, Exton JH: Activation of the ζ isoenzyme of protein kinase C by phosphatidylinositol 3,4,5-triphosphate. *J Biol Chem* 1993;268:13–16.
80. Westel WC, Khan WA, Merchenthaler I et al: Tissue and cellular distribution of the extended family of protein kinase C isoenzymes. *J Cell Biol* 1992;117:121–133.

81. Rybin VO, Steinberg SF: Protein kinase C isoform expression and regulation in the developing rat heart. *Circ Res* 1994;74:299–309.

82. Ron D, Chen CH, Caldwell J et al: Cloning of an intracellular receptor for protein kinase C: a homologue of the beta subunit of G proteins. *Proc Natl Acad Sci U S A* 1994;91:839–843.

83. Hsieh JC, Jurutka PW, Galligan MA et al: Human vitamin D receptor is selectively phosphorylated by protein kinase C on serine 51, a residue crucial to its trans-activation function. *Proc Natl Acad Sci U S A* 1991;88:9315–9319.

84. Ido M, Sekiguchi K, Kikkawa U, Nishizuka Y: Phosphorylation of the EGF receptor from A431 epidermoid carcinoma cells by three distinct types of protein kinase C. *FEBS Lett* 1987;219: 215–218.

85. Sheu FS, Marais RM, Parker PJ et al: Neuron-specific protein F1/GAP-43 shows substrate specificity for the beta subtype of protein kinase C. *Biochem Biophys Res Commun* 1990;171:1236–1243.

86. Bakalyar HA, Reed RR: Identification of a specialized adenylyl cyclase that may mediate odorant detection. *Science* 1990;250:1403–1406.

87. Feinstein PG, Schrader KA, Bakalyar HA et al: Molecular cloning and characterization of a Ca^{2+}/calmodulin-insensitive adenylyl cyclase from rat brain. *Proc Natl Acad Sci U S A* 1991;88: 10173–10177.

88. Gao BN, Gilman AG: Cloning and expression of a widely distributed (type IV) adenylyl cyclase. *Proc Natl Acad Sci U S A* 1991;88:10178–10182.

89. Ishikawa Y, Katsushika S, Chen L et al: Isolation and characterization of a novel cardiac adenylyl-cyclase cDNA. *J Biol Chem* 1992;267:13553–13557.

90. Katsushika S, Chen L, Kawabe J et al: Cloning and characterization of a sixth adenylyl cyclase isoform: types V, VI constitute a subgroup within the mammalian adenylyl cyclase family. *Proc Natl Acad Sci U S A* 1992;89:8774–8778.

91. Premont RT, Chen J, Ma HW et al: Two members of a widely expressed subfamily of hormone-stimulated adenylyl cyclases. *Proc Natl Acad Sci U S A* 1992;89:9809–9813.

92. Paterson JM, Smith SM, Harmar AJ, Antoni FA: Control of a novel adenylyl cyclase by calcineurin. *Biochem Biophys Res Commun* 1995;214:1000–1008.

93. Yoshimura M, Cooper DMF: Cloning and expression of a Ca inhibitable adenylyl cyclase from NCB-20 cells. *Proc Natl Acad Sci U S A* 1992;89:6716–6720.

94. Wallach J, Droste M, Kluxen FW et al: Molecular cloning and expression of a novel type V adenylyl cyclase from rabbit myocardium. *FEBS Lett* 1994;338:257–263.

95. Watson PA, Krupinski J, Kempinski AM, Frankenfield CD: Molecular cloning and characterization of the type VII isoform of mammalian adenylyl cyclase expressed widely in mouse tissues and in S49 mouse lymphoma cells. *J Biol Chem* 1994;269:28893–28898.

96. Villacres EC, Xia Z, Bookbinder LH et al: Cloning, chromosomal mapping, and expression of human fetal brain type I adenylyl cyclase. *Genomics* 1993;16:473–478.

97. Glatt CE, Snyder SH: Cloning and expression of an adenylyl cyclase localized to the corpus striatum. *Nature* 1993;361:536–538.

98. Defer N, Marinx O, Stengel D et al: Molecular cloning of the human type VIII adenylyl cyclase. *FEBS Lett* 1994;351:109–113.

99. Nomura N, Miyajima N, Sazuka T et al: Prediction of the coding sequences of unidentified human genes. I. The coding sequences of 40 new genes (KIAA0001-KIAA0040) deduced by analysis of randomly sampled cDNA clones from human immature myeloid cell line KG-1. *DNA Res* 1994;1: 27–35.

100. Ishikawa Y, Sorota S, Kiuchi K et al: Down-regulation of adenylylcyclase types V, VI mRNA levels in pacing-induced heart failure. *J Clin Invest* 1994;93:2224–2229.

101. Tobise K, Ishikawa Y, Holmer SR et al: Changes in type VI adenylylcyclase isoform expression correlate with a decreased capacity for cyclic AMP generation in the aging ventricle. *Circ Res* 1994;74:596–603.

102. Jacobowitz O, Chen J, Premont RT, Iyengar R: Stimulation of specific types of G_s-stimulated adenylyl cyclases by phorbol ester treatment. *J Biol Chem* 1993;268:3829–3832.

103. Yoshimura M, Cooper DM: Type-specific stimulation of adenylylcyclase by protein kinase C. *J Biol Chem* 1993;268:4604–4607.

104. Choi EJ, Wong ST, Dittman AH, Storm DR: Phorbol ester stimulation of the type I, type III adenylyl cyclases in whole cells. *Biochemistry* 1993;32:1891–1894.

105. Sutkowski EM, Tang W-J, Broome CW et al: Regulation of forskolin interaction with type I, II, V, VI adenylyl cyclases. *Biochemistry* 1994;33:12852–12859.

106. Chen J, Iyengar R: Inhibition of cloned adenylyl cyclases by mutant-activated G_i-alpha and specific suppression of type 2 adenylyl cyclase inhibition by phorbol ester treatment. *J Biol Chem* 1993;268:12253–12256.

107. Impey S, Wayman G, Wu Z, Storm DR: Type I adenylyl cyclase functions as a coincidence detector for control of cyclic AMP response element-mediated transcription: synergistic regulation of transcription by Ca and isoproterenol. *Mol Cell Biol* 1994;14:8272–8281.

108. Taussig R, Quarmby LM, Gilman AG: Regulation of purified type I, type II adenylylcyclases by G protein beta gamma subunits. *J Biol Chem* 1993;268:9–12.

109. Taussig R, Iniguez-Lluhi JA, Gilman AG: Inhibition of adenylyl cyclase by G_i alpha. *Science* 1993; 261:218–221.

110. Morimoto BH, Koshland DEJ: Conditional activation of cAMP signal transduction by protein kinase C: the effect of phorbol esters on adenylyl cyclase in permeabilized and intact cells. *J Biol Chem* 1994;269:4065–4069.

111. Lustig KD, Conklin BR, Herzmark P et al: Type II adenylylcyclase integrates coincident signals from G_s, G_i, and G_q. *J Biol Chem* 1993;268:13900–13905.

112. Jacobowitz O, Iyengar R: Phorbol ester-induced stimulation and phosphorylation of adenylyl cyclase 2. *Proc Natl Acad Sci U S A* 1994;91:10630–10634.

113. Kawabe J, Iwami G, Ebina T et al: Differential activation of adenylylcyclase by protein kinase C isoenzymes. *J Biol Chem* 1994;269:16554–16558.

114. Ebina T, Kawabe J, Ishikawa Y: Conformation-dependent activation of type II adenylyl cyclase by protein kinase C. *J Cell Biochem* 1997; 64:492–498.

115. Zimmermann G, Taussig R: Protein kinase C alters the responsiveness of adenylyl cyclase to G protein α and βγ subunits. *J Biol Chem* 1996; 271:27161–27166.

116. Tang WJ, Krupinski J, Gilman AG: Expression and characterization of calmodulin-activated (type I) adenylyl cyclase. *J Biol Chem* 1991;266:8595–8603.

117. Nakanishi H, Exton JH: Purification and characterization of the ζ isoform of protein kinase from bovine kidney. *J Biol Chem* 1992;267:16347–16354.

118. Kochs G, Hummel R, Meyer D et al: Activation and substrate specificity of the human protein kinase C alpha and zeta isoenzymes. *Eur J Biochem* 1993;216:597–606.

119. Kotani K, Yonezawa K, Hara K et al: Involvement of phosphoinositotide 3-kinase in insulin- or IGF-1-induced membrane ruffling. *EMBO J* 1994;13:2313–2321.

120. Domingesz I, Diaz-Meco MT, Municio MM et al: Evidence for a role of protein kinase C ζ subspecies in maturation of *Xenopus laevis* oocytes. *Mol Cell Biol* 1992;12:3776–3783.

121. Kawabe J, Ebina T, Toya Y et al: Regulation of type V adenylyl cyclase by PMA-sensitive and -insensitive protein kinase C isoenzymes in intact cells. *FEBS Lett* 1996;384:273–276.

122. Yoshimasa T, Bouvier M, Benovic JL et al: Regulation of the adenylyl cyclase signaling pathway: potential role for the phosphorylation of the catalytic unit by protein kinase A and protein kinase C. In: KW McKerns and M Chretien, eds. *Molecular biology of brain and endocrine peptidergic systems*. New York: Plenum, 1988;123–139.

123. Premont RT, Jacobowitz O, Iyengar R: Lowered responsiveness of the catalyst of adenylyl cyclase to stimulation by GS in heterologous desensitization: a role for adenosine 3′,5′-monophosphate-dependent phosphorylation. *Endocrinology* 1992;131:2774–84.

124. Iwami G, Kawabe J, Ebina T et al: Regulation of adenylyl cyclase by protein kinase A. *J Biol Chem* 1995;270:12481–12484.

125. Wang WH, Giebisch G: Dual modulation of renal ATP-sensitive K^+ channel by protein kinases A and C. *Proc Natl Acad Sci U S A* 1991;88:9722–9725.

126. Chen Y, Yu L: Differential regulation by cAMP dependent protein kinase and protein kinase C of the μ opioid receptor coupling to a G protein–activated K channel. *J Biol Chem* 1994;269: 7839–7842.

127. Nair BG, Rashed HM, Patel TB: Epidermal growth factor stimulates rat cardiac adenylate cyclase through a GTP-binding regulatory protein. *Biochem J* 1989;264:563–571.

128. Nair BG, Parikh B, Milligan G, Patel TB: $G_{s\alpha}$ mediates epidermal growth factor elicited stimulation of rat cardiac adenylate cyclase. *J Biol Chem* 1990;265:21317–21322.

129. Nair BG, Rashed HM, Patel TB: Epidermal growth factor produces inotropic and chronotropic effects in rat hearts by increasing cyclic AMP accumulation. *Growth Factors* 1993;8:41–48.

130. Chen Z, Nield HS, Sun H et al: Expression of type V adenylyl cyclase is required for epidermal growth factor-mediated stimulation of cAMP accumulation. *J Biol Chem* 1995;270:27525–27530.]

131. Choi EJ, Xia Z, Storm DR: Stimulation of the type III olfactory adenylyl cyclase by calcium and calmodulin. *Biochemistry* 1992;31:6492–6498.
132. Wayman GA, Impey S, Storm DR: Ca inhibition of type III adenylyl cyclase *in vivo. J Biol Chem* 1995;270:21480–21486.
133. Lisanti MP, Scherer PE, Tang Z, Sargiacomo M: Caveolae, caveolin and caveolin-rich membrane domains: a signaling hypothesis. *Trends Cell Biol* 1994;4:231–235.

*Advances in Second Messenger and
Phosphoprotein Research*, Vol. 32,
edited by Dermot M. F. Cooper
Lippincott–Raven Publishers, Philadelphia © 1998

6

Genetic Characterization of Adenylyl Cyclase Function

Martin J. Cann and Lonny R. Levin

*Department of Pharmacology, Cornell University Medical College,
New York, New York 10021*

The second messenger cyclic adenosine monophosphate (cAMP) induces transient cellular changes by regulating the cAMP-dependent protein kinase A (PKA) and cyclic nucleotide-gated ion channels and long-term modifications by stimulating the cAMP-response element binding protein/cAMP-response element modulation (CREB/CREM) family of transcription factors. Various hormones and neurotransmitters modulate cAMP production by activating G protein–coupled, seven-transmembrane receptors. The guanine nucleotide binding α subunits of stimulatory (G_s) or inhibitory (G_i) G proteins directly modulate the activity of adenylyl cyclase (AC), the enzyme responsible for synthesizing cAMP (1). There are multiple AC isoforms, each displaying distinct regulatory properties and tissue distribution. In a specific region of the brain, dopamine can both increase and decrease cAMP production of an AC isoform by G_s-dependent activation and G_i-mediated inhibition (2). G protein $\beta\gamma$ subunits ($G_{\beta\gamma}$) participate in the guanine nucleotide exchange cycle and the transient activation of α subunits, but they also regulate specific isoforms of AC directly (3). $G_{\beta\gamma}$ stimulation may account for the ability of certain G_i-coupled neurotransmitters to potentiate the effects of G_s-coupled agonists in the brain.

Intracellular cAMP levels are also affected through mechanisms that utilize other signaling pathways to modulate cyclase activity. Calcium (Ca^{2+}) and protein kinase C (PKC), which function in discrete second messenger systems, can independently modulate the activity of various AC isoforms. Ca^{2+}/calmodulin (CaM) activation synergizes with G_s protein stimulation, allowing specific ACs to integrate signals from different second messenger pathways, as in learning and memory. Phospholipase C, which is regulated by its own cohort of G proteins and by tyrosine kinase receptors, can affect the activity of some AC isoforms by causing an increase in intracellular Ca^{2+} (4) and other isoforms by activation of PKC (5). The reasons for these complex regulatory relationships involving components of many signaling pathways and for the variety of differentially modulated isoforms of AC are not yet understood. Identifying the physiologic significance of the different isoforms and the specific roles for their individual regulatory interactions will increase our understanding of how cells respond to their environment. This review will focus on the genetic efforts, using model organisms, to identify functions of the various isoforms of AC.

Elucidating the functional significance of the different forms of AC has proved difficult in mammals. Individual isoforms have distinct modulatory interactions and tissue distribution that suggest possible functional roles; however, the relevance of these unique properties *in vivo* remains unclear. Type III AC (AC3) is the predominant AC expressed in olfactory receptor neurons and was therefore predicted to mediate signal transduction in odorant detection (6); however, its *in vivo* functional significance has not yet been determined. Type V AC (AC5) is the most abundantly expressed cyclase in the corpus striatum (2), the major site of dopamine action in the brain. Its activity is subject to G_s-dependent stimulation and G_i-mediated inhibition (7) consistent with it mediating synaptic responses to dopamine and suggesting it may play a role in the pathology of schizophrenia or Parkinson's disease. However, demonstration that AC5 is involved in dopamine signaling awaits functional characterization. Three closely related isoforms of AC—type II (AC2), type IV (AC4), and type VII (AC7)—are stimulated by $G_{\beta\gamma}$ and may be responsible for the ability of certain G_i-coupled neurotransmitters to increase cAMP levels, but the physiologic significance of this phenomenon is unknown.

To date, only one mammalian AC isozyme has been ascribed a specific physiologic role, and this knowledge is due to genetic characterization of its homolog in *Drosophila melanogaster*. Genetic analysis of AC function in the cellular slime mold *Dictyostelium* has identified novel components of the cAMP signaling pathway that may be conserved in metazoans and demonstrated a biologic significance of βγ regulation of AC. These studies demonstrate the utility of genetic characterization using relevant model organisms and are discussed later.

MODEL SYSTEMS

The use of cAMP as a signaling molecule is nearly universal, being conserved from bacteria to humans. Its roles include modulation of gene expression in bacteria, control of metabolic activity in yeast, and transduction of extracellular signals in higher eukaryotes. The ACs synthesizing cAMP in bacteria and yeast have been studied extensively; however, they appear to be unsuitable for identifying the specific physiologic functions of individual mammalian AC isoforms. Bacterial ACs are unrelated to the known mammalian forms, and the cyclases identified in the yeasts *Saccharomyces cerevisiae* (8) and *Schizosaccharomyces pombe* (9) are structurally distinct from the known metazoan ACs. Additionally, the well-characterized regulatory components responsible for modulating AC activity in mammals are not conserved in either yeast or bacteria.

Dictyostelium

In *Dictyostelium*, cAMP is used as both an extracellular signaling molecule and an intracellular second messenger. Two structurally distinct ACs have been identified in *Dictyostelium* (10). One, ACA, expressed during the aggregation stage of development, shares sequence and structural homology with the metazoan ACs. The other, ACG, is required during spore germination. Although it is structurally distinct from other known ACs, it displays significant primary amino acid sequence homology within its catalytic domain. Studies of the regulation of the more metazoan-like ACA have identified a novel component of the cAMP signaling pathway and demonstrated a biologic significance of G protein $\beta\gamma$ regulation. Additionally, *Dictyostelium* provides a system for genetic selection of mutants useful for investigating the enzymatic activity of AC.

In *Dictyostelium,* nutritional starvation induces a developmental program that involves aggregation of single-cell amoebas into a mobile, multicellular slug and culminates in differentiation into a fruiting body. Cyclic AMP, secreted in oscillatory waves, serves as the extracellular signal for organizing single cells into mounds as well as the intracellular second messenger directing migration toward the organizing center. Each cell's response to extracellular cAMP is mediated by a specific G protein–coupled receptor and includes chemotaxis, increased gene expression, and activation of the aggregation-specific cyclase, ACA. G protein stimulation of ACA is mediated by $\beta\gamma$ subunits, providing the only *in vivo* evidence of a biologic role for $G_{\beta\gamma}$ modulation of AC activity. G protein $\beta\gamma$ subunits are known to mediate signal transduction in the yeast mating pathway, but in this case, $G_{\beta\gamma}$ stimulates a kinase cascade (11). At least five of the mammalian AC isoforms can be directly regulated by $G_{\beta\gamma}$ *in vitro*, but whether this form of regulation occurs *in vivo* has not yet been demonstrated. Interestingly, $G_{\beta\gamma}$ stimulation of ACA is dependent upon a novel, cytosolic protein termed CRAC (for "cytoplasmic regulator of adenylyl cyclase") (12–14), which appears to be functionally analogous to $G_{s\alpha}$-GTP in the $\beta\gamma$ potentiation of mammalian AC activity. Attempts are under way to determine whether a mammalian homolog of CRAC exists (P. N. Devreotes, personal communication).

During aggregation, the cAMP produced by ACA serves as both an intracellular messenger and an extracellular signaling molecule. Because their predicted 12 transmembrane spans are structurally similar to known transporter molecules, it has been suggested that ACA and the known metazoan ACs might secrete cAMP out of the cell. Experiments involving heterologous expression of a novel form of AC from *Dictyostelium* suggest this is not the case. Strains with ACA deleted (*aca*⁻) do not aggregate, but this aggregation-deficient phenotype can be rescued by expression of the germination stage cyclase, ACG (10,15). Although ACG is unresponsive to G protein/CRAC regulation, it has an inherently high basal activity that apparently generates sufficient cAMP to induce the normal developmental program and propagate the aggregation signal. It was further shown that aggregating *aca*/ACG cells secrete cAMP (10), indicating that, at the very least, cAMP secretion is not dependent upon the 12-transmembrane form of cyclase.

ACG is another previously unidentified component in the cAMP signaling pathway. It is predicted to have a unique structure that is reminiscent of membrane-bound guanylyl cyclases with a single transmembrane span and a single catalytic domain (10). ACG appears to be unresponsive to G proteins and forskolin but exhibits a high basal activity. This single-domain cyclase may represent a novel class of ACs that also exist in mammals.

Characterization of the cAMP pathway in *Dictyostelium* has led not only to the identification of novel constituents, such as the single-domain AC (ACG) and the soluble regulator of AC activity (CRAC), but to genetic screens that have isolated mutant ACs that increase our understanding of enzyme catalysis. Two classes of mutant ACs—catalytically inactive and G protein insensitive—were selected by their inability to rescue the aggregation deficiency in the *aca⁻* null strain (16). The various substitutions appeared to define distinct regulatory and catalytic regions within the C_{1a} domain of ACA. Additionally, a constitutively active ACA was isolated by virtue of its ability to rescue the aggregation deficiency of a *crac⁻* null strain (17). The amino acid substitution responsible for activation mapped to the same region as G protein–insensitive substitutions, suggesting it defines a site for regulatory interactions.

Drosophila

The existence of only one AC resembling the known mammalian cyclases significantly limits *Dictyostelium's* use as a model system for determining the specific physiologic functions of individual isoforms. Emerging molecular evidence from our laboratory indicates that the fruit fly *Drosophila melanogaster* contains a family of ACs that appears to reflect the diversity of the mammalian isoforms (M. J. C., V. Iourgenko, B. Kliot, E. Chung, and L. R. L., unpublished observations). Their functions are likely to indicate the roles of their mammalian counterparts because the only AC characterized thus far in the fly, the Rutabaga AC involved in learning and memory, is structurally, biochemically, and functionally conserved with its mammalian homolog, AC1 (18,19). Coupled with its incredibly powerful genetics and extensive array of mutations, the existence of a conserved AC gene family makes *Drosophila* the best model system to investigate the specific functions of the multiple mammalian AC isoforms.

Low-stringency screening of a *Drosophila melanogaster* genomic library with DNA fragments from AC1 (20) and AC2 (21) identified four independent classes of hybridizing phage. Nucleotide sequences of putative exons from three of these classes of phage exhibited more than 67% identity with a highly conserved region within the second catalytic domain of mammalian ACs (18). Specific sequences in this region are absolutely conserved among metazoan ACs but not among guanylyl cyclases (20). These sequences can therefore be used to differentiate between members of the cyclase superfamily, and their presence suggests these *Drosophila* genes encode distinct forms of AC. The predicted amino acid sequence of one was more similar to AC1 than it was to any other mammalian AC, and it was subsequently shown to encode the *Drosophila rutabaga* gene required for learning and memory (18). At the stringency used for library screening, another (DAC-76E) hybridized exclusively with AC2, and partial sequence analysis implicates it as a member of the

AC subclass that includes mammalian AC2, AC4, and AC7 (Fig 1; E. Chung, J. Liao, and L. R. L., unpublished observations). Work in our laboratory has identified at least nine distinct complementary DNAs (cDNAs) that likely encode ACs, and among those characterized, we appear to have found *Drosophila* homologs for mammalian AC3 and AC9 (see Fig. 1; 21a).

The ability of *Drosophila* to learn and retain simple associative tasks enabled the genetic identification of mutants affecting the acquisition and/or storage of information (22,23). The identification of the biochemical activity of one of the initial X chromosome–linked mutants, *dunce*, as a *Drosophila* cAMP phosphodiesterase (24,25) prompted a survey of other learning and memory mutants to determine if they might encode components of the cAMP pathway. *Rutabaga* mutant flies had lower levels of AC activity than wild-type flies and specifically lacked a Ca^{2+}/CaM-stimulated cyclase activity (26). Because these flies expressed normal levels of CaM, it was postulated that the *rutabaga* gene encoded a Ca^{2+}/CaM-responsive AC or a subunit conferring Ca^{2+}/CaM responsiveness to a catalytic subunit of AC.

One of the original genomic cyclase clones mapped to the *rutabaga* locus (26,27), and nucleotide sequence analysis of the corresponding cDNA revealed an open reading frame predicted to encode a protein of 2249 amino acids (18). The amino terminal half of the predicted protein product shares the same overall structure as the previously isolated mammalian ACs (28) and is most closely related to mammalian AC1. This is true not only in the highly conserved catalytic domains but in the less-well-conserved C_{1b} domain as well. Rutabaga AC contains a carboxy terminal extension of more than 1100 amino acids of unknown function. This domain bears no significant homology to any protein in the database and is not required for catalytic activity (M. Sinclair, S. Westman, and L. R. L., unpublished observations). Identification of a function for this domain awaits behavioral characterization *in vivo*.

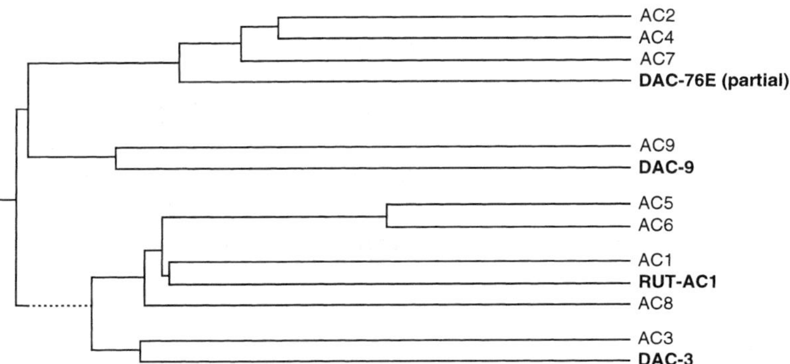

FIG. 1. Phylogenetic analysis of four cloned *Drosophila* ACs. *Drosophila* ACs are shown in bold. Phylogenetic analysis was performed on predicted amino acid sequences of bovine AC1, rat AC2, AC3, AC4, AC5, AC6, and AC8, mouse AC7 and AC9, and Rutabaga-DAC1 from Genbank, and DAC3 (V. Iourgenko, B. Kliot, and L. R. L., in preparation) (21a), and a partial cDNA of DAC-76E (E. Chung, J. Liao, and L. R. L., unpublished observations).

The catalytic activity of Rutabaga AC was assayed by heterologous expression in a human embryonic kidney cell line, HEK-293 (29). Rutabaga AC activity was potently stimulated by forskolin and by G_s protein α subunit (18). Assays of crude fly homogenates detected a specific loss of the CaM-responsive cyclase activity in flies homozygous for the original *rut1* mutant allele (26,30). The basal activity of heterologously expressed Rutabaga AC was decreased by inclusion of EGTA in the assay, indicating it had been stimulated by endogenous Ca^{2+}. Addition of CaM further stimulated AC activity fivefold, confirming that Rutabaga AC is Ca^{2+}/CaM responsive (18). Rutabaga AC activity displayed the characteristic biphasic Ca^{2+} dependence (31); it was activated over a broad range of low Ca^{2+} concentrations (maximal activation between 1 and 30 μM) but was inhibited by higher concentrations.

Current models for learning and memory have posited at least one molecular component in the system to be sensitive to the spatial and temporal responses of distinct neurotransmitter pathways (32). The ability to be activated by two second messenger pathways and the memory deficit in Rutabaga mutants suggest that the Rutabaga AC performs this integrative function in flies. It appears that this function is conserved in AC1, the mammalian homolog of Rutabaga. Like Rutabaga, AC1 is synergistically activated by G proteins and Ca^{2+}/CaM (33), expressed in the regions of the brain involved in learning and memory (2,34), and mice lacking AC1 are deficient in spatial learning (19).

Understanding the role of the Rutabaga AC in learning and memory requires knowledge of the genetic defect responsible for the *rutabaga* mutant phenotype. Comparison of genomic DNA sequences from homozygous *rut1* flies and from the Canton-S parent strain identified a single-point mutation that resulted in an arginine substituted for the glycine at amino acid 1026. This glycine is conserved both in cytoplasmic domains of the Rutabaga AC and in each of the known metazoan ACs and guanylyl cyclases. The G1026R mutation, introduced into the wild-type cDNA by oligonucleotide-directed mutagenesis, abrogated AC activity in transfected HEK-293 cells (18), indicating that the biochemical and phenotypic defects seen *in vivo* are due to complete loss of Rutabaga catalytic activity.

Dunce, another learning and memory gene, is preferentially expressed in mushroom bodies (35), the region of the insect brain independently shown to be responsible for integrating and storing sensory information (36). RNA *in situ* hybridization and immunocytochemistry revealed a strikingly similar pattern of expression for *rutabaga* AC (30). Mammalian AC1 is preferentially expressed in the hippocampus (34), the functionally analogous structure to the fly mushroom body, further demonstrating the conservation between fly and mammalian AC homologs.

Although not detectable by conventional methods, Rutabaga AC is also expressed in the ovaries. Mutations in the *dunce* cAMP phosphodiesterase cause female sterility. Isolation of recessive suppressors of the *dunce* female sterile phenotype yielded new *rutabaga* alleles (37). Because its phenotype could be suppressed by diminished AC activity, *dunce* female sterility is caused by elevated cAMP. In contrast, the learning and memory deficiency associated with either *dunce* or *rutabaga* mutations do not appear to be simply attributable to either elevated or depressed cAMP levels. The *rutabaga dunce* double mutant is still deficient for learning and memory (26), indicating that the biochemical requirement for cAMP is more likely temporal, and perhaps spatial. This supports the theory that signal transduction is defined by the frequency of second messenger generation as opposed to its absolute amplitude.

In an "enhancer detector" experiment (38), a collection of P-element insertions resulting in β-galactosidase expression in mushroom bodies was isolated (30); several of these insertions contained P elements inserted in the *rutabaga* locus (18). Several of these lines showed significantly reduced learning in an odor avoidance paradigm (39) and failed to complement the *rut1* allele (30). *Rutabaga* expression in each of the mutant P-element lines, as detected by both RNA *in situ* hybridization and immunohistochemistry using affinity-purified anti-Rutabaga antibody, was reduced. In addition, the proximity of P-element insertion to the *rutabaga* transcriptional start site correlated with the severity of learning and memory deficit, lack of expression, and diminished Ca^{2+}/CaM-responsive AC activity in mutant flies (18,30). The P-element-inserted lines therefore represent new *rutabaga* alleles with differing degrees of penetrance that can be used to gain further insight into the role of AC in learning and memory.

Ultimately, the physiologic relevance of specific modes of AC regulation can be investigated *in vivo*. Substitution of phenylalanine-503 with arginine in AC1 renders it insensitive to Ca^{2+}/CaM modulation (40). We can introduce the analogous substitution into Rutabaga AC and determine whether it is also resistant to Ca^{2+}/CaM regulation. This mutationally compromised AC can then be introduced by P-element-mediated transformation into *rutabaga* mutant flies and tested for its ability to rescue their learning and memory defect. Failure to rescue would confirm that Ca^{2+}/CaM activation of AC is crucial for learning and memory. Alternatively, heterologous expression of mammalian homologs and other AC isoforms can identify that specific regulatory interactions are important.

The genetic analysis of the *Drosophila* AC1 homolog revealed its specific functional role in the fly (18,26). Subsequent genetic experiments in mice confirmed that this function is conserved between flies and mammals. Homologous recombination was used to generate a strain of mice deficient for AC1 that displayed a learning deficit in the Morris water maze test (19). These studies demonstrated the structural, biochemical, and functional conservation of the *rutabaga* AC with its mammalian homolog AC1, and they suggest that genetic analysis of other fly cyclases will reveal the functional significance of each of the individual mammalian AC isoforms.

As stated earlier, *Drosophila* appears to express homologs for most, if not all, of the mammalian AC isoforms (M. J. C., V. Iourgenko, B. Kliot, E. Chung, and L. R. L., unpublished observations). Molecular cloning of each *Drosophila* AC and characterization of its biochemical properties and pattern of expression will demonstrate its relatedness to the known mammalian isoforms. Chromosomal *in situ* localization will identify its physical position relative to the genetic map and indicate whether any of the extensive array of *Drosophila* mutants carries a defect in that particular cyclase. Genetic analysis of the mutant phenotype would specify the physiologic role of that AC in *Drosophila* that can provide an indication of the function of its mammalian counterpart.

ROLES OF PROTEIN KINASE A IN *DROSOPHILA* DEVELOPMENT

At present, one can study *Drosophila* deficient in the downstream effectors of the cAMP signal transduction pathway to predict possible phenotypes of mutant AC isoforms. The major effector of cAMP is PKA, which has been extensively studied biochemically but has only recently been subjected to genetic analysis. The discovery of one predominant gene for the *Drosophila* catalytic (C) subunit, DC0 (41,42), has permitted genetic analysis of the ramifications of the loss of PKA. It is these PKA-deficient phenotypes that will provide clues to the possible physiologic roles of AC in *Drosophila*.

Two genes encoding regulatory (R) subunits have been identified in *Drosophila* (42), but these have not yet yielded clues to the possible *in vivo* roles of cAMP. The RI subunit gene produces at least three major transcripts that are homologous to the mammalian type I R subunit. Two of the alternatively spliced transcripts produce putative R subunits with N-terminal truncations of 57 or 87 amino acids, but these are of unknown physiologic relevance (42). The predominant C subunit, DC0, shows 82% overall identity to mouse C subunit (41,42), is expressed at all developmental stages, and appears to be responsible for the majority, if not all, of the PKA activity in adult flies (43). Efforts to uncover related kinases identified two genes, designated DC1 and DC2, which show 45% and 49% identity to mouse C subunit, respectively, and may encode additional C subunits of *Drosophila* PKA (42). As demonstrated *in vitro*, DC2 encodes an active kinase that is regulated by R subunit (44), but whether DC1 or DC2 are cAMP-regulated kinases *in vivo* has been difficult to assess because DC0 null mutations are lethal. Various hypomorphic DC0 alleles with differing penetrance cause a variety of phenotypes throughout *Drosophila* development. It is these phenotypes that will be summarized in the remainder of this chapter.

Oogenesis

The first suggestion that cAMP is important for oogenesis derives from the *dunce* learning and memory mutant. Certain mutations in the *dunce* cAMP phosphodiesterase cause female sterility, and these can be suppressed by mutations inactivating *rutabaga* AC (26,37). The functionally relevant AC is not likely to be *rutabaga*; *rut1* flies do not show any defect in oogenesis. Females expressing only a weak DC0 mutant allele are also sterile, and their phenotypes suggest two distinct functions for PKA during oogenesis.

The egg chamber in *Drosophila* consists of an oocyte surrounded by 15 nurse cells. Selective transport of organelles, RNAs, and proteins from nurse cells to oocyte occurs through visible gaps in the plasma membranes. This transport, which is essential for the viability of the oocyte and the earliest patterning events in the pre-blastoderm embryo, is dependent on the cytoskeletal architecture. DC0 protein is closely associated with germ cell membranes, and flies expressing only a weak DC0 mutant allele sometimes form abnormal, multinucleated nurse cells (45). It is thought that PKA phosphorylation maintains the cortical cytoskeletal infrastructure responsible for the intercellular bridges; absence of this PKA phosphorylation would lead to multinucleation of the nurse cells. Temporal regulation of PKA activity by cAMP does not appear to be essential in this process because heterologous expression of a constitutively active mutant mouse C subunit (mc*) can rescue the DC0-deficient phenotype (D. Kalderon, personal communication). However, the membrane localization of DC0 suggests that it is locally acting cAMP, presumably produced by an as yet unidentified AC that is important for PKA activation. There is precedent for a role of cAMP in maintaining cell–cell interactions. In lymphocyte adhesion, fusion mediated by the integrin LAF-1 or by the immunoglobulin superfamily member ICAM is altered by agents that modulate cAMP synthesis or PKA activity (46,47).

Besides maintaining the structural integrity of the oocyte, PKA appears to be important during oogenesis for the specification of the anteroposterior axis of the developing blastoderm. During the early stages of oogenesis, the microtubule network facilitates the nonspecific entry of messenger RNA (mRNA) and proteins into the immature oocyte. As oogenesis progresses, a structural rearrangement in the microtubule network determines the specific transport of macromolecules, such as the mRNAs *bicoid* (*bcd*) and *oskar* (*osk*), to the oocyte poles. During embryogenesis, bicoid protein forms a gradient from anterior to posterior in the cellular blastoderm and activates anterior gap genes, and oskar accumulates at the posterior pole and positions the cells that are fated to become germ cells in the adult (48). The localization of *osk* and *bcd* to the poles of the mature oocyte depends on the establishment of polarity in the microtubule cytoskeleton during oogenesis. This change in the cytoskeletal scaffold appears to require PKA and is thought to be initiated by a transient cAMP pulse. As a result, DC0 mutants show visual abnormalities in the organization of the microtubule network, and they fail to localize the microtubule-ependent motor protein kinesin to the posterior pole of the oocyte (49). Interestingly, the DC0 mutant alleles that disrupt the normal polarization of *osk* and *bcd* are mechanistically distinct from the alleles causing abnormal cell–cell interactions, which suggests that two different AC isoforms could be responsible for generating cAMP to activate the single C subunit.

Embryonic Patterning

Roles for maternally encoded PKA in embryogenesis are multifarious, judging from the variety of phenotypes of the unhatched embryos. The range of phenotypes varies from minor deformities in the embryonic cuticle to gross morphologic changes (43). No particular region of the embryo appears to be preferentially affected, with abnormalities being noted in anterior, thoracic, abdominal, and terminal structures.

Although lacking genetic evidence, a specific role for PKA and AC in the patterning of the dorsal–ventral axis in the *Drosophila* embryo has been hypothesized. Toll protein, although ubiquitous in the embryonic plasma membrane, is activated by a ventral-specific signal to stimulate the Pelle serine/threonine kinase. Pelle kinase activity, which remains restricted to the ventral side of the embryo, liberates the Dorsal transcription factor from its cytosolic binding protein, Cactus, allowing Dorsal to enter the nucleus and specify a ventral fate (50). The activity of a reporter gene driven by a *dorsal* specific promoter is dependent upon the nuclear localization of introduced dorsal protein. Dorsal contains a consensus PKA phosphorylation site, and cotransfection of *dorsal* with cDNAs encoding either the Toll receptor or DC0 significantly increases reporter activity (51), which suggests that PKA stimulates nuclear localization of Dorsal. However, there is little direct evidence that PKA regulation of dorsal is relevant in the embryo. Embryos in which the Dorsal PKA consensus phosphorylation site is mutated do not show a deficiency in Dorsal nuclear localization but do have a weakly dorsalized phenotype (D. Kalderon, personal communication). This may indicate that PKA maintains the transcriptional efficiency of Dorsal. The strongest genetic evidence supporting this role for PKA comes from lymphocyte activation in mammals. The proteins involved in dorsal–ventral patterning in *Drosophila* are related to those in the signaling pathway that triggers transcription of lymphocyte specific genes. Toll is homologous to the interleukin-1 (IL-1) receptor, whereas Cactus and Dorsal are homologous to IkB and NF-kB, respectively. IL-1 activation of NF-kB transcription is modulated by PKA (52,53), suggesting that cAMP may also play a role in the dorsal–ventral signaling pathway in *Drosophila*.

This discussion has so far provided evidence for a role for PKA in the earliest stages of *Drosophila* development, from oogenesis to the dorsal–ventral and anteroposterior patterning events. Is there evidence for other roles for PKA in later development? The use of weaker mutant DC0 alleles has allowed a more extensive, albeit incomplete, study of PKA mutant phenotypes in later stages of development.

Larval Patterning

Females heterozygous for two different, weakly mutant DC0 alleles show reduced fertility and lay only a small number of eggs that hatch. Hatched larvae have a requirement for zygotically expressed DC0, as opposed to the maternally encoded DC0 required for embryogenesis. Most DC0-deficient larvae have no obvious phenotypic abnormalities but do not molt and die within a few days; only those that express the weakest DC0 alleles survive to pupate.

A definitive role for PKA during larval development has been established by analyzing the phenotype of DC0 mosaic flies. DC0 mosaics contain clones of cells homozygous for PKA deficiency that are produced by inducing mitotic recombination in phenotypically wild-type flies heterozygous for the null DC0 allele. Mosaic induction early in larval development, during the first instar, revealed that PKA modulates the Hedgehog (Hh) signal transduction pathway establishing positional identity. Hedgehog was first identified as a segment polarity gene in the embryo involved in controlling gene transcription in cells anterior to its own expression. Hedgehog is now known to be a diffusible morphogen controlling cell fate and differentiation in a variety of tissues throughout embryonic and larval development. One system in particular in which Hh signaling is involved is the specification of cell fate in larval imaginal discs. Imaginal discs are epithelial sacs that invaginate from the surface of the 5-hour-old embryo and differentiate from the first to third instar larvae to form adult appendages, such as legs, antennae, eyes, and wings. In the leg and wing imaginal discs, Hh expressed in the posterior compartment diffuses anteriorly to induce a narrow band of expression of growth factor–like molecules, including *decapentaplegic* (*dpp*), of the transforming growth factor-β family, and *wingless* (*wg*), of the Wnt family. Decapentaplegic and Wg dictate cell fate throughout the imaginal disc. In the developing eye imaginal disc, Hh expressed in the posterior compartment induces a small anterior band of Dpp expression. Decapentaplegic induces anterior differentiation and stimulates further Hh expression, causing the band of differentiating cells, the morphogenetic furrow, to progress across the eye disc from posterior to anterior. Two other components play integral roles in this signaling pathway. Patched (Ptc) is a transmembrane protein and an Hh receptor (54,55). It functions to antagonize the expression of *wg* and *dpp* and sequesters Hh by limiting its diffusion into the anterior compartment (56). Smoothened (Smo) is a seven-transmembrane-span protein that activates *dpp* and *wg* transcription in the posterior compartment in response to a Hh signal (57,58). Smoothened has been demonstrated to form a complex with Hh and Ptc (55).

Protein kinase A mutations affect appendage development by interfering with normal cell differentiation in the imaginal discs (59–64). The loss of PKA mimics ectopic Hh expression or loss of the Ptc receptor and results in ectopic anterior expression of *dpp* and *wg* in the absence of the inductive signal from Hh protein. Some of the possibilities for Hh and PKA signaling that include the involvement of an AC are depicted in Fig. 2. In model A, Ptc receptor stimulation by Hh stimulates Smo activity. Smoothened inhibits AC, possibly by a G_i-mediated mechanism, down-regulating PKA, which represses expression of *dpp* and *wg*; Hh therefore up-regulates expression by antagonizing a high endogenous AC activity. Although this model is consistent with the biochemical nature of the cAMP signaling pathway, it is not supported by the genetic data (62). First, Hh activates *dpp* and *wg* expression in the presence of a constitutively active C subunit, mc*, indicating that Hh does not act solely to inhibit the PKA activity. Second, mc* does not compensate for loss of Ptc, implying that Ptc and PKA are in distinct pathways. Model B is similar to A except that Hh acts through Ptc both to stimulate Smo activity and to inhibit AC activity by an Smo-independent mechanism. Model B predicts that Ptc modulates AC activity even though it does not appear to be a member of the seven-transmembrane receptor family; model B therefore appears unlikely biochemically, in addition to being inconsistent with the genetic observations described earlier. In model C, which agrees with both biochemical and genetic data, Ptc and an AC act in parallel pathways to repress *dpp* and *wg* expression. Hh relieves this repression by stimulating Smo via Ptc

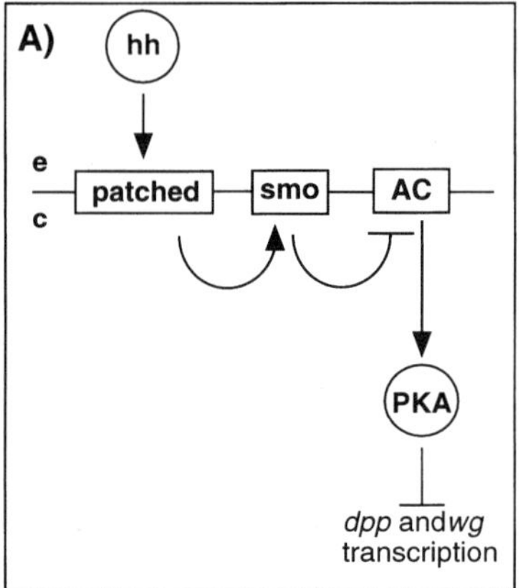
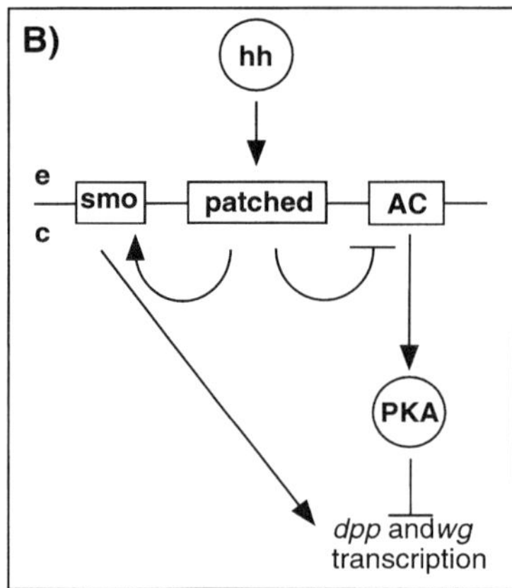
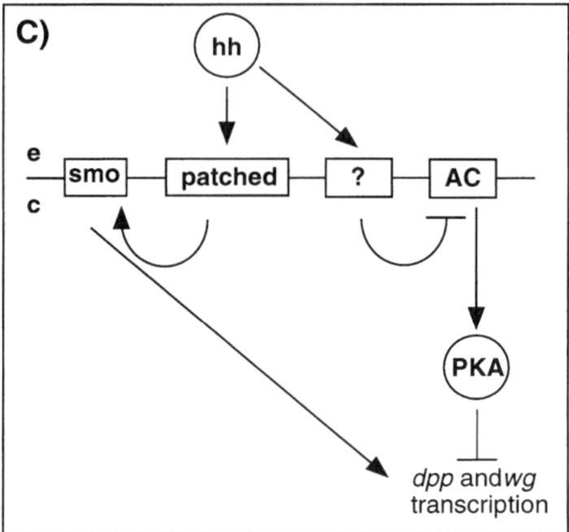

FIG. 2. Three possible mechanisms for Hh and PKA signal transduction. Refer to text for explanation. "e" denotes extracellular and "c" denotes the cytosol. Lines with arrow heads indicate positive interactions; lines with bars indicate negative interactions.

in conjunction with down-regulating the activity of AC through a second, unidentified Hh receptor. This unidentified gene could encode a second G protein–linked seven-transmembrane receptor that modulates AC. Alternatively, there may not be a receptor activating AC; PKA could be activated by an AC isoform with high intrinsic basal activity. In this case, regulation of cAMP production would be achieved by spatial and temporal control of AC expression. Other possibilities exist, some of which do not include a role for AC. The molecular and genetic characterization of an AC isoform expressed during larval development will aid in understanding the roles of PKA, Hh, Smo, and Ptc in this signaling pathway.

The roles of PKA in oogenesis, embryogenesis, and larval development involve responses to very different signals, implying biological roles for multiple isoforms of AC showing distinct regulatory properties and responding differentially to molecular cues. There are also potential roles for the cAMP signaling pathway in the adult fly that have been unexplored here. In summary, a number of AC isoforms appear to be required throughout *Drosophila* development, in a strict spatial-temporal pattern, to regulate the activity of the cAMP pathway.

CONCLUSION

The mammalian isoforms of AC possess distinct regulatory properties and exhibit differential patterns of expression. These biochemical and structural differences are presumably essential for their unique biologic roles. A complete understanding of the process of signal transduction therefore requires elucidation of the physiological significance of the individual isoforms of AC and of the *in vivo* relevance of their specific modes of regulation. These can both be accomplished by genetic analysis of the growing family of AC isoforms in *Drosophila*.

ACKNOWLEDGMENTS

We would like to thank Drs. Daniel Kalderon and Peter N. Devreotes for communicating results prior to publication, and Drs. Daniel Kalderon and Jochen Buck, as well as members of the Buck and Levin laboratories, for helpful suggestions on the preparation of this chapter. Work cited from this laboratory is supported by a grant from the National Institutes of Health (GM52891).

REFERENCES

1. Taussig R, Tang WJ, Hepler JR, Gilman AG: Distinct patterns of bidirectional regulation of mammalian adenylyl cyclases. *J Biol Chem* 1994;269:6093–6100.
2. Glatt CE, Snyder SH: Cloning and expression of an adenylyl cyclase localized to the corpus striatum [see comments]. *Nature* 1993;361:536–538.
3. Tang WJ, Gilman AG: Type-specific regulation of adenylyl cyclase by G protein beta gamma subunits. *Science* 1991;254:1500–1503.
4. Felder CC, Kanterman RY, Ma AL, Axelrod J: A transfected m1 muscarinic acetylcholine receptor stimulates adenylate cyclase via phosphatidylinositol hydrolysis. *J Biol Chem* 1989;264:20356–20362.
5. Federman AD, Conklin BR, Schrader KA et al: Hormonal stimulation of adenylyl cyclase through Gi-protein βγ subunits. *Nature* 1992;356:159–161.
6. Bakalyar HA, Reed RR: Identification of a specialized adenylyl cyclase that may mediate odorant detection. *Science* 1990;250:1403–1406.
7. Taussig R, Iniguez-Lluhl JA, Gilman AG: Inhibition of adenylyl cyclase by Gia. *Science* 1993;261:218–221.
8. Kataoka T, Broek D, Wigler M: DNA sequence and characterization of the *S. cerevisiae* gene encoding adenylate cyclase. *Cell* 1985;43:493–505.
9. Young D, Riggs M, Field J et al: The adenylyl cyclase gene from *Schizosaccharomyces pombe*. *Proc Natl Acad Sci U S A* 1989;86:7989–7993.
10. Pitt GS, Milona N, Borleis J et al: Structurally distinct and stage-specific adenylyl cyclase genes play different roles in *Dictyostelium* development. *Cell* 1992;69:305–315.
11. Whiteway M, Hougan L, Dignard D et al: The STE4 and STE18 genes of yeast encode potential β and γ subunits of the mating factor receptor-coupled G protein. *Cell* 1989;56:467–477.
12. Lilly P, Wu L, Welker DL, Devreotes PN: A G-protein beta-subunit is essential for *Dictyostelium* development. *Genes Dev* 1993;7:986–995.

13. Lilly PJ, Devreotes PN: Identification of CRAC, a cytosolic regulator required for guanine nucleotide stimulation of adenylyl cyclase in *Dictyostelium. J Biol Chem* 1994;269:14123–14129.
14. Insall R, Kuspa A, Lilly PJ et al: CRAC, a cytosolic protein containing a pleckstrin homology domain, is required for receptor and G protein–mediated activation of adenylyl cyclase in *Dictyostelium. J Cell Biol* 1994;126:1537–1545.
15. Pitt GS, Brandt R, Lin KC et al: Extracellular cAMP is sufficient to restore developmental gene expression and morphogenesis in *Dictyostelium* cells lacking the aggregation adenylyl cyclases (ACA). *Genes Dev* 1993;7:2172–2180.
16. Parent CA, Devreotes PN: Isolation of inactive and G protein–resistant adenylyl cyclase mutants using random mutagenesis. *J Biol Chem* 1995;270:22693–22696.
17. Parent CA, Devreotes PN: Constitutively active adenylyl cyclase mutant requires neither G proteins nor cytosolic regulators. *J Biol Chem* 1996;271:18333–18336.
18. Levin LR, Han PL, Hwang PM et al: The *Drosophila* learning and memory gene rutabaga encodes a Ca^{2+}/Calmodulin-responsive adenylyl cyclase. *Cell* 1992;68:479–489.
19. Wu ZL, Thomas SA, Villacres EC et al: Altered behavior and long-term potentiation in type I adenylyl cyclase mutant mice. *Proc Natl Acad Sci U S A* 1995;92:220–224.
20. Krupinski J, Coussen F, Bakalyar HA et al: Adenylyl cyclase amino acid sequence: possible channel- or transporter-like structure. *Science* 1989;244:1558–1564.
21. Feinstein PG, Schrader KA, Bakalyar HA et al: Molecular cloning and characterization of a Ca^{2+}/calmodulin-insensitive adenylyl cyclase from rat brain. *Proc Natl Acad Sci U S A* 1991;88:10173–10177.
21a. Iourgenko V, Kliot B, Cann MJ, Levin LR: Cloning and characterization of a drosophila adenylyl cyclase homologous to mammalian type IX. *FEBS Letts* 1997 (in press).
22. Dudai Y, Jan Y-N, Byers D et al: *Dunce,* a mutant of *Drosophila* deficient in learning. *Proc Natl Acad Sci U S A* 1976;73:1684–1688.
23. Quinn WG, Harris WA, Benzer S: Conditioned behavior in *Drosophila melanogaster. Proc Natl Acad Sci U S A* 1974;71:708–712.
24. Chen CN, Denome S, Davis RL: Molecular analysis of cDNA clones and the corresponding genomic coding sequences of the *Drosophila dunce+* gene, the structural gene for cAMP phosphodiesterase. *Proc Natl Acad Sci U S A* 1986;83:9313–9317.
25. Qiu YH, Chen CN, Malone T et al: Characterization of the memory gene *dunce* of *Drosophila melanogaster. J Mol Biol* 1991;222:553–565.
26. Livingstone MS, Sziber PP, Quinn WG: Loss of calcium/calmodulin responsiveness in adenylate cyclase of *rutabaga,* a *Drosophila* learning mutant. *Cell* 1984;37:205–215.
27. Livingstone MS: Genetic dissection of *Drosophila* adenylate cyclase. *Proc Natl Acad Sci U S A* 1985;82:5992–5996.
28. Tang W-J, Gilman AG: Adenylyl cyclases. *Cell* 1992;70:869–872.
29. Gorman CM, Gies DR, McCray G: Transient production of proteins using an adenovirus transformed cell line. *DNA Protein Eng Techniques* 1990;2:3–10.
30. Han PL, Levin LR, Reed RR, Davis RL: Preferential expression of the *Drosophila rutabaga* gene in mushroom bodies, neural centers for learning in insects. *Neuron* 1992;9:619–627.
31. Brostrom CO, Huang Y-C, Breckenbridge BM, Wolff DJ: Identification of a calcium-binding protein as a calcium-dependent regulator of brain adenylate cyclase. *Proc Natl Acad Sci U S A* 1975;72:64–68.
32. Abrams TW, Kandel ER: Is contiguity detection in classical conditioning a system or a cellular property? Learning in *Aplysia* suggests a possible molecular site. *Trends Neurosci* 1988;11:128–135.
33. Tang WJ, Krupinski J, Gilman AG: Expression and characterization of calmodulin-activated (type I) adenylyl cyclase. *J Biol Chem* 1991;266:8595–8603.
34. Xia Z, Refsdal CD, Merchant KM et al: Distribution of mRNA for the calmodulin-sensitive adenylate cyclase in rat brain: expression in areas associated with learning and memory. *Neuron* 1991;6:431–443.
35. Nighorn A, Healy MJ, Davis RL: The cyclic AMP phosphodiesterase encoded by the *Drosophila dunce* gene is concentrated in the mushroom body neuropil. *Neuron* 1991;6:455–467.
36. de Belle JS, Heisenberg M: Associative odor learning in *Drosophila* abolished by chemical ablation of mushroom bodies. *Science* 1994;263:692–695.
37. Bellen HJ, Gregory BK, Olsson CL, Kiger JA Jr: Two *Drosophila* learning mutants, *dunce* and *rutabaga,* provide evidence of a maternal role for cAMP on embryogenesis. *Dev Biol* 1987;121:432–444.
38. O'Kane CJ, Gehring W: Detection *in situ* of genomic regulatory elements in *Drosophila. Proc Natl Acad Sci U S A* 1987;84:9123–9127.

39. Tully T, Quinn W: Classical conditioning and retention in normal and mutant *Drosophila melanogaster. J Comp Physiol* 1985;157:263–277.
40. Wu Z, Wong ST, Storm DR: Modification of the calcium and calmodulin sensitivity of the type I adenylyl cyclase by mutagenesis of its calmodulin binding domain. *J Biol Chem* 1993;268: 23766–23768.
41. Foster JL, Higgins GC, Jackson FR: Cloning, sequence, and expression of the *Drosophila* cAMP-dependent protein kinase catalytic subunit gene. *J Biol Chem* 1988;263:1676–1681.
42. Kalderon D, Rubin GM: Isolation and characterization of *Drosophila* cAMP-dependent protein kinase genes. *Genes Dev* 1988;2:1539–1556.
43. Lane ME, Kalderon D: Genetic investigation of cAMP-dependent protein kinase function in *Drosophila* development. *Genes Dev* 1993;7:1229–1243.
44. Melendez A, Li W, Kalderon D: Activity, expression and function of a second *Drosophila* protein kinase A catalytic subunit gene. *Genetics* 1995;141:1507–1520.
45. Lane ME, Kalderon D: Localization and functions of protein kinase A during *Drosophila* oogenesis. *Mech Dev* 1995;49:191–200.
46. Haverstick DM, Gray LS: Lymphocyte adhesion mediated by lymphocyte function-associated antigen-1. I. Long term augmentation by transient increases in intracellular cAMP. *J Immunol* 1992; 149:389–396.
47. Haverstick DM, Gray LS: Lymphocyte adhesion mediated by lymphocyte function associated antigen-1. II. Interaction between phorbol ester- and cAMP-sensitive pathways. *J Immunol* 1992;149: 397–402.
48. St Johnston D, Nusslein-Volhard C: The origin of pattern and polarity in the *Drosophila* embryo. *Cell* 1992;68:201–219.
49. Lane ME, Kalderon D: RNA localization along the anteroposterior axis of the *Drosophila* oocyte requires PKA-mediated signal transduction to direct normal microtubule organization. *Genes Dev* 1994;8:2986–2995.
50. Grosshans J, Bergmann A, Haffter P, Nusslein-Volhard C: Activation of the kinase Pelle by Tube in the dorsoventral signal transduction pathway of *Drosophila* embryo. *Nature* 1994;372:500–501.
51. Norris JL, Manley JL: Selective nuclear transport of the *Drosophila* morphogen dorsal can be established by a signaling pathway involving the transmembrane protein Toll and protein kinase A. *Genes Dev* 1992;6:1657–1667.
52. Shirakawa F, Yamashita U, Chedid M, Mizel SB: Cyclic AMP-an intracellular second messenger for interleukin-1. *Proc Natl Acad Sci U S A* 1988;85:8201–8205.
53. Shirakawa F, Chedid M, Suttleo J et al: Interleukin-1 and cyclic AMP induce K immunoglobulin light-chain expression via activation of an NF-kB-like DNA binding protein. *Mol Cell Biol* 1989; 9:959–964.
54. Marigo V, Davey RA, Zuo Y et al: Biochemical evidence that patched is the Hedgehog receptor. *Nature* 1996;384:176–179.
55. Stone DM, Hynes M, Armanini M et al: The tumour-suppressor gene patched encodes a candidate receptor for Sonic Hedgehog. *Nature* 1996;384:129–134.
56. Chen Y, Struhl G: Dual roles for patched in sequestering and transducing Hedgehog. *Cell* 1996;87:553–563.
57. Alcedo J, Ayzenzon M, Von Ohlen T et al: The *Drosophila* smoothened gene encodes a seven-pass membrane protein, a putative receptor for the hedgehog signal. *Cell* 1996;86:221–232.
58. Van den Heuvel M, Ingham PW: Smoothened encodes a receptor-like serpentine protein required for hedgehog signaling. *Nature* 1996; 382:547–551.
59. Heberlein U, Singh CM, Luk AY, Donohoe TJ: Growth and differentiation in the *Drosophila* eye coordinated by *hedgehog. Nature* 1995;373:709–711.
60. Jiang J, Struhl G: Protein kinase A and hedgehog signaling in *Drosophila* limb development. *Cell* 1995;80:563–572.
61. Lepage T, Cohen SM, Diaz-Benjumea FJ, Parkhurst SM: Signal transduction by cAMP-dependent protein kinase A in *Drosophila* limb patterning. *Nature* 1995;373:711–715.
62. Li W, Ohlmeyer JT, Lane ME, Kalderon D: Function of protein kinase A in hedgehog signal transduction and *Drosophila* imaginal disc development. *Cell* 1995;80:553–562.
63. Pan D, Rubin GM: cAMP-dependent protein kinase and *hedgehog* act antagonistically in regulating decapentaplegic transcription in *Drosophila* imaginal discs. *Cell* 1995;80:543–552.
64. Strutt DI, Wiersdorff V, Mlodzik M: Regulation of furrow progression in the *Drosophila* eye by cAMP-dependent protein kinase A. *Nature* 1995;373:705–709.

*Advances in Second Messenger and
Phosphoprotein Research,* Vol. 32,
edited by Dermot M. F. Cooper
Lippincott–Raven Publishers, Philadelphia © 1998

7

Class III Adenylyl Cyclases: Regulation and Underlying Mechanisms

Wei-Jen Tang, Shuizhong Yan, and Chester L. Drum

*Department of Pharmacological and Physiological Sciences, University of Chicago,
Chicago, Illinois 60637*

Cyclic adenosine monophosphate (cAMP) is a key second messenger. It activates
several target molecules, including cAMP-dependent protein kinases, cyclic nucleotide–gated ion channels, cAMP receptors (in *Dictyostelium*), and catabolic activating protein (in *Escherichia coli*). Activation of the cAMP-modulated target proteins controls diverse phenomena involving metabolism, gene transcription,
olfaction, heart rate, learning and memory, drug addiction, and cell differentiation.
Defects in components that are involved in cAMP-mediated pathways are linked to
several human diseases, including diabetes, certain thyroid and pituitary tumors, and
mental retardation (1,2).

The level of intracellular cAMP is determined by the rate of synthesis and the rate
of clearance. Clearance of intracellular cAMP is regulated either by cyclic nucleotide phosphodiesterase or by cAMP efflux mechanisms that secrete or transport
cAMP out of cells. Five classes of phosphodiesterases regulating cAMP degradation
are encoded by 11 genes, and their enzymatic activity can be regulated by a diversity
of signals, including calcium (via Ca^{2+}/calmodulin) and cyclic guanosine
monophosphate (cGMP) (3,4). Cyclic AMP efflux, a regulatory mechanism that exists from *E. coli* to human, is less clear in the molecular identity of the involved components (5,6).

Cyclic AMP is synthesized by adenylyl cyclases (ACs), enzymes that convert
adenosine triphosphate (ATP) to cAMP. Three classes of ACs have been cloned
based on the conservation of their catalytic domains: class I, ACs from Enterobacte-
ria, including *E. coli*; type II, toxin ACs, including calmodulin-activated toxins from
Berdetella pertussis and *Bacillus anthrocis*; class III, ACs homologous from bacteria to human. Class III ACs are the focus of this review. (Excellent reviews of
classes I and II ACs can be found in refs. 7 and 8.)

REGULATION OF CLASS III ADENYLYL CYCLASES

Class III ACs have four distinct molecular designs (Fig. 1). First are the integral membrane proteins that contain two hydrophobic stretches (~20 kDa each), each consisting of six putative transmembrane helices, and two homologous cytoplasmic domains (C_1 and C_2 domains, 30–40 kDa each). This design includes nine mammalian ACs (type I to type IX [AC1 to AC9]), fruit fly *rutabaga* (Dm *rutabaga*), and slime mold (Dd ACA) (9–22). Second are the integral membrane proteins that have one transmembrane helix. This class is found in the parasites *Trypanosoma brucei* (Tb ESAG, expression site-associated gene) and *Leishmania donovani* (Ld-RAC, receptor-AC) and in the slime mold (*Dictyostelium discoideum*, Dd ACG) (22–24). Third are the peripheral membrane proteins from yeast (*Schizosaccharomyces*

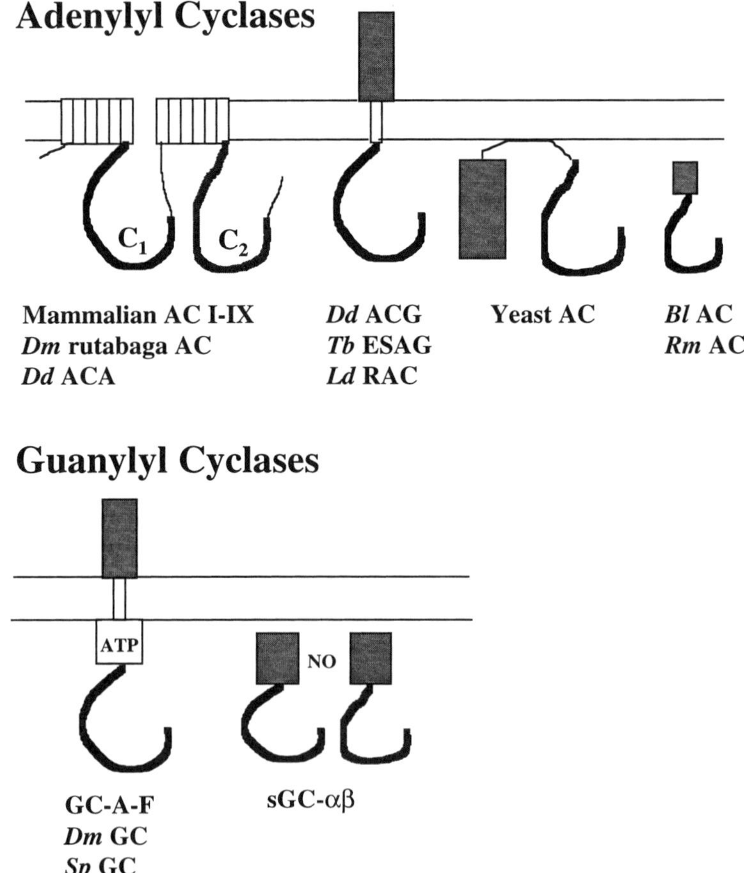

FIG. 1. Molecular diversity of adenylyl and guanylyl cyclases. Bold lines indicate the cyclase domain. ATP, ATP-binding regulatory domain in the membrane-bound guanylyl cyclases; NO, nitric oxide. See text for description.

pombe, Saccharomyces cerevisiae, and *Saccharomyces klyuyveri*) (25–27). Fourth are the soluble enzymes from the bacteria *Brevibacterium liquefaciens* (Bl AC) and *Rhizobium meliloti* (Rm AC) (28,29). The properties of these four designs of enzymes are summarized in Table 1. Class III ACs also share the domains that are present in guanylyl cyclases, enzymes that convert guanosine triphosphate (GTP) to cGMP (see Fig. 1). This family of guanylyl cyclases includes integral membrane proteins that have one transmembrane helix (GC-A to GC-F, guanylyl cyclases from fruit fly and sea urchin), and heterodimeric soluble proteins (sGC-α and sGC-β) (30–41).

Adenylyl cyclase activity can be regulated diversely by extracellular and intracellular stimuli. The enzyme activity of ACs with 12-transmembrane helices is regulated mostly through serpentine receptors that couple to G proteins. Of the four G protein classes, G_s, $G_{i/o}$, and G_q can modulate AC activity (42,43). Signals that activate G_s result in the release of the activated α subunit of G_s, which can activate all mammalian and *Drosophila* ACs with 12-transmembrane helices (42). In olfactory epithelium, odorant-bound receptors activate the olfactory-specific G protein α subunit G_{olf}, which then stimulates olfactory-specific AC3 (10). Signals that activate $G_{i,o,z}$ can inhibit AC5 and AC6 activity by release of the GTP-bound α subunit of G_i, G_o, and G_z (44,45). The concomitant release of G protein $\beta\gamma$ from $G_{i/o}$ seems to play the major role in inhibiting type I enzyme (45,46). On the contrary, signals that activate $G_{i,o}$ could result in stimulation of type II, IV, and VII AC activity by release of the G protein $\beta\gamma$ (12,46,47). G protein $\beta\gamma$ is also implicated in the regulation of the CRAC (cytoplasmic regulator of adenylyl cyclase)-activated activity of ACA from *Dictyostelium* (48,49). Homologous enzymes that are regulated by different G proteins (i.e., G protein α and $\beta\gamma$) also exist in ACs from lower eukaryotes. Adenylyl cyclase from *S. cerevisiae* is activated by ras, a small GTP-binding protein, whereas that from *S. pombe* is stimulated by Gpa2, a yeast homolog of the G protein α subunit (50,51).

A more complex pattern of regulation can be achieved by signals that activate the G_q family (42,43). Activation of the G_q family can increase the enzyme activity of phospholipase C-β, resulting in the increase of intracellular Ca^{2+} by inositol triphosphate and activation of protein kinase C (PKC) by diacylglycerol. These signals can regulate AC activity via four distinct mechanisms: by Ca^{2+} ion (inhibition of type V and VI enzymes) (15,52,53), by Ca^{2+}-calmodulin (activation of type I, VIII, and *rutabaga* enzymes) (14,21,54), by Ca^{2+}-calcineurin (inhibition of AC9) (55), and by phosphorylation of protein kinase C (PKC) (resulting in either activation or inhibition of type II, IV, and V enzymes) (56,57). Regulation of AC activity by phosphorylation is not limited to PKC. Activation of cAMP-dependent protein kinase can in turn inhibit the activity of type V and VI enzyme, resulting in feedback inhibition (58).

Adenylyl cyclase could bind extracellular stimuli directly. Adenylyl cyclases with a single transmembrane helix have 40–80 kDa extracellular domains. By analogy to the membrane-bound guanylyl cyclases, this extracellular domain might bind extracellular ligands and increase cAMP production. Consistent with this notion, a fragment of chicken blood protein, α^D-globin is shown to activate AC activity from *T.*

TABLE 1. *Properties of class III adenylyl cyclases*

	Type*	Amino acid residues	Expression	Regulators[†]			
				$G_{s\alpha}$	$\beta\gamma$	Fsk	Others
12-transmembrane proteins	I	1134	Brain	+	−	+	$G_{o,i\alpha}(-)$,Ca^{2+}-CaM(+)
	II	1090	Brain, lung	+	+	+	PKC(+)
	III	1144	Olfactory	+	0	+	
	IV	1064	Widely expressed	+	+	+	PKC
	V	1184	Heart, brain, others	+	0	+	$G_{i,o,z\alpha}(-)$,PKA(−), PKC(+),$Ca^{2+}(-)$
	VI	1165	Heart, brain, others	+	0	+	$G_{i,o,z\alpha}(-)$,PKC(+),$Ca^{2+}(-)$
	VII	1099	Lung, spleen, heart, brain	+	+	+	
	VIII	1248	Brain	+	?	+	Ca^{2+}-CaM(+)
	IX	1353	Widely expressed	+	0	+/0	Calcineurin(−)
	Dm *rutabaga*	2249	Mushroom body	+	0	+	Ca^{2+}-CaM(+)
	Dd ACA	1407	During aggregation	0	+	0	CRAC(+)
Single-transmembrane protein	Dd ACG	858	Fruiting body	?	?	0	?
	Tb ESAG 4	1229	Bloodstream form	?	?	0	α^D-globin fragment[‡]
	Ld RAC	1380	Promastigote	?	?	0	?
Peripheral membrane protein	Sc CYR	2026	Constitutive	?	?	0	RAS,CAP
	Sp CYR	1693	Constitutive	?	?	0	Gpa2
Soluble proteins	Bl AC	403	Constitutive	?	?	0	Pyruvate
	Rm AC	193	Constitutive	?	?	0	?

*Adenylyl cyclase types I–IX are mammalian (9–17); *rutabaga* is from *Drosophila* (17); ACA and ACG are from *Dictyostelium* (22); Tb ESAG 4 is from *Trypanosoma* (24); Ld RAC is from *Leishmania* (23); CYR is from *S. cerevisiae* and *S. pombe* (26,27); and Bl and Rm AC are from *Brevibacterium* and *Rhizobium*, respectively (28,29).

[†]+, stimulation; −, inhibition; 0, no effect; +/0, small stimulation in mammalian cells and no stimulation in Sf9 cells (19).

[‡]Activate adenylyl cyclase from *T. cruzi* (59).

cruzi (59). The natural ligands of ACs with large, potentially ligand-binding domains remain to be determined for *Dictyostelium* (ACG), *T. brucei*, and *L. donovani*. Adenylyl cyclases could also be regulated by small intracellular ligands directly. Enzyme from *B. liquefaciens* is regulated by the metabolic intermediate pyruvate. Mammalian AC is inhibited by adenosine analogs; however, the physiologic significance of this inhibition remains elusive (60).

PROPERTIES OF THE CATALYTIC DOMAINS OF CLASS III ADENYLYL CYCLASE

A 10–15-kDa domain (designated the cyclase domain) is conserved broadly among 21 different ACs (class III) and all known guanylyl cyclases (see Fig. 1). Phylogenetic analysis indicates that these domains can be divided into four groups (Fig. 2A): group I is from guanylyl cyclases; group II is the second cytoplasmic domain of ACs with 12-transmembrane helices and the cyclase domain of ACG; group III includes the cyclase domains of ACs from yeast, *T. brucei, L. donovani,* and *B. liquefaciens*; and group IV is the cyclase domain of AC from *R. meliloti*. It is worth noting that although AC C_2 and ACA C_2 are closely related, AC C_1 and ACA C_1 are less related to each other than to the first and second groups, respectively. Sequence analysis of the C_1 domain reveals that mammalian and *Drosophila* ACs can be divided into six groups; their regulation is also closely related to their groupings (Fig. 2B; Table 1).

The cyclase domains consist of two relatively conserved domains (A and B) joined by a highly variable linker (Fig. 3). The length of the linker sequence can be very short (zero in Bl AC) or up to 96 amino acid residues long (in ACA C_1 domain). There is a short insertion (1–13 amino acids) at the variable region of the C_2 domain of mammalian ACs but none at the analogous region of the C_1 domain. The linker regions are most likely solvent exposed and are likely to be involved in the regulation of enzyme activity. This region could be involved in binding of G protein $\beta\gamma$ because a peptide consisting of sequences from the linker region of the C_2 domain of AC2 blocks the G protein $\beta\gamma$–mediated activities (61).

The dimer of cyclase domains appears to be the functional unit for catalysis. The ACs with 12-transmembrane helices have two naturally linked cyclase domains (C_1 and C_2). The C_1 or C_2 domain alone has little enzyme activity; however, mixing the two domains results in a $G_{s\alpha}$ and forskolin-activated enzyme activity (62,63). Soluble guanylyl cyclase is a heterodimer (α and β), and both subunits are necessary for nitric oxide–regulated enzyme activity (30). Membrane-bound guanylyl cyclases have only one cyclase domain, and they exist as a dimer or a tetramer (64–66). Catalytically inactive, mutated forms of guanylyl cyclases can form a complex with wild-type guanylyl cyclase and inactivate it, indicating again that the functional unit is formed by at least two cyclase domains (67). Interestingly, other proteins that have one cyclase domain, including ACs from yeast and *B. liquefaciens*, also exist as a dimer (68,69).

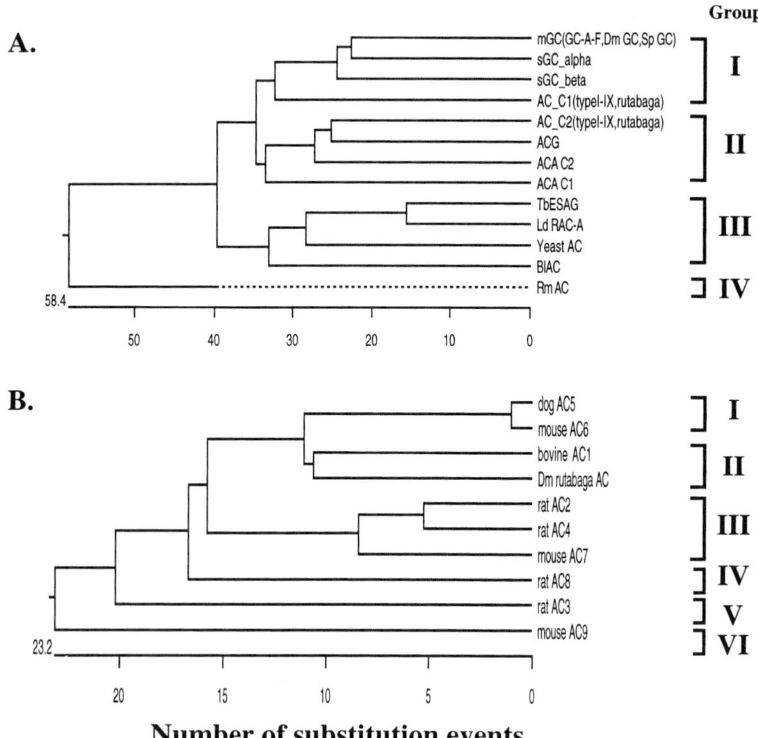

Number of substitution events

FIG. 2. Phylogeny of cyclase domains from adenylyl and guanylyl cyclases (**A**) and mammalian and fly adenylyl cyclases (**B**). The protein sequences analyzed included adenylyl cyclases from mammalian and *D. melanogaster* (types I–IX, *rutabaga*; AC1 [GenBank accession number M25579], AC2 [M80550], AC3 [M55075], AC4 [M80633], AC5 [M88649], AC6 [M94968], AC7 [U12919], AC8 [L26986], AC9 [Z50190], and *Rutabaga* [M81887]), *Dictyostelium* (ACA [M87279] and ACG [M87278]), yeast (*S. cerevisiae* [M12057], *S. pombe* [M24942], and *S. klyuyveri* [X56042]), *Trypanosoma* (Tb ESAG; X52118, X52120, and X52121), *Leishmania* (LdRAC; U17042, U17043), *Brevibacterium* (BlAC; X57541), *Rhizobium* (RmAC; M35096), membrane-bound forms of guanylyl cyclases (mGC; mammalian GC-A [J05677], GC-B [M26896], GC-C [M55636], GC-D [L37203], GC-E [L36029], and GC-F [L36030], *D. melanogaster* [X72800, L35598], and sea urchin [M22444]), and soluble guanylyl cyclases α (sGC_alpha; mammalian α1 [M57405] and α2 [X63282], *D. melanogaster* [U27117]) and β (sGC_beta; mammalian β1 [M22562] and β2 [M57507], and *D. melanogaster* [U27123]). The sequence alignment was performed using DNASTAR MegAlign Clustal method with a structural residue weight table; consensus sequences were first obtained within the family of cyclases and those sequences were used to compare among different families of the enzymes.

How do the dimeric cyclase domains form their nucleotide binding sites? The homodimer of cyclase domains, including adenylyl and guanylyl cyclases with a single transmembrane helix, should have at least two distinct nucleotide binding sites. Kinetic analysis shows that the forskolin-activated mammalian ACs deviate from the Michaelis–Menton equation with respect to Mg^{2+}-ATP (70). Furthermore, mutations at either C_1 or C_2 of AC1 can affect the K_m value (70). The evidence suggests that the heterodimer of cyclase domains from mammalian AC has more than one nu-

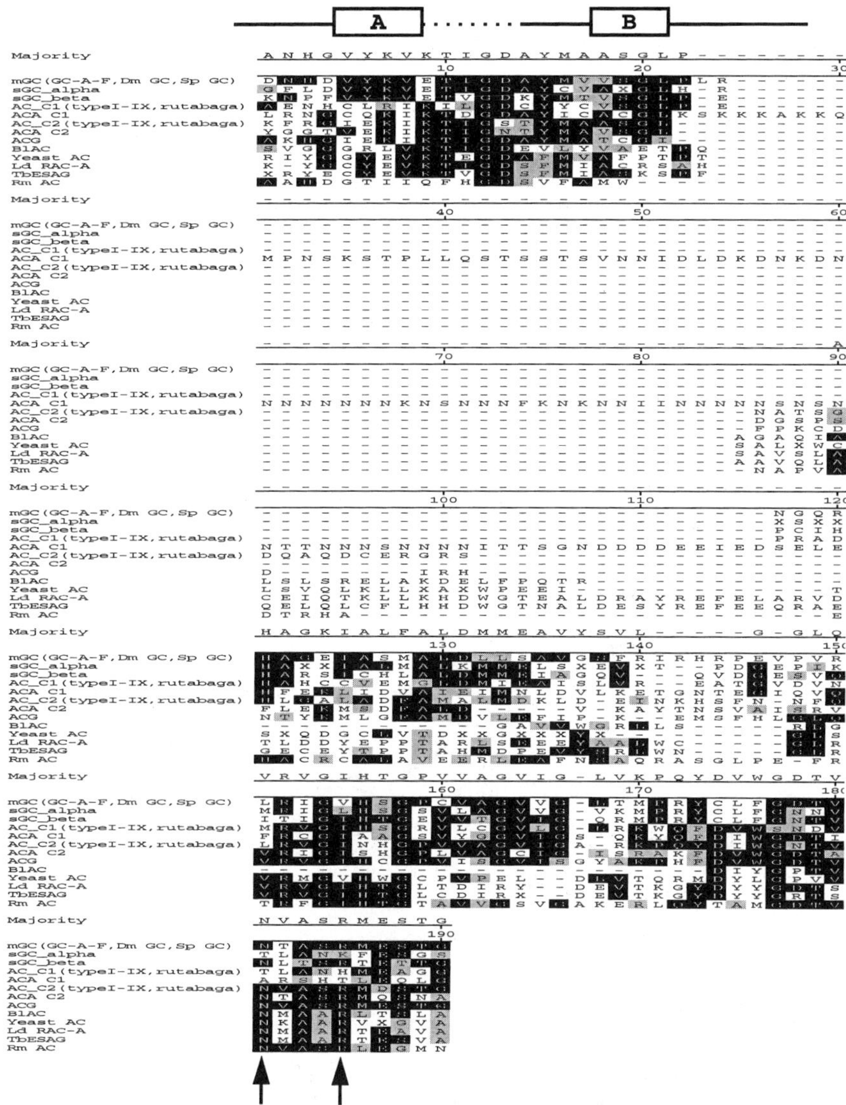

FIG. 3. Sequence comparison of the cyclase domains from adenylyl and guanylyl cyclases. Diagram (top) shows the relationship of region A and B, which is connected by a highly variable region (dotted line). Sequences shaded black are absolutely conserved amino acid residues and those shaded gray are highly conserved. Regions A and B are the two segments that are shaded mostly black or gray and thus are highly conserved. The analysis was performed as for Fig. 2. *Arrows* show two conserved amino acid residues that are crucial for catalysis (see text for description).

cleotide binding site. It remains to be determined whether there are more than two nucleotide binding sites in the dimeric cyclase domains and whether the nucleotide binding sites are formed within or between two cyclase domains (Fig. 4A).

Within the cyclase domains, catalysis may occur at one or both of the nucleotide binding sites. Analysis using the point mutations at AC1 suggests that only one of the cyclase domains (C_2 but not C_1) is the catalytic domain. This hypothesis is further supported by genetic analysis of the C_1 and C_2 domains in *E. coli* (71). Expres-

A. Substrate binding of cyclase domains

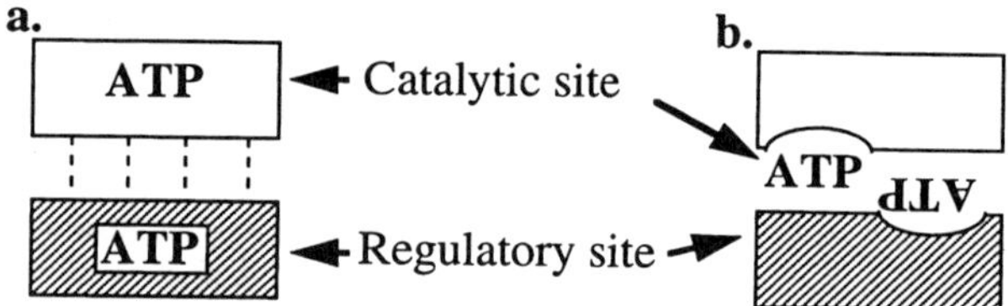

B. Catalytic site

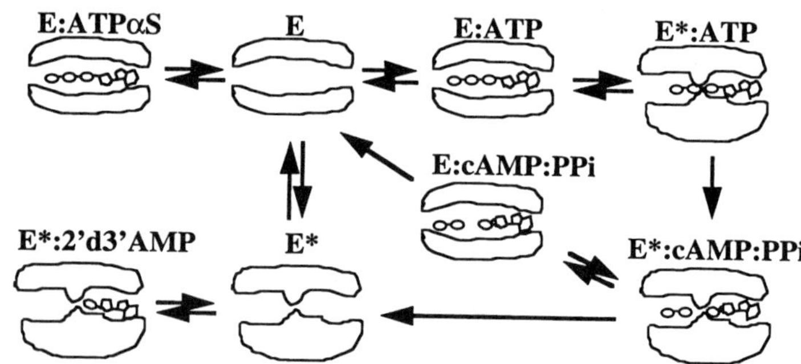

C. Regulatory site

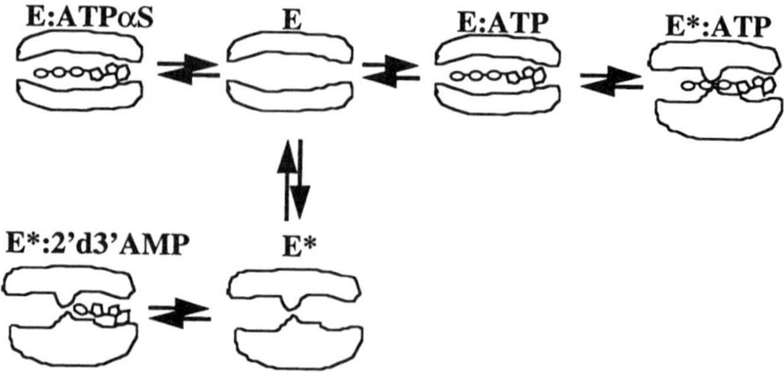

FIG. 4. Proposed model for catalysis. (**A**) Organization of ATP binding in the cyclase domains. (**B,C**) The proposed kinetic cycle of catalytic and regulatory domain. See text for description.

sion of the C_2 but not the C_1 domain in *E. coli* can drive cAMP-mediated maltose catabolism. The inequivalence of cyclase domains in catalysis may not be restricted to the heterodimer of cyclase domains. In certain assay conditions, the homodimer of cyclase domains from membrane-bound guanylyl cyclases exhibits kinetic features deviant from the Michaelis–Menton equation.

What amino acid substitutions inactivate the C_1 domain of AC? Sequence analysis of all the cyclase domains reveals that two residues (arginine and asparagine) are conserved among one of the cyclase domains from the heterodimers (the catalytic C_2 domains of ACs and the β subunits of guanylyl cyclases) and all of those from the homodimers (Fig. 3, arrows). These two residues are not conserved in the C_1 domains of ACs or the α subunit of soluble guanylyl cyclases. Mutational analysis has shown that these two residues are crucial for catalysis (S.-Z. Yan, Z.-H. Hwang, and W.-J. Tang, unpublished data) (70). It would be interesting to test whether the conversion of these two residues could restore the enzymatic activity of C_1 domains. The proposed role of these residues in catalysis also predicts that the β subunit of soluble guanylyl cyclase is the catalytic domain, whereas the α subunit is involved in regulatory function; this hypothesis remains to be evaluated.

What are the functions for the noncatalytic C_1 domain of ACs? Although the C_1 domain itself appears to be catalytically inactive, its association with the C_2 domain is required for high enzymatic activity. We speculate that the C_1 domain binds the substrate and is involved in regulating the enzymatic activity by transferring the binding energy of the substrate to the C_2 domain. It has been shown that mammalian phosphofructokinase and hexokinase have a similar design wherein one of two homologous domains is left catalytically inactive (72). The noncatalytic domain could have arisen by gene duplication, with subsequent mutations of residues that are crucial for catalysis. The result of having one catalytic and one noncatalytic domain in a functional unit is to increase greatly the regulation of these enzymes compared to their single-domain yeast and bacterial counterparts.

UNDERLYING MECHANISMS FOR COMPLICATED REGULATION OF CLASS III ADENYLYL CYCLASES

To understand the mechanisms of regulation for class III AC, the kinetic steps in the enzyme cycle need to be determined. Although AC from *B. liquefaciens* was crystallized more than 20 years ago, no molecular structure of any class III cyclase domain has been determined (68). Furthermore, no direct biochemical analysis has been done to address the properties of the nucleotide binding sites. Thus the proposed enzyme cycle is based only on the following kinetic and mutational analyses. Adenylyl cyclase is largely inactive in the absence of stimulators, despite the presumption that it can bind substrate readily (K_m for metal-ATP is about 50–500 μM, at least 10-fold lower than intracellular ATP concentration). There is little inhibition of enzymatic activity by cAMP and pyrophosphate (PPi), implying that release of product is not the rate-limiting step.

Adenosine analogs that specifically inhibit AC activity are termed P-site inhibitors because of the requirement of an intact purine ring. Deoxylation at 2' or 5' OH groups and phosphorylation at 3' OH groups (tri-P>di-P>mono-P) of P-site inhibitors increase their potency (73,74). P-site inhibitors resemble cAMP, yet product inhibition of AC is weak. They resemble substrate but the mechanism for inhibition is not competitive. The inhibition of AC activity by P-site inhibitors is activity dependent: the higher the enzyme activity of AC, the more potent the P-site inhibitors become (75,76). Mutations in the noncatalytic C_1 domain dramatically reduce the sensitivity of AC to P-site inhibitors without affecting the K_m value, so do those at the catalytic C_2 domain (70). This excludes the possibility of a single site of interaction for P-site inhibitors.

Kinetic and mutational analyses have led us to hypothesize that the enzyme is regulated by an open/closed transition of the catalytic domain (Fig. 4B) (77,78). At the catalytic site, substrate is bound in the open state. The conformational change leading to the closed state brings crucial residues to bear on the α-phosphate and 3' OH groups of ATP, where catalysis must occur. Such a conformational change might occur between the conserved A and B segments, induced by a conformational shift of the linker between them (see Fig. 3). After catalysis, the open state could reform either prior to or after release of product. By analogy, the noncatalytic site would also have an open/closed transition (Fig. 4C). The rate of formation of the closed state at either catalytic or regulatory sites could limit the rate of reaction, as could the rate for returning to the open conformation from the nucleotide-free closed state. The activators of AC would facilitate these rate-limiting steps, whereas the inhibitors would hinder them. The activators can also enhance the enzyme activity by facilitating the interaction between the C_1 and C_2 domains, adding another mechanism for regulation (62,63).

In our hypothesis, P-site inhibitors and substrate bind to different conformations of the same site. We also hypothesize that P-site inhibitors bind to both nucleotide binding sites (the catalytic and regulatory sites). Whereas substrate can bind only to the open conformation, P-site inhibitors have preferential affinity for the closed state. Although P-site inhibitors and substrate bind the same site, inhibition of AC by P-site inhibitors is not competitive with respect to substrate (see Appendix). Our model also predicts that the sensitivity of P-site inhibitors would depend on the duration of the empty closed state and the activators that promote the formation of the closed state, thus enhancing the enzyme activity. Our model predicts that activators would also increase the sensitivity of the enzymes to P-site inhibitors, fitting well with the observation from kinetic analysis. Point mutations in AC1 could upset the balance of open/closed states at either C_1 or C_2 domains, resulting in a significant reduction of their response to P-site inhibitors (70).

Catalytic velocity of adenylyl and guanylyl cyclases is fit well with the Michael–Menton equation under most assay conditions, indicating that only one nucleotide binding site operates kinetically. However, our model predicts that there are two nucleotide binding sites at the functional units of cyclase domains and these two sites are not equal in the heterodimers. One possibility is that binding of the nucleotides is ordered, first to the noncatalytic site and then to the catalytic site. To evaluate this model, a reasonable quantity of pure protein from the two cyclase domains of heterodimers is needed. We have expressed and purified the C_1 and C_2 domains, and experiments to address the nucleotide binding sites are in progress (62).

With the advent of multiple isoforms of AC, the regulatory and physiologic diversity of cAMP production has grown to an unprecedented level of complexity. The emerging ability to study catalytic and regulatory functions within mechanistically based models results in a variety of testable theories. Within this framework we describe a spectrum of AC forms and propose a model of catalysis based on a growing body of experimental evidence. With the gradual, conceptual synthesis of physiological and biochemical data, the ultimate roles of cAMP signaling may one day be resolved.

REFERENCES

1. Lyons J, Landis CA, Harsh G et al: Two G protein oncogenes in human endocrine tumors. *Science* 1990;249:655–659.
2. Petrij F, Giles RH, Dauwerse HG et al: Rubinstein-Taybi syndrome caused by mutations in the transcriptional co-activator CBP. *Nature* 1995;376:348–351.
3. Beavo JA, Conti M, Heaslip RJ: Multiple cyclic nucleotide phosphodiesterases. *Mol Pharmacol* 1994;46:399–405.
4. Corti M, Nemoz G, Sette C, Vicini E: Recent progress in understanding the hormonal regulation of phosphodiesterases. *Endocr Rev* 1995;16:370–389.
5. Brunton LL, Heasley LE: cAMP export and its regulation by prostaglandin A1. *Methods Enzymol* 1988;159:83–93.
6. Barber R, Butcher RW: The egress of cyclic AMP from metazoan cells. *Adv Cyclic Nucl Res* 1983; 15:119–138.
7. Danchin A: Phylogeny of adenylyl cyclases. *Adv Second Messenger Phosphoprotein Res* 1993; 27:109–162.
8. Peterkofsky A, Reizer A, Reizer J et al: Bacterial adenylyl cyclases. *Prog Nucleic Acid Res Mol Biol* 1993;44:31–65.
9. Feinstein PG, Schrader KA, Bakalyar HA et al: Molecular cloning and characterization of a Ca^{2+}/calmodulin-insensitive adenylyl cyclase from rat brain. *Proc Natl Acad Sci U S A* 1991;88: 10173–10177.
10. Bakalyar HA, Reed RR: Identification of a specialized adenylyl cyclase that may mediate odorant detection. *Science* 1990;250:1403–1406.
11. Krupinski J, Coussen F, Bakalyar HA et al: Adenylyl cyclase amino acid sequence: possible channel- or transporter-like structure. *Science* 1989;244:1558–1564.
12. Gao B, Gilman AG: Cloning and expression of a widely distributed (type IV) adenylyl cyclase. *Proc Natl Acad Sci U S A* 1991;88:10178–10182.
13. Katsushika S, Chen L, Kawabe J et al: Cloning and characterization of a sixth adenylyl cyclase isoform: types V and VI constitute a subgroup within the mammalian adenylyl cyclase family. *Proc Natl Acad Sci U S A* 1992;89:8774–8778.
14. Cali JJ, Zwaagstra JC, Mons N et al: Type VIII adenylyl cyclase. A Ca^{2+}/calmodulin-stimulated enzyme expressed in discrete regions of rat brain. *J Biol Chem* 1994;269:12190–12195.
15. Yoshimura M, Cooper DM: Cloning and expression of a Ca^{2+}-inhibitable adenylyl cyclase from NCB-20 cells. *Proc Natl Acad Sci U S A* 1992;89:6716–6720.
16. Premont RT, Chen J, Ma HW et al: Two members of a widely expressed subfamily of hormone-stimulated adenylyl cyclases. *Proc Natl Acad Sci U S A* 1992;89:9809–9813.
17. Ishikawa Y, Katsushika S, Chen L et al: Isolation and characterization of a novel cardiac adenylylcyclase cDNA. *J Biol Chem* 1992;267:13553–13557.
18. Watson PA, Krupinski J, Kempinski AM, Frankenfield CD: Molecular cloning and characterization of the type VII isoform of mammalian adenylyl cyclase expressed widely in mouse tissues and in S49 mouse lymphoma cells. *J Biol Chem* 1994;269:28893–28898.
19. Premont RT, Matsuoka I, Mattei M-G et al: Identification and characterization of a widely expressed form of adenylyl cyclase. *J Biol Chem* 1996;271:13900–13907.

20. Paterson JM, Smith SM, Harmar AJ, Antoni FA: Control of a novel adenylyl cyclase by calcineurin. *Biochem Biophys Res Commun* 1995;214:1000–1008.
21. Levin LR, Han PL, Hwang PM et al: The *Drosophila* learning and memory gene rutabaga encodes a Ca^{2+}/calmodulin-responsive adenylyl cyclase. *Cell* 1992;68:479–489.
22. Pitt GS, Milona N, Borleis J et al: Structurally distinct and stage-specific adenylyl cyclase genes play different roles in *Dictyostelium* development. *Cell* 1992;69:305–315.
23. Sanchez MA, Zeoli D, Klamo EM et al: A family of putative receptor-adenylate cyclases from *Leishmania donovani*. *J Biol Chem* 1995;270:17551–17558.
24. Alexandre S, Paindavoine P, Tebabi P et al: Differential expression of a family of putative adenylate/guanylate cyclase genes in *Trypanosoma brucei*. *Mol Biochem Parasitol* 1990;43:279–288.
25. Young D, O'Neill K, Broek D, Wigler M: The adenylyl cyclase–encoding gene from *Saccharomyces klyuyveri*. *Gene* 1991;102:129–132.
26. Young D, Riggs M, Field J et al: The adenylyl cyclase gene from *Schizosaccharomyces pombe*. *Proc Natl Acad Sci U S A* 1989;86:7989–7993.
27. Kataoka T, Broek D, Wigler M: DNA sequence and characterization of the *S. cerevisiae* gene encoding adenylate cyclase. *Cell* 1985;43:493–505.
28. Peters EP, Wilderspin AF, Wood SP et al: A pyruvate-stimulated adenylate cyclase has a sequence related to the fes/fps oncogenes and to eukaryotic cyclases. *Mol Microbiol* 1991;5:1175–1181.
29. Beuve A, Boesten B, Crasnier M et al: *Rhizobium meliloti* adenylate cyclase is related to eucaryotic adenylate and guanylate cyclases. *J Bacteriol* 1990;172:2614–2621.
30. Nakane M, Arai K, Saheki S et al: Molecular cloning and expression of cDNAs coding for soluble guanylate cyclase from rat lung. *J Biol Chem* 1990;265:16841–16845.
31. McNeil L, Chinkers M, Forte M: Identification, characterization, and developmental regulation of a receptor guanylyl cyclase expressed during early stages of *Drosophila* development. *J Biol Chem* 1995;270:7189–7196.
32. Yang, RB, Foster DC, Garbers DL, Fulle HJ: Two membrane forms of guanylyl cyclase found in the eye. *Proc Natl Acad Sci U S A* 1995;92:602–606.
33. Gigliotti S, Cavaliere V, Manzi A et al: A membrane guanylate cyclase *Drosophila* homolog gene exhibits maternal and zygotic expression. *Dev Biol* 1993;159:450–461.
34. Schulz S, Green CK, Yuen PS, Garbers DL: Guanylyl cyclase is a heat-stable enterotoxin receptor. *Cell* 1990;63:941–948.
35. Yuen PS, Potter LR, Garbers DL: A new form of guanylyl cyclase is preferentially expressed in rat kidney. *Biochemistry* 1990;29:10872–10878.
36. Yamaguchi M, Rutledge LJ, Garbers DL: The primary structure of the rat guanylyl cyclase A/atrial natriuretic peptide receptor gene. *J Biol Chem* 1990;265:20414–20420.
37. Koesling D, Harteneck C, Humbert P et al: The primary structure of the larger subunit of soluble guanylyl cyclase from bovine lung: homology between the two subunits of the enzyme. *FEBS Lett* 1990;266:128–132.
38. Schulz S, Singh S, Bellet RA et al: The primary structure of a plasma membrane guanylate cyclase demonstrates diversity within this new receptor family. *Cell* 1989;58:1155–1162.
39. Thorpe DS, Garbers DL: The membrane form of guanylate cyclase. Homology with a subunit of the cytoplasmic form of the enzyme. *J Biol Chem* 1989;264:6545–6549.
40. Nakane M, Saheki S, Kuno T et al: Molecular cloning of a cDNA coding for 70 kilodalton subunit of soluble guanylate cyclase from rat lung. *Biochem Biophys Res Commun* 1988;157:1139–1147.
41. Shah, S Hyde DR: Two *Drosophila* genes that encode the alpha and beta subunits of the brain soluble guanylyl cyclase. *J Biol Chem* 1995;270:15368–15376.
42. Sunahara RK, Dessauer CW, Gilman AG: Complexity and diversity of mammalian adenylyl cyclase. *Annu Rev Pharmacol Toxicol* 1996;36:461–480.
43. Taussig R, Gilman AG: Regulation of adenylyl cyclases. *J Biol Chem* 1995;270:1–4.
44. Kozasa T, Gilman AG: Purification of recombinant G proteins from Sf9 cells by hexahistidine tagging of associated subunits. Characterization of α_{12} and inhibition of adenylyl cyclase by α_z. *J Biol Chem* 1995;270:1734–1741.
45. Taussig R, Tang WJ, Hepler JR, Gilman AG: Distinct patterns of bidirectional regulation of mammalian adenylyl cyclases. *J Biol Chem* 1994;269:6093–6100.
46. Tang WJ, Gilman AG: Type-specific regulation of adenylyl cyclase by G protein βγ subunits. *Science* 1991;254:1500–1503.
47. Yoshimura M, Ikeda H, Tabakoff B: μ-opioid receptors inhibit dopamine stimulated activity of type V adenylyl cyclase but enhance the dopamine stimulated activity of type VII adenylyl cyclase. *Mol Pharmacol* 1996;50:43–51.

48. Lilly PJ, Devreotes PN: Chemoattractant and GTP gamma S-mediated stimulation of adenylyl cyclase in *Dictyostelium* requires translocation of CRAC to membranes. *J Cell Biol* 1995;129:1659–1665.

49. Parent CA, Devreotes PN: Molecular genetics of signal transduction in *Dictyostelium*. *Ann Rev Biochem* 1996;65:411–440.

50. Isshiki T, Mochizuki N, Maeda T, Yamamoto M: Characterization of a fission yeast gene, gpa2, that encodes a G alpha subunit involved in the monitoring of nutrition. *Genes Dev* 1992;6:2455–2462.

51. Toda T, Uno I, Ishikawa T et al: In yeast, RAS proteins are controlling elements of adenylate cyclase. *Cell* 1985;40:27–36.

52. Chiono M, Mahey R, Tate G, Cooper DM: Capacitative Ca^{2+} entry exclusively inhibits cAMP synthesis in C6–2B glioma cells. Evidence that physiologically evoked Ca^{2+} entry regulates Ca(2+)-inhibitable adenylyl cyclase in non-excitable cells. *J Biol Chem* 1995;270:1149–1155.

53. Yu HJ, Ma H, Green RD: Calcium entry via L-type calcium channels acts as a negative regulator of adenylyl cyclase activity and cyclic AMP levels in cardiac myocytes. *Mol Pharmacol* 1993;4:689–693.

54. Tang WJ, Krupinski J, Gilman AG: Expression and characterization of calmodulin-activated (type I) adenylylcyclase. *J Biol Chem* 1991;266:8595–8603.

55. Antoni FA, Barnard RJO, Shipston MJ et al: Calcineurin feedback inhibition of agonist-evoked cAMP formation. *J Biol Chem* 1995;270:28055–28061.

56. Kawabe J, Iwami G, Ebina T et al: Differential activation of adenylyl cyclase by protein kinase C isoenzymes. *J Biol Chem* 1994;269:16554–16558.

57. Jacobowitz, O Iyengar R: Phorbol ester-induced stimulation and phosphorylation of adenylyl cyclase 2. *Proc Natl Acad Sci U S A* 1994;91:10630–10634.

58. Iwami G, Kawabe J-I, Ebina T et al: Regulation of adenylyl cyclase by protein kinase A. *J Biol Chem* 1995;270:12481–12484.

59. Fraidenraich D, Pena C, Isola EL et al: Stimulation of Trypanosoma cruzi adenylyl cyclase by an αD-globin fragment from Triatoma hindgut: effect on differentiation of epimastigote to trypomastigote forms. *Proc Natl Acad Sci U S A* 1993;90:10140–10144.

60. Bushfield M, Shoshani I, Johnson RA: Tissue levels, source, and regulation of 3'-AMP: an intracellular inhibitor of adenylyl cyclases. *Mol Pharmacol* 1990;38:848–853.

61. Chen J, Devivo M, Dingus J et al: A region of adenylyl cyclase 2 critical for regulation of G protein βγ subunits. *Science* 1995;268:1166–1169.

62. Yan S-Z, Hahn D, Huang Z-H, Tang W-J: Two cytoplasmic domains of mammalian adenylyl cyclase form a Gsa- and forskolin-activated enzymes *in vitro*. *J Biol Chem* 1996;271:10941–10945.

63. Whisnant RE, Gilman AG, Dessauer CW: Interaction of the two cytoplasmic domains of mammalian adenylyl cyclase. *Proc Natl Acad Sci U S A* 1996;93:6621–6625.

64. Lowe DG: Human natriuretic peptide receptor-A guanylyl cyclase is self-associated prior to hormone binding. *Biochemistry* 1992;31:10421–10425.

65. Chinkers M, Wilson EM: Ligand-independent oligomerization of natriuretic peptide receptors. Identification of heteromeric receptors and a dominant negative mutant. *J Biol Chem* 1992;267:18589–18597.

66. Wilson EM, Chinkers M: Identification of sequences mediating guanylyl cyclase dimerization. *Biochemistry* 1995;34:4696–4701.

67. Thompson DK, Garbers DL: Dominant negative mutations of the guanylyl cyclase-A receptor. *J Biol Chem* 1995;270:425–430.

68. Takai K, Kurashina Y, Suzuki-Hori C et al: Adenylate cyclase from *Brevibacterium liquefaciens*: I. Purification, crystallization, and some properties. *J Biol Chem* 1974;249:1965–1972.

69. Mitts MR, Grant DB, Heideman W: Adenylate cyclase in Saccharomyces cerevisiae is a peripheral membrane protein. *Mol Cell Biol* 1990;10:3873–3883.

70. Tang WJ, Stanzel M, Gilman AG: Truncation and alanine-scanning mutants of type I adenylyl cyclase. *Biochemistry* 1995;34:14563–14572.

71. Tang WJ, Gilman AG: Construction of a soluble adenylyl cyclase activated by Gsa and forskolin. *Science* 1995;268:1769–1772.

72. Fothergill-Gilmore LA, Michels PA: Evolution of glycolysis. [Review]. *Prog Biophys Mol Biol* 1993;59:105–235.

73. Desaubry L, Shoshani I, Johnson RA: 2',5'-Dideoxyadenosine 3'-polyphosphates are potent inhibitors of adenylyl cyclases. *J Biol Chem* 1996;271:2380–2382.

74. Johnson RA, Yeung SM, Stubner D et al: Cation and structural requirements for P site–mediated inhibition of adenylate cyclase. *Mol Pharmacol* 1989;35:681–688.

75. Florio, VA, Ross EM: Regulation of the catalytic component of adenylate cyclase. Potentiative interaction of stimulatory ligands and 2′,5′-dideoxyadenosine. *Mol Pharmacol* 1983;24:195–202.
76. Johnson RA, Shoshani I: Kinetics of "P"-site-mediated inhibition of adenylyl cyclase and the requirements for substrate. *J Biol Chem* 1990;265:11595–11600.
77. Schulz GE: *Curr Opin Struct Biol* 1991;1:883.
78. Cox S, Radzio-Andzelm E, Taylor SS: Domain movements in protein kinase. *Curr Opin Struct Biol* 1994;4:893–901.

APPENDIX

The steady-state equation derived from the model of the kinetic cycle in the catalytic site of adenylyl cyclases.

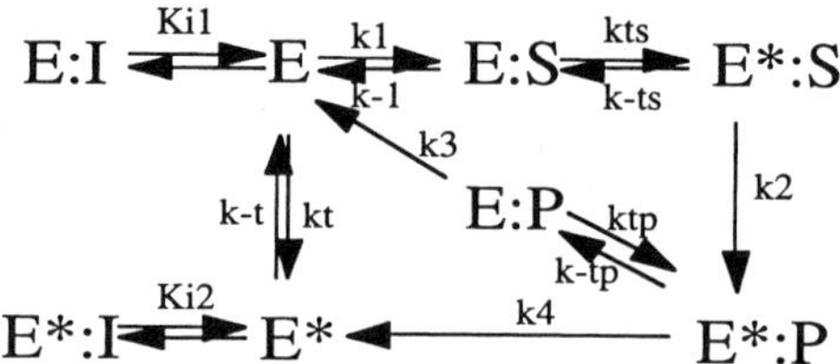

The model assumes that catalysis and release of product are fast and irreversible. The symbols are: v = velocity, V_m = maximal enzyme velocity, s = substrate, i = inhibitor, E = open state, E^* = closed state, K_{i1}, K_{i2} = dissociation constants for inhibitors 1 and 2. Using the King–Altman method, the velocity for the catalytic site of adenylyl cyclase at steady state is predicted to be:

$$\frac{v}{V_m} = \frac{A[s]}{B + C[s] + \dfrac{[i_1]}{K_{i1}} \bullet D + \dfrac{[i_2]}{K_{i2}} \bullet (E + F[s])}$$

1. If the inhibitor binds the open state (E) only, the equation for catalytic rate (velocity) at the steady state will be:

$$\frac{v}{V_m} = \frac{A[s]}{C[s] + \left(B + \dfrac{[i]}{K_i} \bullet D\right)}$$

Increase of substrate concentration could reverse the effect of the inhibitor; thus the inhibitor is competitive with respect to substrate.

2. If the inhibitor binds the closed state (E^*) only, the equation for catalytic rate at the steady state will be:

$$\frac{v}{V_{\mathrm{m}}} = \frac{A[s]}{\left(C + F \bullet \dfrac{[i]}{K_i}\right)[s] + \left(B + \dfrac{[i]}{K_i} \bullet E\right)}$$

If the inhibitor binds both the open and closed states, the equation for catalytic rate at the steady state will be:

$$\frac{v}{V_{\mathrm{m}}} = \frac{A[s]}{\left(C + F \bullet \dfrac{[i]}{K_i}\right)[s] + \left(B + \dfrac{[i]}{K_i} \bullet (D + E)\right)}$$

In both cases, the inhibitor could reduce the enzyme activity even at very high substrate concentration; thus, it is not competitive with respective to substrate.

Advances in Second Messenger and Phosphoprotein Research, Vol. 32, edited by Dermot M. F. Cooper
Lippincott–Raven Publishers, Philadelphia © 1998

8

Calcium Control of Adenylyl Cyclase: The Calcineurin Connection

Ferenc A. Antoni, Susan M. Smith, James Simpson, Roberta Rosie, George Fink, and Janice M. Paterson

MRC Brain Metabolism Unit, University of Edinburgh, Edinburgh, EH8 9JZ, Scotland, United Kingdom

In multicellular organisms, the reception and transduction of intercellular signals such as hormones and neurotransmitters into the cell interior are fundamental to homeostatic adaptation. The first recognized intracellular messengers were calcium ions and cyclic adenosine $3'5'$ monophosphate (cAMP), which is synthesized from adenosine triphosphate (ATP) by adenylyl cyclase (AC). It is not overtly exaggerated to suggest that by the mid-1980s the respective roles of these intracellular messengers had been analyzed to an extent that the topic was graying into boredom through familiarity, with tyrosine kinases and protein phosphatases stealing most of the limelight.

The lull in cAMP signaling has been radically dispersed by the introduction of reverse genetics into the field (1). As a result, mammalian AC has been reborn as an expanding group of highly complex proteins currently numbering nine isotypes. Each enzyme appears to function as a uniquely designed molecular signal processor (2–4). Notionally, structural divergence means functionally distinct execution of the generic function of cAMP synthesis from ATP. A major challenge for current efforts is to elucidate the distinctive functional properties of the AC subfamilies—a task at least partially accomplished—and to correlate them with structural features—a less well-developed area mainly because of the technical difficulties encountered in site-directed mutagenesis of ACs (5,6).

The control of ACs is summarized in Table 1, and the basic structural features have been highlighted in earlier reviews (7,8). The key point is that the cAMP signal in virtually all types of cell is potentially subject to dynamic control by a variety of intracellular messenger systems and that the unique functional properties of AC isotypes are fundamental in bringing about this dynamism.

This point is illustrated in Fig. 1A, B, which shows a cell that expresses two main types of AC, types II and IX (AC2 and AC9) (note that early publications [9,10] re-

TABLE 1. *Summary of the control of mammalian adenylyl cyclases by various regulators*

Regulator	Action	Adenylyl cyclase isotype	Comment
$G_{s\alpha}$	Stimulation	All	Some cyclases are less responsive than others, but see ref. 4
$G_{i\alpha}$	Inhibition	All except AC2	The exact reason for AC_2 insensitivity toward $G_{i\alpha}$ is controversial (13,14)
$G_{\beta\gamma}$	Inhibition	AC1, AC(8)?	This effect is observed only upon stimulation by $G_{s\alpha}$ or Ca^{2+}/ calmodulin
	Stimulation	AC2	Requires $G_{s\alpha}$ or forskolin; kinase C may also be sufficient
Kinase C	Stimulation	AC2, AC7	Very pronounced enhancement; synergy with $G_{s\alpha}$ or forskolin; just detectable with $G_{\beta\gamma}$
	Stimulation	AC1, AC4, AC5, AC6	Demonstrable but modest increases of activity in intact cells; requires forskolin or active $G_{s\alpha}$
Ca^{2+}	Stimulation	AC1, AC8	Requires calmodulin; synergy with $G_{s\alpha}$ or forskolin
	Inhibition	AC5, AC6, AC9	Apparently direct action; evident upon stimulation by $G_{s\alpha}$; less pronounced after forskolin or GTP
	Inhibition	AC9	Mediated by calcineurin; evident upon stimulation by $G_{s\alpha}$; not apparent after forskolin

fer to this cyclase as no. 10), and is equipped with voltage-regulated calcium channels (typically L-type channels) as well as intracellular Ca^{2+} stores. The stoichiometry of receptors, G_s, and cyclases is such that activation by a receptor $(R[G_s])$ conservatively coupling to G_s will cause a unit increase in cAMP synthesis occurring primarily through AC9. The increase in cAMP causes depolarization of the membrane potential and evokes calcium entry through membrane ion channels. The rise of intracellular calcium suppresses cAMP levels by inhibiting the synthesis of cAMP and by enhancing its breakdown through activation of Ca^{2+}/calmodulin-(CaM) activated cyclic nucleotide phosphodiesterase (PDE). This arrangement may result in cytoplasmic Ca^{2+}/cAMP oscillations the parameters of which depend on the time constants of the three main components described earlier: activated cyclase, calcium channel, and PDE as modeled by Cooper and co-workers (2).

If AC9 alone is present (Fig. 1B), co-activation of G_s and G_i/G_q by a promiscuous receptor such as $PACAP_1$ (see ref. 11 for review) or by two receptors such as CRF_1 (G_s coupled) and $V_{1\beta}$ (G_i/G_q coupled) (12) would produce a lower cAMP response when compared with activation of a G_s-coupled receptor alone, because inhibition by $G_{i\alpha}$ would ensue. Moreover, the enhanced mobilization of intracellular Ca^{2+} from IP3-sensitive stores would also inhibit AC9 and thus favor a rapid decay of the cAMP response.

By contrast, if AC2 is present (Fig. 1C), the cAMP response to the combined activation of G_s and G_i/G_q will differ in at least two important aspects. First, it will be greater than the response to the G_s-coupled agonist alone; second, it may be largely

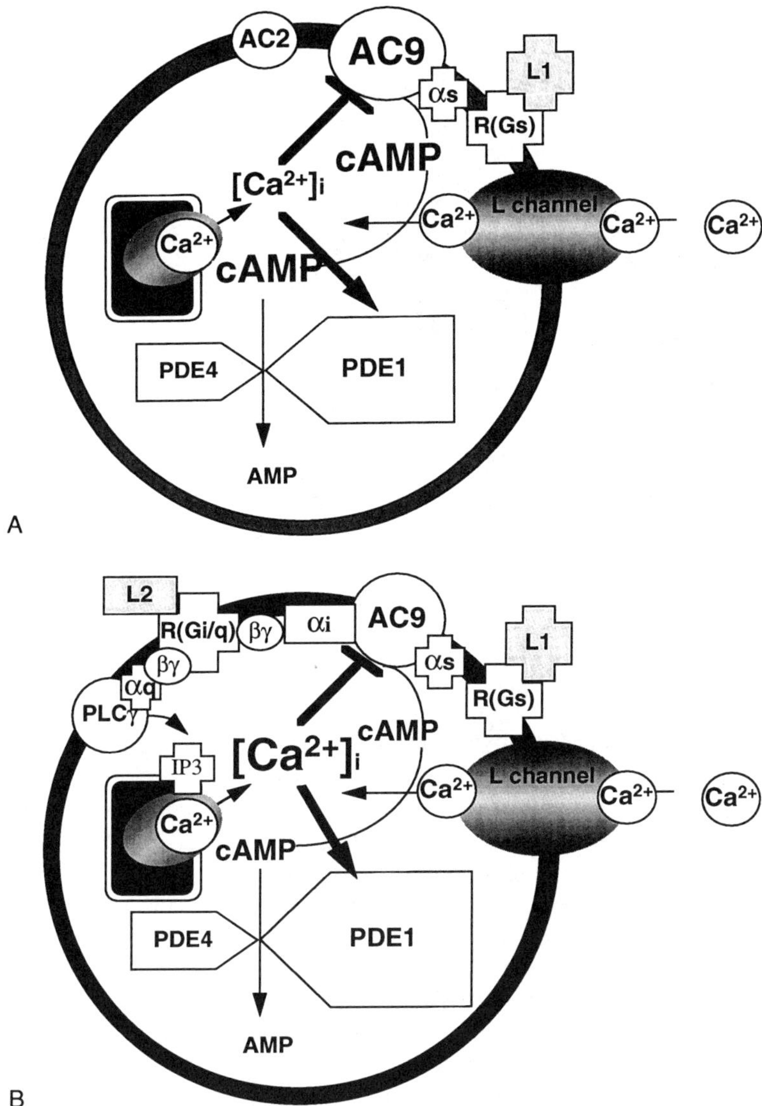

FIG 1. Generation of different types of cAMP signal with two different types of adenylyl cyclase. Stimulatory influences are shown either as thick arrows or items in plus signs. Inhibition is symbolized by lines ending in *T* or by letters enclosed in rectangles (minus sign). The sizes of the font and boxes correlate with changes in concentration or rates of activity. αs, αi, αq, and βγ, the respective subunits of heterotrimeric G proteins; L1 and L2, receptor ligands; R(G$_s$), receptor showing coupling to G$_s$ only; R(G$_{i/q}$), receptor coupling to both G$_i$ and G$_q$; PDE1, Ca²⁺/CaM-activated cylic nucleotide phosphodiesterase; PDE4, cAMP-specific, Ca²⁺-independent cyclic nucleotide phosphodiesterase. (*Figure continues.*)

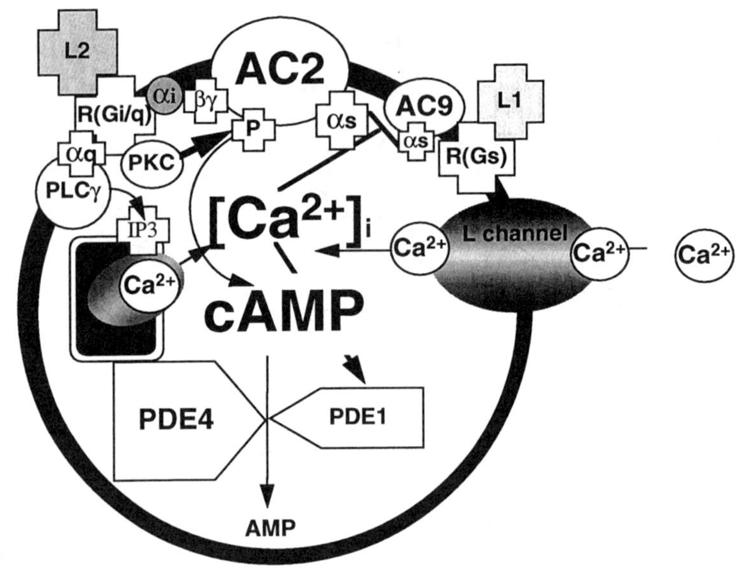

FIG. 1. *Continued.*

resistant to feedback inhibition by intracellular Ca^{2+}. The enhancement of the cAMP response is brought about by a synergistic interaction between $G_{s\alpha}$ and phosphorylation by protein kinase C (13,14). The mechanism of this synergy is not fully understood but appears to involve an increase in the affinity of AC2 for $G_{s\alpha}$ upon phosphorylation by kinase C (see Chapters 4 and 5) and an apparent resistance of AC2 toward the inhibitory action of $G_{i\alpha}$ (13,14). Thus, the bulk of cAMP will arise from AC2, which is not subject to inhibition by Ca^{2+} or $G_{i\alpha}$.

In addition, emerging evidence on PDEs suggests that the resulting high levels of cAMP may reprofile PDE activity. In some systems, cAMP-specific, Ca^{2+}-independent PDE activity is enhanced, whereas Ca^{2+}/CaM-dependent PDE activity is inhibited by cAMP-dependent phosphorylation (15). Thus the synthesis as well as the breakdown of cAMP may be uncoupled from intracellular free Ca^{2+} levels by defined patterns of acute stimuli. The net result is a cAMP signal that is essentially resistant to inhibition by Ca^{2+} ions, thus uncoupling cAMP levels from Ca^{2+} entry mechanisms, or intracellular Ca^{2+}-binding proteins (calbindin, parvalbumin, CaM) that may feed into the cAMP cascade by modulating intracellular Ca^{2+} transients (12,16,17).

This and various other models (2,18) that have been proposed on the basis of studies in transfected cells or cell lines have yet to be demonstrated in physiologically relevant contexts. Furthermore, the model in Fig. 1 does not take into account the possibility of topographic segregation of AC signaling units within a single cell (2,19) and the well-known spatiotemporal inhomogeneities of the intracellular Ca^{2+} signal (2). A prime challenge for the understanding of the physiologic significance and the potential therapeutic exploitation of the structural and functional diversity of

AC is to understand the mechanisms and adaptive pressures that govern the expression and function of AC isotypes. These issues will be highlighted by presenting the state of the art on the novel AC isotype AC9.

FUNCTIONAL PROPERTIES OF AC9

Deductions on the Basis of Primary Amino Acid Structure

Adenylyl cyclase 9 was discovered by homology based polymerase chain reaction (PCR) of mouse brain messenger RNA (mRNA) (10,20) as well as through a functionally driven analysis of AC mRNAs from AtT20 mouse pituitary corticotrope tumor cells (9,21). The primary structure of this enzyme places it firmly in the family of ACs; however, it is also clear that AC9 is quite distinct from currently known isotypes and forms a novel subfamily (9,20) (Fig. 2). Its most remarkable features are the long cytoplasmic loops (Fig. 3) that are more than 100 amino acids longer than in previously published cyclases. A further potential member of the AC9 subfamily is a sequence deposited in GenBank (Z46958) (http://www2.ncbi.nlm.gov/genbank; NCBI, Bethesda, MD); however, this has not been reported so far to give rise to a functionally active cyclase.

Data from our laboratory (9,21; also see following discussion) suggest that AC9 is inhibited by intracellular Ca^{2+} and that this effect is mediated by the Ca^{2+}/CaM-activated Ser/Thr protein phosphatase calcineurin (protein phosphatase 2B) (22). Control of cal-

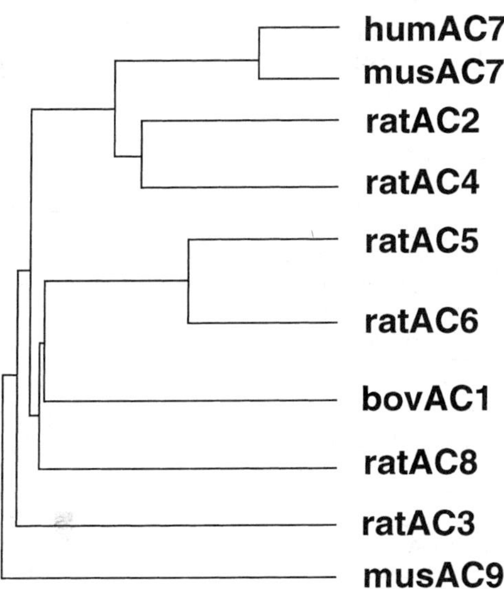

FIG. 2. Dendrogram of mammalian AC amino acid sequences. (The alignments were generated with the PileUp program of the GCG, University of Wisconsin package provided through the MRC Human Genome Mapping Project.)

cineurin activity by Ca^{2+} is through two proteins—a tightly bound E-F-hand subunit and CaM, the association of which to calcineurin is Ca^{2+} dependent (22).

It is important to note that the protein phosphatase activity of calcineurin is inhibited by nanomolar concentrations of a bimolecular complex formed between the immunosuppressant agent FK506 (tacrolimus) and FK-binding protein 12 (FKBP12) (23). FKBP12 is a ubiquitously present small chaperone enzyme that preserves the conformation of a number of intrinsic membrane proteins (24). Notably, calcineurin is not a substrate for FKBP12 under physiologic conditions; it only interacts with FKBP12 in the presence of FK506 (25,26), although this view has been challenged recently (27,28). FK506 is a bacterial toxin and the inhibition by FK506/FKBP12 of calcineurin, an enzyme fundamental for adaptation in fungi (27), is the result of opportunistic microbial evolution (29). In clinical therapy, FK506 is a widely used immunosuppressant because calcineurin is pivotal for the activation of T-lymphocytes (25,26). It is also known that FK506 and the mechanistically related immunosuppressant cyclosporin A have serious toxic side effects precisely because of the widespread occurrence of their cellular targets in the body (26).

The crystal structure of a calcineurin/FK506/FKBP12 complex has been recently solved (30,31), which has revealed that FK506/FKBP12 binds to the substrate-binding site of calcineurin. Moreover, evidence from other laboratories suggests that calcineurin associates with a number of cellular proteins. Some of these, such as AKAP79 (A-kinase anchoring protein 79) (32), are thought to act as molecular scaffolds under the plasma membrane and thus may play a role in the substrate targeting of calcineurin. Another calcineurin ligand, the transcription factor NFAT1 (nuclear factor of activated T-lymphocytes), is a prominent calcineurin substrate in T-lymphocytes (33,34).

Closer examination of the primary structure of AC9 (see Fig. 3) indicates that between residues 503 and 611 it contains a domain similar in sequence to FKBP12.

FIG. 3. Apparent homology of the primary structure of AC9 with FK506 binding proteins. All mammalian ACs reported so far conform to this structural scheme: A short N-terminal cytoplasmic segment is followed by six hydrophobic transmembrane segments (M1). The next long cytoplasmic segment (C_1) can be divided into two sections: C_{1a} (thicker lines), important for catalytic activity and highly conserved among ACs, and C_{1b}, which is nonconserved and is thought to be important for isotype-specific regulation. This is followed by a quasi-duplicate of the first segments M2, C_{2a}, C_{2b}, which have characteristics similar to their 1 indexed counterparts. The lower part of the figure shows the justification of the alignment of mouse AC9 (residues 503–611) with FKBP12 and FKBP13 isotypes. Alignment with yeast FKBP12 was initially carried out using GeneJockeyII and subsequently optimized by hand using the conserved substitutions (1) C, (2) S T P A G, (3) N D E Q, (4) H R K, (5) M I L V, (6) F Y W. Dots indicate at least one identity between AC9 and the FKBPS shown; lines denote conserved amino acid substitutions. The FKBP12 species in the top section are all potent inhibitors of calcineurin with submicromolar IC_{50}-s when complexed with FK506, whereas the three FKBP13s in the lower section bind FK506 but have no appreciable calcineurin-inhibiting activity. Note that if the 80s loop of human FKBP13 is exchanged for the 80s loop of human FKBP12, the resulting hybrid protein inhibits calcineurin with a potency similar to that of FKBP12 (35,36). Moreover, exchange of the single lysine residue for isoleucine in the 80s loop of FKBP13 dramatically enhances the calcineurin inhibitory potency of FKBP13, and the whole of the 40s loop of FKBP12 may be exchanged for that of FKBP13 without appreciable loss of calcineurin inhibition (35,36).

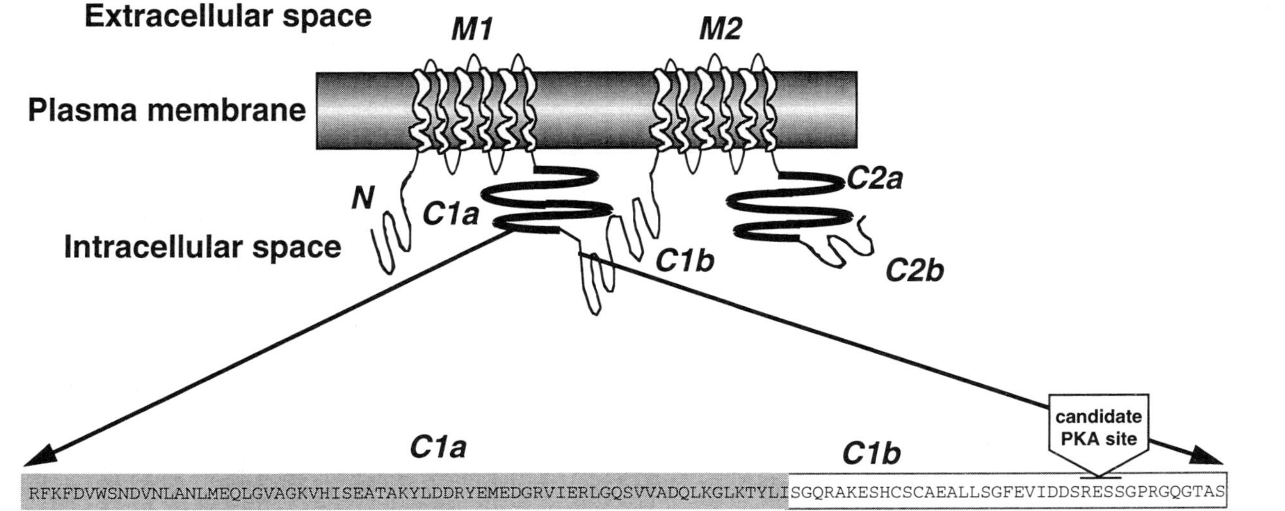

Extracellular space
M1
M2
Plasma membrane
N
C1a
C2a
Intracellular space
C1b
C2b
candidate PKA site
C1a
C1b
RFKFDVWSNDVNLANLMEQLGVAGKVHISEATAKYLDDRYEMEDGRVIERLGQSVVADQLKGLKTYLI
SGQRAKESHCSCAEALLSGFEVIDDSRESSGPRGQGTAS

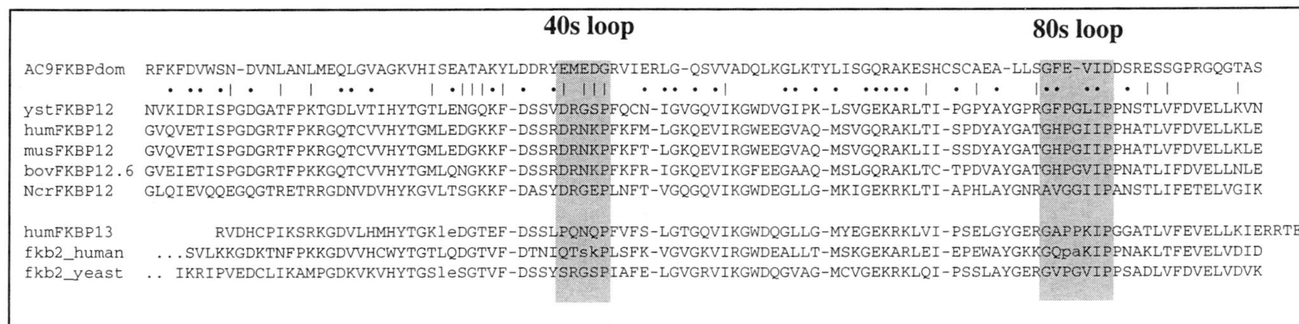

40s loop   80s loop

AC9FKBPdom   RFKFDVWSN-DVNLANLMEQLGVAGKVHISEATAKYLDDRYEMEDGRVIERLG-QSVVADQLKGLKTYLISGQRAKESHCSCAEA-LLSGFE-VIDDSRESSGPRGQGTAS
              • • • • | • | | • • • | • | | | | • | • • | | | • • • • | • | •   • | • •   • | | |
ystFKBP12   NVKIDRISPGDGATFPKTGDLVTIHYTGTLENGQKF-DSSVDRGSPFQCN-IGVGQVIKGWDVGIPK-LSVGEKARLTI-PGPYAYGPRGFPGLIPPNSTLVFDVELLKVN
humFKBP12   GVQVETISPGDGRTFPKRGQTCVVHYTGMLEDGKKF-DSSRDRNKPFKFM-LGKQEVIRGWEEGVAQ-MSVGQRAKLTI-SPDYAYGATGHPGIIPPHATLVFDVELLKLE
musFKBP12   GVQVETISPGDGRTFPKRGQTCVVHYTGMLEDGKKF-DSSRDRNKPFKFT-LGKQEVIRGWEEGVAQ-MSVGQRAKLII-SSDYAYGATGHPGIIPPHATLVFDVELLKLE
bovFKBP12.6 GVEIETISPGDGRTFPKKGQTCVVHYTGMLQNGKKF-DSSRDRNKPFKFR-IGKQEVIKGFEEGAAQ-MSLGQRAKLTC-TPDVAYGATGHPGVIPPNATLIFDVELLNLE
NcrFKBP12   GLQIEVQQEGQGTRETRRGDNVDVHYKGVLTSGKKF-DASYDRGEPLNFT-VGQGQVIKGWDEGLLG-MKIGEKRKLTI-APHLAYGNRAVGGIIPANSTLIFETELVGIK

humFKBP13   RVDHCPIKSRKGDVLHMHYTGKleDGTEF-DSSLPQNQPFVFS-LGTGQVIKGWDQGLLG-MYEGEKRKLVI-PSELGYGERGAPPKIPGGATLVFEVELLKIERRTEL
fkb2_human   ...SVLKKGDKTNFPKKGDVVHCWYTGTLQDGTVF-DTNIQTskPLSFK-VGVGKVIRGWDEALLT-MSKGEKARLEI-EPEWAYGKKGQpaKIPPNAKLTFEVELVDID
fkb2_yeast   .. IKRIPVEDCLIKAMPGDKVKVHYTGSleSGTVF-DSSYSRGSPIAFE-LGVGRVIKGWDQGVAG-MCVGEKRKLQI-PSSLAYGERGVPGVIPPSADLVFDVELVDVK

This analogy is intriguing for several reasons. First, the degree of primary sequence similarity between $AC9_{503-611}$ and FKBP12/13 proteins is roughly 50%. The relationship of structure and function in FKBP12 has been analyzed extensively by site-directed mutagenesis (35–38), and it is known that the amino acid segment absolutely required for the interaction of FK506/FKBP12 with calcineurin is the 80s loop (see Fig. 3). The critical amino acid residues in the alignment of FKBP12 with AC9 shown in Fig. 3 are compatible with this function (35–38).

Second, the FKBP-like domain of AC9 is found in a segment of the cyclase primary sequence where the highly conserved cytoplasmic C_{1a} domain is joined to the highly variable C_{1b} domain. This region has been shown to confer unique properties with respect to regulation by Ca^{2+}/CaM on at least two other Ca^{2+}-regulated ACs. In AC1, a putative Ca-binding site (39) that is requisite for the stimulation of catalytic activity by Ca^{2+}/CaM (40) is located at the C-terminal end of the proposed FKBP-like domain.[1] A splice variant of AC8 that has a 66-amino-acid deletion a few amino acids downstream from a region corresponding to the putative CaM-binding site in AC1 has profoundly different (fourfold) sensitivity toward Ca^{2+}/CaM (41) than the full-length variant. It would thus appear that the C_{1a}–C_{1b} border region is an important site of allosteric control in ACs regulated by Ca^{2+}.

Third, a phosphorylated glutathione-S-transferase (GST)–AC9 fusion protein derived from the FKBP-like domain of AC9 is dephosphorylated by calcineurin (42; and see later discussion).

Collectively, these observations on the primary structure of AC9 support the notion derived from purely pharmacologic analyses that AC9 is a calcineurin-regulated AC.

Pharmacology in Intact Cells

When stably expressed in human embryonic kidney (HEK-293) cells, AC9 appears to have high "basal" activity as estimated in the presence of blockers of PDE but shows little appreciable stimulation by endogenously expressed receptors for 41-residue corticotrophin releasing factor (CRF41) (Fig. 4). Stimulation by forskolin, a drug known to act directly on AC, is evident (see Fig. 4) but is much smaller than expected on basis of the very marked increase in unstimulated cAMP accumulation. Although these data are not readily interpretable, a reproducible feature is that calcium depletion of the cells caused a marked increase of both basal and forskolin-stimulated cAMP synthesis (see Fig. 4), and reintroduction of Ca^{2+} caused a concentration-dependent inhibition. The levels of intracellular free Ca^{2+} under the conditions employed do not exceed 1 μM as measured by fura2 fluorescence

---

[1]AC1 and AC9 were aligned on the basis of the homologies of the C_{1a} domains. Intriguingly, residues 487–491 in AC1 are identical to the 80's (FGPLI) loop of yeast FKBP12, and the C_{1a}–C_{1b} border in AC1 also has a relatively high degree of sequence similarity with the FKBP 12/13 family of proteins (F. A. Antoni, unpublished data). This raises the possibility that AC1 may also interact with calcineurin.

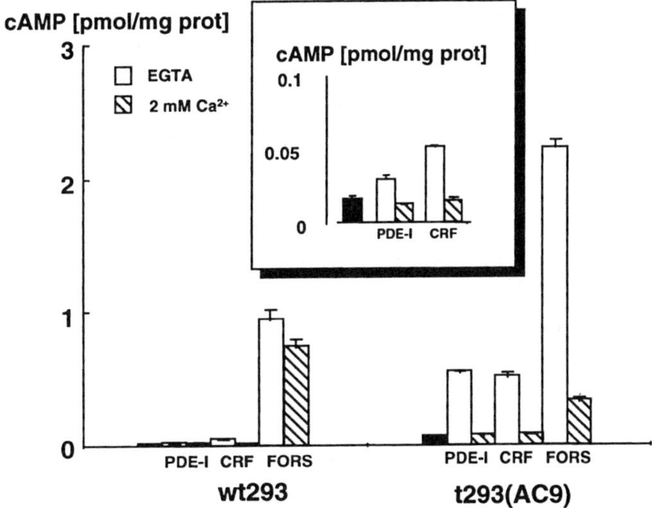

FIG. 4. Inhibition of AC9 by Ca^{2+} in stably transfected HEK-293 cells. HEK-293 cells were transfected by electroporation with a full-length mouse AC9 cDNA (9) in pcDNA3 and subjected to selection in geneticin. All eight cell clones selected and tested had the same properties with respect to cAMP responses in the presence of PDE blockers. Cells in suspension were preincubated with Ca^{2+} free Hanks balanced salt solution (BSS) containing 5 μM A23187 for 20 minutes at 37°C, PDE-I (1 mM IBMX and 0.1 mM rolipram) were added at time 0, corticotropin releasing factor (CRF, 3 nM) or forskolin (FORS, 10 μM) were added at 5 minutes. Some cells also received $CaCl_2$ (2 mM final) at this time and the incubation was terminated at 20 minutes by adding HCl to a final concentration of 0.1 M. cAMP was measured by RIA after acetylation. The black columns show the levels of cAMP in cells before the addition of PDE-I. The inset shows the changes of cAMP levels in the wild-type 293 cells on an enlarged scale. Data shown are means ± SEM, n = 4/group, representative of three similar experiments. Note that forskolin-induced cAMP levels were much more susceptible to inhibition by Ca^{2+} in transfected cells than in wild-type cells. The average intracellular levels of Ca^{2+} in HEK-293 cells under similar conditions have been estimated to be in the low-micromolar range under similar conditions (40).

(40,43); hence, these data support the finding from AtT20 cells that AC9 is a Ca^{2+} inhibitable AC controlled by physiologically relevant concentrations of intracellular free Ca^{2+}.

With respect to the lack of stimulation of AC9 by CRF41 and the high "basal" levels of cAMP, stable overexpression of AC2 in HEK-293 cells in the same expression vector (pcDNA3) resulted in cAMP levels that were high basally, markedly stimulated by phorbol esters, but once more, unresponsive to stimulation by CRF41 (A. Leitch and F.A. Antoni, unpublished data). By contrast, CRF41 produces robust increases of cAMP in wild-type HEK-293 cells (9,21) (see Fig. 4). Although the possibility that receptors for CRF41 are lost from the transfected cells cannot be excluded, the plausible explanation of these findings is that the dramatic increase of cyclase catalytic subunit expression, which can be estimated to be around 100–200-fold on the basis of radioimmunoassay measurements, strips $G_{s\alpha}$ from heterotrimeric G_s and thus activates a significant proportion of the catalytic moiety. This may be masked by long-term adaptation of PDE activity as documented in other systems (44) and revealed by blockers of PDE routinely applied in assays of cAMP production. Activation by $G_{s\alpha}$ in the absence of agonist would also explain why in stably transfected HEK-293 cells forskolin stimulation of AC9 is readily suppressed by Ca^{2+} (see Fig. 4), whereas in AtT20 cells Ca^{2+} inhibition was negligible with forskolin as the stimulus (9,21).

When expressed transiently in HEK-293 cells or endogenously in the wild-type AtT20 cell line, where it is the predominant isotype, AC9 appears to have further interesting properties (9,21). Inhibition by intracellular Ca^{2+} can be readily demonstrated and is apparently mediated by the Ca^{2+}/CaM-activated protein phosphatase calcineurin (22). Briefly, agonist-induced cAMP accumulation in these systems is increased by immunosuppressants such as FK506 and cyclosporin A. Multifaceted pharmacologic analysis indicated that the actions of immunosuppressants are due to the inhibition of calcineurin activity (9,21) and that the site of calcineurin action is AC. In AtT20 cells, cAMP production is very potently inhibited by the endogenous increase of Ca^{2+}, as intracellular Ca^{2+} chelators such as BAPTA-AM and fura 2-AM dose-dependently enhance the cAMP response to CRF41 (21) (F.A. Antoni, unpublished data). The rise of intracellular Ca^{2+} in this system is unequivocally dependent on calcium entry through L-type Ca^{2+} channels (45), whereas intracellular Ca^{2+} pools are not demonstrable (46). The latter feature has greatly facilitated the analysis of immunosuppressant action on the cAMP cascade. Immunosuppressant drugs have profound Ca^{2+}-mobilizing actions in a variety of systems, which are due to their effects on intracellular Ca^{2+} channels and are not entirely attributable to the inhibition of calcineurin (28,47,48).

Taken together, the pharmacologic analysis of AC9-mediated cAMP synthesis indicates that AC9 is active in the phosphorylated form and is progressively suppressed upon dephosphorylation by Ca^{2+}/calcineurin.

Analysis of AC9–Ca^{2+} Interaction in Cell Free Systems

Studies by Premont and co-workers (20) on overexpressed AC9 in insect Sf9 and monkey embryonic kidney cells revealed the basic features of AC9 to be robust activation by $G_{s\alpha}$, surprisingly weak activation by forskolin, and no apparent effect of Ca^{2+}, Ca^{2+}/CaM, or $G_{\beta\gamma}$. It is perhaps worth noting that the effects of Ca^{2+} on AC9 expressed in the human cell line were not reported and, hence, the data are not directly comparable with the studies shown later, as major species- and cell-specific variations in the post-translational control of signaling proteins are known to occur.

In AtT20 cells that express predominantly AC9 as the endogenous cyclase, the CRF41 stimulation of AC activity in crude membrane preparations is largely GTP dependent. The responses obtained were low: Only a threefold increase of cAMP levels could be observed within 20 minutes. By comparison, 20–30-fold increases of cAMP production are elicited within the same time period by CRF41 in intact cells, suggesting that components of the coupling mechanism may be lost during homogenization. Although the crude membrane preparation contains only 10% or less of the

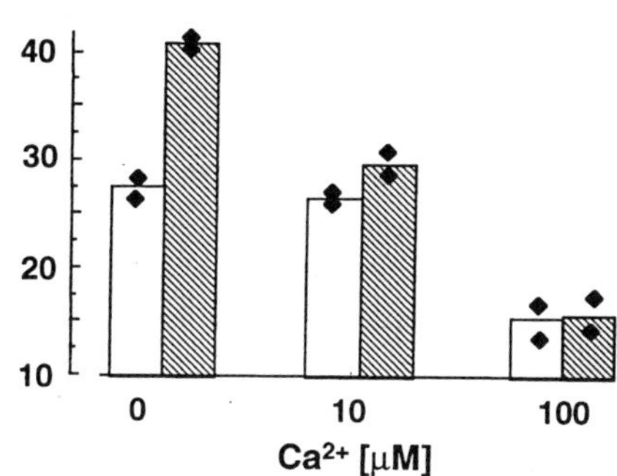

FIG. 5. Effect of $CaCl_2$ on cAMP synthesis evoked by CRF41 in AtT20 cell membranes. AtT20 cells were detached from culture plates with PBS/0.1% EDTA, pelleted by centrifugation and homogenized in buffer containing protease inhibitors (49). A crude membrane pellet was prepared by centrifugation, resuspended and reacted with 2 mM ATP in 30 mM TRIS-HCl, pH 7.1 in the presence of 1 mM $MgCl_2$, creatine phosphokinase/phosphocreatine, 1 mM IBMX, and varying amounts of GTP at 30°C. The reaction was stopped at 20 minutes by adding HCl to a final concentration of 0.1 M. Cyclic AMP was measured by RIA after 25-fold dilution in 0.1 HCl and acetylation. Open bars, basal; striped bars, 10nM CFR.

total cellular calcineurin activity (49; J. Simpson and F. A. Antoni, unpublished data), addition of Ca^{2+} caused a concentration-dependent inhibition of cAMP synthesis with an $IC_{50} < 10$ μM (Fig. 5). This inhibition was not blocked by a complex of FK506/FKBP12, which blocked 90% of the Ca^{2+}/CaM-dependent protein phosphatase activity in AtT20 extracts prepared by hypotonic lysis (50; J. Simpson and F. A. Antoni, unpublished data).

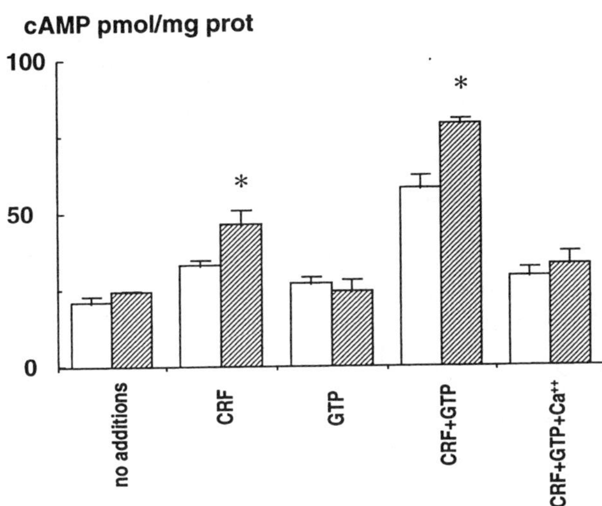

FIG. 6. The effect of preincubation of AtT20 cells with ethanol vehicle (open bars) or 1 μM each of the immunosuppressants FK506 and cyclosporin A for 30 minutes (striped bars) at 37°C on CRF41 (15 nM) stimulation of membrane AC activity that was measured as described in the legend for Fig. 5. Data are means ± SE, $n = 3$/group, 1-way ANOVA and Newman-Keuls test. Effect of GTP on CRF41 stimulation was significant ($p < 0.05$). The asterisk indicates a significant effect of immunosuppressant treatment on the CRF response. Note that 30 μM $CaCl_2$ blocked CRF41 stimulation in both groups. Furthermore, separate experiments showed that $CaCl_2$ at this concentration had no significant effect on basal cAMP synthesis in the presence of GTP.

Pretreatment of intact cells with maximally effective concentrations of FK506 and cyclosporin A clearly enhanced GTP-dependent stimulation of cAMP synthesis by CRF41 in membrane preparations (Fig. 6), which could be completely suppressed by 30 μM CaCl$_2$. Similar results were obtained by stimulating cAMP synthesis with the nonhydrolyzable GTP analog GTP-γ-S (J. Simpson and F. A. Antoni, unpublished data), suggesting that the effects of immunosuppressants and Ca^{2+} are distal to the CRF41 receptor.

These data, taken together with the results from HEK-293 cells stably transfected with AC9, suggest that Ca^{2+} inhibition of AC9 could occur by direct interaction with the catalytic moiety as proposed for AC5 and AC6 (2), as well as through dephosphorylation by calcineurin. Further studies are clearly required to explore these issues and their relevance for physiologic control.

The major problem in trying to reconstruct the regulation of AC9 by calcineurin in a cell free system is the lack of knowledge regarding the phosphorylation of AC9 by protein kinase(s). Is this a constitutive phosphorylation, or is it triggered by agonist stimulation?

In some experiments, addition of the catalytic subunit of protein kinase A from bovine heart (Sigma P2645) to crude membrane preparations produced an enhancement of the response to CRF41, which, however, still remained much smaller than the response in intact cells. The effect of protein kinase A catalytic subunit on CRF41 stimulation of AC and the presence of a candidate protein kinase A phosphorylation motif in the FKBP-homology domain of AC9 (see Fig. 3) prompted us to examine this question in further detail.

Our preliminary findings (42) show that a fusion protein of GST-AC9$_{511-668}$ is phosphorylated by protein kinase A (Sigma, or type Ia recombinant subunit produced in *E. coli*; courtesy of R. Clegg, Hannah Institute, Ayr, UK). Roughly 10% of GST-AC9$_{511-668}$ is phosphorylated with no incorporation of label into GST alone. The relatively low level of phosphorylation suggests that protein kinase A is not the physiologic kinase for this site; alternatively, expression as a GST fusion protein may result in a suboptimal conformation of the phosphorylation site.

More important, ^{32}P-GST-AC9$_{511-668}$ is dephosphorylated by calcineurin (both from bovine brain and from AtT20 cell extracts) in a Ca^{2+}/CaM-dependent manner. The rate of dephosphorylation is comparable to that of the RII substrate peptide (51) and is blocked by the FK506/FKBP12 complex (J. Simpson, J. M. Paterson, and F. A. Antoni, unpublished data). Further analysis of this process and the identification of the protein kinase(s) that phosphorylate AC9 are in progress.

Summary

In summary, analysis of the functional properties of AC9 has revealed that the activity of this AC is connected to a signaling pathway inhibited by calcineurin. The exact mechanism of calcineurin action remains to be characterized, but preliminary data with an AC9-derived GST-fusion protein suggest a direct action of calcineurin in the cytoplasmic loop where a putative FKBP12-like domain and, hence, potential calcineurin docking site could be discerned. Demonstration of the inhibition of AC9 by Ca^{2+} at the level of the catalytic subunit independently of calcineurin in membrane preparations raises the possibility of differential control, depending on the local concentrations of Ca^{2+} and its target enzymes. An intricate feedback regulatory circuit is thus formed that is potentially coupled to a variety of Ca^{2+}-dependent signaling processes that are effective both in the short term (e.g., electrical activity and mobilization of intracellular Ca^{2+} stores) and in the long term (e.g., regulation of the expression of Ca^{2+}-binding proteins and components of the CaM/calcineurin protein phosphatase complex) (Fig. 7).

TISSUE DISTRIBUTION OF AC9 AND FUNCTIONAL IMPLICATIONS

The tissue distribution profiles of ACs are highly distinctive (52–54) and provide important indications with respect to functional significance. Although the analysis of the tissue distribution of AC9 and its potential splice variants is far from complete, it is clear that this is an enzyme widely distributed in the body (Fig. 8) (20). It is possibly the most common isotype described so far; note, however, that there are clear indications that levels of protein do not closely correlate with the mRNA signal, particularly in skeletal muscle (20; J. Simpson, unpublished data). Northern analysis with various probes suggests a single mRNA species of approximately 8.5 kb, which is considerably larger than the coding sequence. Chromosomal localization using a mouse cDNA probe indicated that a single gene is found in the human genome on chromosome 16 (20).

The most interesting clues with respect to the physiologic function of AC9 may be derived from the distribution of the mRNA in the brain. By far the highest intensity *in*

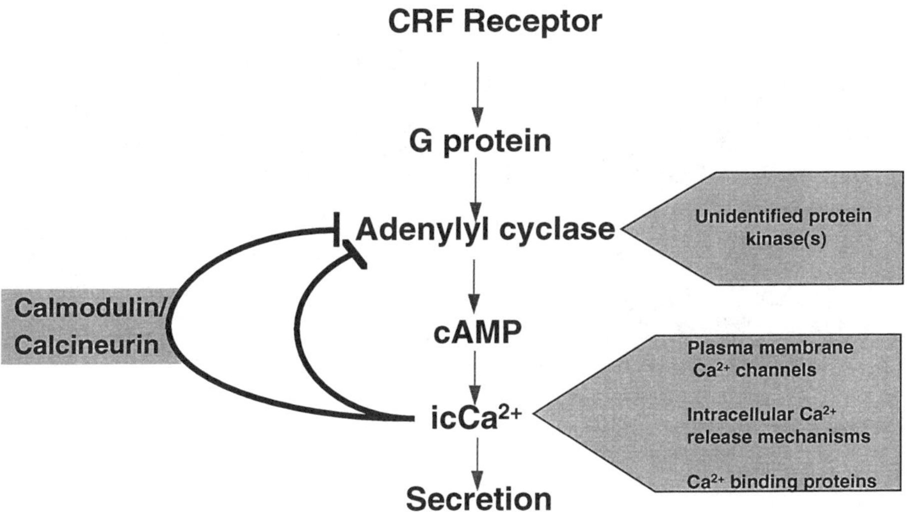

FIG. 7. Scheme of Ca^{2+}/calcineurin feedback inhibition of AC9.

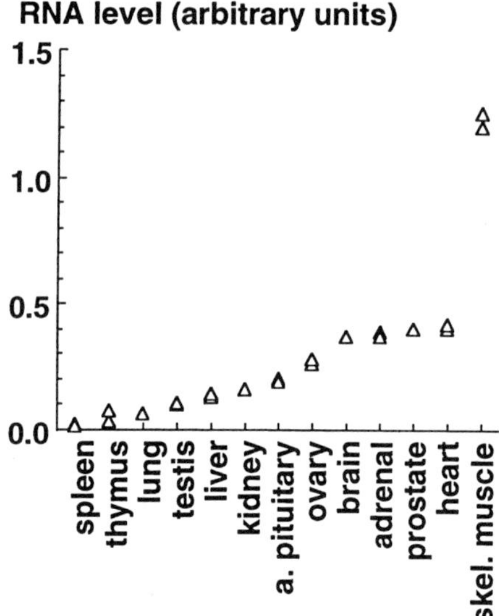

FIG. 8. Levels of AC9 mRNA in mouse tissues. Adult balb/c mice were killed by cervical dislocation and various tissues were rapidly excised and processed for RNAse protection assay as described (21). Data are phosphorimager-generated density units, standardized with β-actin mRNA run in each sample as internal standard. Points are individual values; all samples were assayed in duplicate.

situ hybridization signal was detected in the pyramidal neurons of the hippocampus (Fig. 9). Preliminary data by radioimmunoassay and immunoblot analysis also confirm this observation with respect to protein expression, with the obvious limitation that the tissue dissection does not resolve the pyramidal cells from the surrounding tissue.

It is important to note that AC9 is the first Ca^{2+}-inhibitable AC mRNA present in the hippocampus at levels comparable to AC1 and AC2. Of further note is that the

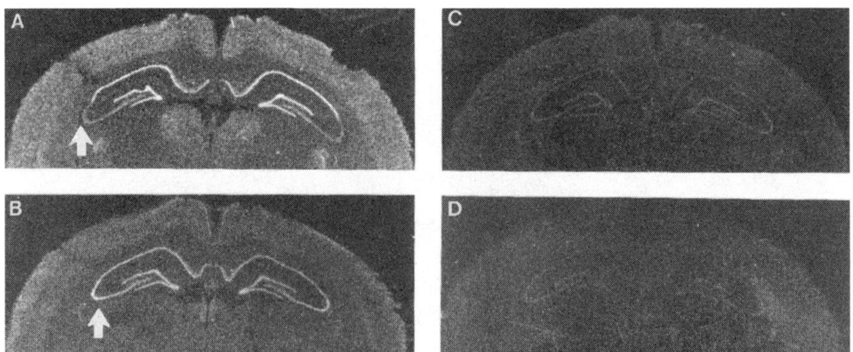

FIG. 9. Distribution of AC2 (**A**) and AC9 (**B**) mRNA in rat brain. *In situ* hybridization was carried out using species-specific ^{35}S-labeled antisense riboprobes derived from isotype-specific segments of the respective cyclases (9). (**C,D**) The respective sense-reacted sections that serve as background. Most notable is the relatively low labeling for AC2 in the CA3 region of the hippocampal pyramidal cell layer (*arrow*), whereas the signal for AC9 is somewhat more intense in this region when compared with other parts of the pyramidal cell layer. Also note the intense staining for AC9 in the dentate gyrus where AC1 is also expressed at high levels (52).

CA3 field contains a relatively higher level of AC9 mRNA than the CA1 field or the dentate gyrus, whereas the AC2 mRNA, which is also highly expressed throughout the hippocampal formation, is barely above background in the CA3 field (see Fig. 9A). If AC9 functions in hippocampal pyramidal neurons with characteristics similar to those found in AtT20 cells (21) (i.e., it is inhibited by Ca^{2+}/calcineurin), then current hypotheses of long-term potentiation and Ca^{2+}-induced cAMP synthesis in hippocampal neurons will require further analysis to accommodate calcineurin-mediated inhibition of cAMP synthesis. Data on calcineurin distribution show that it is highly abundant in the hippocampal formation (55,56) and has been implicated both in long-term depression (57,58) and in long-term potentiation (59), but none of these experiments were controlled for changes in cAMP metabolism.

Of further interest is the presence of AC9 mRNA in the supraoptic nucleus (Fig. 10), which contains vasopressin- and oxytocin-synthesizing neuroendocrine motoneurons. The dorsal pattern of distribution of the mRNA signal in the supraoptic nucleus suggests that it is preferentially expressed by oxytocinergic neurons (60). This may be of functional importance because the electrical activity pattern of these cells is generated by Ca^{2+}-mediated control that involves calbindin and features a prominent afterhyperpolarization potential (61). It is therefore apparent that Ca^{2+} synchronizes membrane currents in these cells, and in view of the prominent effects of cAMP-dependent phosphorylation on the activity of ion channels close control of cAMP synthesis is likely to be an important component of this process. Potential activators of cAMP in the supraoptic nucleus are pituitary adenylate cyclase–activating peptide (PACAP) and CRF41, which are synthesized by the same group of neurons that appear to express AC9 (62,63); furthermore, supraoptic neurons express receptors for both of these neuropeptides (64,65).

Of further note is the relatively high expression of AC9 in several peripheral endocrine tissues such as adrenal cortex, ovary, and testis. As these steroid-synthesizing glands undergo dynamic cycles of cell proliferation/differentiation that is regulated by trophic hormones of the adenohypophysis, a potential role of AC9 could be the Ca^{2+}/calcineurin-dependent coupling of the cell cycle to cAMP synthesis.

Adenylyl cyclase 9 is also abundant in the anterior pituitary gland where the synthesis of cAMP stimulated by CRF41 is under inhibitory control by intracellular Ca^{2+} (12,66). Furthermore, in adenohypophysial corticotrophs, opposition of Ca^{2+} and cAMP appears to be the pivot of the inhibitory effect of adrenal corticosteroids on hormone secretion (12).

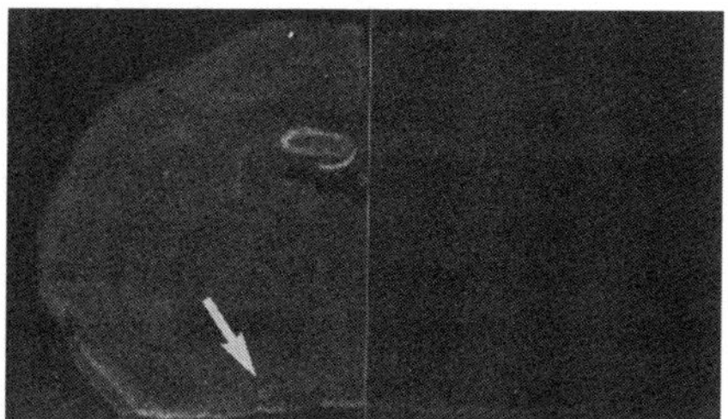

FIG. 10. Distribution of AC9 mRNA in the brain: rat hypothalamus. Although the levels of AC9 mRNA in the hypothalamus appear to be relatively low, note the selective labeling in the dorsal part of the supraoptic nucleus (*arrow*, left-hand panel) with an antisense ^{35}S-riboprobe; the sense ^{35}S-riboprobe labeling in an anatomically comparable section is shown on the right-hand panel.

IS AC9 UNIQUE?

Regulation of cAMP accumulation by intracellular free Ca^{2+} is a relatively common finding (2,12,67). However, the physiologic significance of this process is only currently beginning to be appreciated as vastly more is known about the cellular handling of Ca^{2+} as well as the diversity of the underlying molecular machinery. Thus it is not surprising that the understanding of the mechanism(s) by which Ca^{2+} at low micromolar, and in some cases submicromolar, concentrations modulates cAMP synthesis is far from complete.

With respect to calcineurin, there is only one further study available (68), which analyzed ACTH- and angiotensin II–mediated cAMP production in bovine adrenal glomerulosa and COS7 cells. This study employed pharmacologic tools and showed that the enhancement of cAMP synthesis in both cell systems by protein kinase C could be blocked by immunosuppressant inhibitors of calcineurin and prevented by rapamycin, an antagonist of the calcineurin-inhibiting action of FK506. The AC involved has not been identified, although accounts of COS7 cyclase isotyping (10,69) suggest that the most likely candidate is AC7 (70).

FUTURE DIRECTIONS

A wide variety of approaches will have to be taken to explore the physiologic and pathophysiologic consequences of AC diversity. On the one hand, purely genetic approaches such as generation of transgenic animals and screening of the human population for genetic polymorphisms will have to be undertaken; on the other hand, reagents to probe the role of a given AC (e.g., active site-directed antisera in a single cell) need to be developed.

All these approaches are wrought with difficulties. The first null mutant mice for AC1 (71) showed deficits in certain aspects of neurotransmission and learning, but it was not possible to establish a direct correlation of these changes with a specific cellular locus or context of AC1 function. If the cAMP signal is regulated by redundantly organized intracellular feedback circuits (see Fig. 7), then it will be exceedingly difficult to detect alterations using the genetic-knockout/phenotype-screening approach. Screening of the human gene pool, an increasing proportion of which is characterized with respect to associated diseases, may well prove to be the more fruitful and cost-effective approach.

The transcriptional control of AC expression and the exploration of gene structure have not progressed very far; only AC1 and AC3 have been partially characterized in this respect. This is certainly an area likely to undergo considerable expansion in the future.

Cell line models endogenously and predominantly expressing a certain isotype of AC could also contribute to the exploration of the unique properties of individual isotypes. Cell lines predominantly expressing AC6 (72), AC7 (70), and AC9 (21) have been reported in the literature. In contrast, stable overexpression of a given isotype may yield atypical results (see earlier and summary in ref. 4), although the studies reported so far were all carried out in the HEK-293 cell line, which may not be an optimal host for this type of approach.

Development of isotype-specific reagents would greatly enhance the understanding of cellular control mechanisms involving ACs. A potentially fruitful approach could be the inducible expression *in vitro* and *in vivo* of negative dominant mutants or miniproteins based on the sequence of putative autoinhibitory and regulatory domains in ACs (6,73,74). Specific antisera directed against the active sites of cyclase isotypes are also potentially valuable tools but have not been reported so far.

The exploitation of AC diversity in clinical therapy is in its infancy (75,76). Therapeutic agents targeting AC isotypes potentially offer considerable advantages. When used in combination with compounds affecting receptors, interference at multiple sites of the same signal transduction cascade is possible and dosage requirements for the individual components of treatment regimes may be reduced to minimize unwanted side effects. Similarly, influence of functionally concerted pathways such as activation of AC and inhibition of calcineurin hold the promise of immunosuppressive treatment with diminished toxic actions (75,76).

ACKNOWLEDGMENTS

We thank Professor Tony Harmar for the AC dendrogram, Dr. Roger Clegg for recombinant protein kinase A catalytic subunit and helpful discussions, and the Medical Research Council for continued financial support.

REFERENCES

1. Krupinski J, Coussen F, Bakalyar HA et al: AC amino acid sequence: possible channel- or transporter-like structure. *Science* 1989;244:1558–1564.
2. Cooper DMF, Mons N, Karpen JW: Adenylyl cyclases and the interaction between calcium and cAMP signalling. *Nature* 1995;374:421–424.
3. Antoni FA: Calcium regulation of adenylyl cyclase—relevance for endocrine control. *Trends Endocrinol Metab* 1996;8:7–14.
4. Sunahara RK, Dessauer CW, Gilman AG: Complexity and diversity of mammalian adenylyl cyclases. *Ann Rev Biochem* 1996;36:461–480.
5. Tang WJ, Stanzel M, Gilman AG: Truncation and alanine-scanning mutants of type-I adenylyl-cyclase. *Biochemistry* 1995;34:14563–14572.
6. Levin LR, Reed RR: Identification of functional domains of adenylyl cyclase using *in vivo* chimeras. *J Biol Chem* 1995;270:7573–7579.
7. Taussig R, Gilman AG: Mammalian membrane-bound adenylyl cyclases. *J Biol Chem* 1995;270:1–4.
8. Iyengar R: Multiple families of G_s-regulated adenylyl cyclases. In: Brown BL, Dobson PRM, eds. *Advances in second messenger and phosphoprotein research.* New York: Raven Press, 1993:27–36.
9. Paterson JM, Smith SM, Harmar AJ, Antoni FA: Control of a novel adenylyl cyclase by calcineurin. *Biochem Biophys Res Commun* 1995;214:1000–1008.
10. Premont RT: Identification of adenylyl cyclases by amplification using degenerate primers. *Methods Enzymol* 1994;238:116–127.
11. Harmar AJ, Lutz EM: Multiple receptors for PACAP and VIP. *Trends Pharmacol Sci* 1994;15:97–99.
12. Antoni FA: Calcium checks cyclic AMP: mechanism of corticosteroid feedback in adenohypophysial corticotrophs. *J Neuroendocrinol* 1996;8:659–672.
13. Taussig R, Iniguez-Lluhi JA, Gilman AG: Inhibition of adenylyl cyclase by $G_{i\alpha}$. *Science* 1993;261:219–221.

14. Jacobowitz O, Iyengar R: Phorbol ester-induced stimulation and phosphorylation of adenylyl cyclase 2. *Proc Natl Acad Sci U S A* 1994;91:10630–10634.

15. Ang KL, Antoni FA: Rolipram-inhibitable cyclic nucleotide phosphodiesterase (PDE) in rat adenohypophysis: Potential functional role in corticotrophs and somatotrophs. *Br J Pharmacol* 1996; 119:361.

16. Baimbridge K, Celio M, Rogers J: Calcium binding proteins in the nervous system. *Trends Neurosci* 1992;15:303–308.

17. Epstein PN, Ribar TJ, Decker GL et al: Elevated β-cell calmodulin produces a unique insulin secretory defect in transgenic mice. *Endocrinology* 1992;130:1387–1393.

18. Wayman GA, Hinds TR, Storm DR: Hormone stimulation of type-III adenylyl-cyclase induces Ca^{2+} oscillations in HEK-293 cells. *J Biol Chem* 1995;270:24108–24115.

19. Jurevicius J, Fischmeister R: cAMP compartmentation is responsible for a focal activation of cardiac Ca^{2+} channels by β-adrenergic agonists. *Proc Natl Acad Sci U S A* 1996;93:295–299.

20. Premont RT, Matsuoka I, Mattei M-G et al: Identification and characterization of a widely expressed from of adenylyl cyclase. *J Biol Chem* 1996;271:13900–13907.

21. Antoni FA, Barnard RJO, Shipston MJ et al: Calcineurin feedback inhibition of agonist-evoked cAMP formation. *J Biol Chem* 1995;270:28055–28061.

22. Klee CB, Cohen P: The calmodulin regulated protein phosphatase. In: Cohen P, Klee CB, eds. *Calmodulin*. Amsterdam: Elsevier, 1988:225–245.

23. Liu J, Farmer Jr JD, Lane WS et al: Calcineurin is a common target of cyclophylin-cyclosporin A and FKBP-FK506 complexes. *Cell* 1991;66:807–815.

24. Fischer G: Peptidyl-prolyl cis/trans isomerases and their effectors. *Angewandte chemie—international edition in english.* 1994;33:1415–1436.

25. Schreiber SL: Immunophilin-sensitive protein phosphatase action in cell signaling pathways. *Cell* 1992;70:365–368.

26. Sigal NH, Dumont FJ: Cyclosporin A, FK-506, and rapamycin: pharmacologic probes of lymphocyte signal transduction. *Ann Rev Immunol* 1992;10:519–560.

27. Cardenas M, Hemenway C, Muir RS et al: Immunophilins interact with calcineurin in the the absence of exogenous immunosuppressive ligands. *EMBO J* 1994;13:5944–5957.

28. Cameron AM, Steiner JP, Roskams AJ et al: Calcineurin associated with the inositol 1,4,5-trisphosphate receptor-FKBP12 complex modulates Ca^{2+} flux. *Cell* 1995;83:463–472.

29. Pahl A, Keller U: FK-506-binding proteins from *Streptomycetes* producing immunosuppressive macrolactones of the FK-506 type. *J Bacteriol* 1992;174:5888–5894.

30. Kissinger CR, Parge HE, Knighton DR et al: Crystal-structures of human calcineurin and the human FKBP12-FK506-calcineurin complex. *Nature* 1995;378:641–644.

31. Griffith JP, Kim JL, Kim EE et al: X-ray structure of calcineurin inhibited by the immunophilin immunosuppressant FKBP12-FK506 complex. *Cell* 1995;82:507–522.

32. Coghlan VM, Perrino BA, Howard M et al: Association of protein-kinase-A and protein-phosphatase-2B with a common anchoring protein. *Science* 1995;267:108–111.

33. Wesselborg S, Fruman DA, Sagoo JK et al: Identification of a physical interaction between calcineurin and nuclear factor of activated T-cells (NFATP). *J Biol Chem* 1996;271:1274–1277.

34. Loh C, Shaw KTY, Carew J et al: Calcineurin binds the transcription factor NFAT1 and reversibly regulates its activity. *J Biol Chem* 1996;271:10884–10891.

35. Rosen MK, Yang D, Martin PK, Schreiber SL: Activation of an inactive immunophilin by mutagenesis. *J Am Chem Soc* 1993;115:821–822.

36. Yang D, Rosen MK, Schreiber SL: A composite FKBP12-FK506 surface that contacts calcineurin. *J Am Chem Soc* 1993;115:819–820.

37. Aldape RA, Futer O, DeCenzo MT et al: Charged surface residues of FKBP12 participate in formation of the FKBP12-FK506-calcineurin complex. *J Biol Chem* 1992;267:16029–16032.

38. Braun W, Kallen J, Mikol V et al: Three-dimensional structure and actions of immunosuppressants and their immunophilins. *FASEB J* 1995;9:63–72.

39. Vorherr T, Knopfel L, Hofmann F et al: The calmodulin-binding domain of nitric oxide synthase and adenylyl cyclase. *Biochemistry (U S A)* 1993;32:6081–6088.

40. Wu Z, Wong ST, Strom DR: Modification of the calcium and calmodulin sensitivity of the type I adenylyl cyclase by mutagenesis of its calmodulin binding domain. *J Biol Chem* 1993;286:23766–23768.

41. Cali JJ, Parekh RS, Krupinski J: Splice variants of type-VIII adenylyl-cyclase—differences in glycosylation and regulation by Ca^{2+}/calmodulin. *J Biol Chem* 1996;271:1089–1095.

42. Paterson J, Simpson J, Antoni F: Dephosphorylation of the C1b segment of adenylyl cyclase type 9 by calcineurin. *Biochem Soc Trans* 1996;24:583S.

43. Cooper DMF, Yoshimura M, Zhang Y, Chiono M, Mahey: Capacitative Ca^{2+} entry regulates Ca^{2+}-sensitive adenylyl cyclases. *Biochem J* 1994;297:437–440.

44. Nemoz G, Sette C, Hess M et al: Activation of cyclic-nucleotide phosphodiesterases in FRTL-5 thyroid- cells expressing a constitutively active G_s-alpha. *Mol Endocrinol* 1995;9:1279–1287.

45. Antoni FA, Hoyland J, Woods MD, Mason WT: Glucocorticoid inhibition of stimulus-evoked adrenocorticotrophin release caused by suppression of intracellular calcium signals. *J Endocrinol* 1992;133:R13-R16.

46. Fiekers JF, Konopka LM: Spontaneous transients of $[Ca^{2+}]_{(i)}$ depend on external calcium and the activation of L-type voltage-gated calcium channels in a clonal pituitary cell-line (AtT-20) of cultured mouse corticotropes. *Cell Calcium* 1996;19:327–336.

47. Collins JH: Sequence-analysis of the ryanodine receptor—possible association with a 12K, FK506-binding immunophilin protein-kinase-C inhibitor. *Biochem Biophys Res Commun* 1991;178:1288–1290.

48. Kaftan E, Marks AR, Ehrlich BE: Effects of rapamycin on ryanodine receptor Ca^{2+}-release channels from cardiac-muscle. *Circ Res* 1996;78:990–997.

49. Antoni FA, Shipston MJ, Smith SM: Inhibitory role for calcineurin in stimulus secretion coupling revealed by FK506 and cyclosporin A in pituitary corticotrope tumor cells. *Biochem Biophys Res Commun* 1993;194:226–233.

50. Antoni FA, Shipston MJ, Woods MD et al: Secretagogue glucocorticoid interactions in the control of anterior pituitary adrenocorticotrophin (ACTH) secretion. In: Mornex R, Jaffiol C, Leclère J, eds. *Progress in endocrinology: proceedings of the Ninth International Congress of Endocrinology, Nice 1992.* Carnforth, UK: Parthenon Publishers, 1993:530–535.

51. Blumenthal DK, Takio K, Hansen RS, Krebs EG: Dephosphorylation of cAMP-dependent protein kinase regulatory subunit (TypeII) by calmodulin-dependent protein phosphatase. *J Biol Chem* 1986;261:8140–8145.

52. Xia ZG, Refsdal CD, Merchant KM et al: Distribution of messenger-RNA for the calmodulin-sensitive adenylate cyclase in rat brain: expression in areas associated with learning and memory. *Neuron* 1991;6:431–443.

53. Mons N, Cooper DMF: Adenylate cyclases: critical foci in neuronal signaling. *Trends Neurosci* 1995;18:536–542.

54. Mons N, Cooper DMF: Selective expression of one Ca^{2+}-inhibitable adenylyl cyclase in dopaminergically innervated rat brain regions. *Mol Brain Res* 1994;22:236–244.

55. Buttini M, Luyten M, Limonta S, Boddeke H: Distribution of calcineurin A beta isoenzyme mRNAs in rat brain: comparison with the distribution of calcineurin A alpha isoenzyme mRNAs. *Neurosci Res Commun* 1994;14:9–16.

56. Polli JW, Billingsley ML, Kincaid RL: Expression of the calmodulin-dependent protein phosphatase, calcineurin, in rat brain: developmental patterns and the role of nigrostriatal innervation. *Dev Brain Res* 1991;63:105–119.

57. Mulkey RM, Endo S, Shenolikar S, Malenka RC: Involvement of a calcineurin/inhibitor-1 phosphatase cascade in hippocampal long-term depression. *Nature* 1994;369:486–488.

58. Hodgkiss JP, Kelly JS: Only *de novo* long-term depression (LTD) in the rat hippocampus *in vitro* is blocked by the same low concentration of FK506 that blocks LTD in the visual cortex. *Brain Res* 1995;705:241–246.

59. Lu YF, Hayashi Y, Moriwaki A et al: FK506 a Ca^{2+}/calmodulin-dependent phosphatase inhibitor inhibits the induction of long-term potentiation in the rat hippocampus. *Neurosci Lett* 1996;205:103–106.

60. Burlet A, Tonon MC, Tankosic P et al: Comparative immunocytochemical localization of corticotropin releasing-factor (CRF-41) and neurohypophyseal peptides in the brain of Brattleboro and Long-Evans rats. *Neuroendocrinology* 1983;37:64–72.

61. Li Z, Decavel C, Hatton GI: Calbindin-D_{28K}: role in determining intrinsically generated firing patterns in rat supraoptic neurones. *J Physiol* 1995;488:601–608.

62. Köves K, Görcs TJ, Arimura A: Colocalization of PACAP, but not of VIP, with oxytocin in the hypothalamic magnocellular neurons of colchicine treated and pituitary-stalk sectioned rats. *Endocrine* 1994;2:1169–1175.

63. Antoni FA, Palkovits M, Makara GB et al: Immunoreactive ovine corticotropin-releasing factor (oCRF-LI) in the hypothalamo-hypophyseal tract of the rat. In: Usdin E, Axelrod J, Kvetnansky R,

eds. *Stress: the role of catecholamines and other neurotransmitters*. New York: Gordon and Breach, 1984:233–241.

64. Rivest S, Laflamme N, Nappi RE: Immune challenge and immobilization stress induce transcription of the gene encoding the CRF receptor in selective nuclei of the rat hypothalamus. *J Neurosci* 1995; 15:2680–2695.

65. Hill JM, Harris A, Hiltonclarke DI: Regional distribution of guanine nucleotide-sensitive and guanine nucleotide-insensitive vasoactive-intestinal-peptide receptors in rat brain. *Neuroscience* 1992;48:925–932.

66. Antoni FA: Vasopressin and the endocrine response to stress. *Neth J Zool* 1995;45:98–102.

67. Cooper DMF, Brooker G: Ca^{2+}-inhibited adenylyl cyclase in cardiac tissue. *Trends Pharmacol Sci* 1993;14:34–36.

68. Baukal AJ, Hunyady L, Catt KJ, Balla T: Evidence for participation of calcineurin in potentiation of agonist-stimulated cyclic AMP formation by the calcium mobilizing hormone, angiotensin II. *J Biol Chem* 1994;269:24546–24549.

69. Tang WJ, Krupinski J, Gilman AG: Expression and characterisation of calmodulin-activated (type I) adenylyl cyclase. *J Biol Chem* 1991;266:8595–8603.

70. Watson PA, Krupinski J, Kempinski AM, Frankenfield CD: Molecular cloning and characterization of the type VII isoform of mammalian adenylyl cyclase expressed widely in mouse tissues and in S49 mouse lymphoma cells. *J Biol Chem* 1994;269:28893–28898.

71. Wu ZL, Thomas SA, Villacres EC et al: Altered behavior and long-term potentiation in type-I adenylyl-cyclase mutant mice. *Proc Natl Acad Sci U S A* 1995;92:220–224.

72. DeBernardi MA, Munshi R, Yoshimura M et al: Predominant expression of type-VI cyclase in C6–2B rat glioma cells may account for inhibition of cyclic AMP accumulation by calcium. *Biochem J* 1993;293:325–328.

73. Kawabe J, Ebina T, Ismail S et al: A novel peptide inhibitor of adenylyl cyclase. *J Biol Chem* 1994;269:24906–24911.

74. Chen J, DeVivo M, Dingus J et al: A region of adenylyl cyclase 2 critical for regulation by G protein bg subunits. *Science* 1995;268:1166–1169.

75. Kerwin Jr JF: Adenylate cyclase subtypes as molecular drug targets. *Annu Rep Med Chem* 1994; 29:287–296.

76. Furukawa Y, Matsumori A, Hirozane T et al: Immunomodulation by an adenylate-cyclase activator, nkh477, *in vivo* and *in vitro*. *Clin Immunol Immunopathol* 1996;79:25–35.

Advances in Second Messenger and Phosphoprotein Research, Vol. 32, edited by Dermot M. F. Cooper Lippincott–Raven Publishers, Philadelphia © 1998

9

Adenylyl Cyclases and Alcohol

Boris Tabakoff and Paula L. Hoffman

Department of Pharmacology, University of Colorado Health Sciences Center, Denver, Colorado 80262

Ethanol is a two-carbon molecule that seemingly holds little structural information. This observation, together with reports of ethanol's actions on a myriad of physiologic and biochemical events, had led researchers to consider ethanol as a rather nonspecific drug that acts by perturbing the order of membrane lipids and secondarily affecting the function of the proteins residing in those lipids (1). The "lipid perturbation hypothesis" of ethanol's actions continues to attract the attention of researchers, but several caveats have generated caution in accepting too literally the proposition that ethanol's actions are a result of intercalation with membrane bulk lipids. For instance, the perturbation of bulk lipid order by physiologically relevant concentrations of ethanol is exceedingly small (2) and the concentration of ethanol within membranes at the time of the proposed perturbation of the membrane-bound proteins is not compatible with the actual concentration of ethanol necessary to perturb the proteins via disruption of the membrane bulk lipid milieu (1,3). In addition, the perturbation of the action of certain membrane proteins by ethanol and other alcohols is quite different in character from perturbation of these proteins by other lipid-soluble anesthetics (4–6), although the membrane concentrations of such anesthetics, or the disordering of membrane lipids by the other anesthetics, are equivalent to those produced by ethanol. It has also been recognized recently that ethanol, at concentrations of 10–50 mM, has a selective and specific effect on certain proteins, which implies the presence of proteins that can be considered as "receptive elements" on which ethanol preferentially acts. The 10–50 mM concentrations of ethanol correspond to levels, when present in human brain, that result in mild to significant intoxication but are well below the level of ethanol necessary to produce anesthesia (7).

The adenylyl cyclase (AC) signal transduction system can be considered as one of the "receptive elements" for ethanol (8), because significant perturbations in the function of this system have been reported at and below 50 mM concentrations of ethanol, and chronic exposure of cells in culture, or whole animals, to ethanol produces notable adaptive changes in the function of the AC system. Since the components of the AC signal transduction system (receptors, G proteins, catalytic components) are membrane-bound proteins, one should not ignore the lipid-related aspects of ethanol's actions, but the following discussion will illustrate that the character of particular G proteins and isoforms of AC predisposes to ethanol-induced perturbations of the cyclic adenosine monophosphate (cAMP)-generating systems. Additionally, we will discuss emerging evidence of an association between the inherited differences in the characteristics of an individual's AC signal transduction system and a predisposition to alcoholism.

ACUTE ACTIONS OF ETHANOL

The initial demonstrations of ethanol's actions on the AC signaling system were published in 1970 by Gorman and Bitensky (9). These data demonstrated an activation of the liver membrane AC system by ethanol. Since that time, additional studies utilizing membrane preparations from a number of organs or cell types have demonstrated that ethanol will enhance the activation of AC catalytic activity by agonists acting at G_s-coupled receptors.

Studies of the acute actions of ethanol on brain AC activity, and on AC activity in cells in culture, provide insights into the mechanisms by which ethanol perturbs AC activity. In membrane preparations of cerebral cortical tissue (10) or striatal tissue (4) from mice, concentrations of ethanol up to 500 mM had little effect on basal AC activity. On the other hand, 50 mM ethanol significantly enhanced AC activity if guanine nucleotides (GppNHp) (in the presence or absence of G_s-coupled receptor agonists) were present in the incubation mixtures. These demonstrations indicated that G proteins play an important role in the expression of ethanol's actions on AC. The studies of Bode and Molinoff (11) focused attention on the G_s heterotrimeric complex and particularly on the α subunit of G_s (α_s) as a necessary component in the stimulation of AC by ethanol. In membrane preparations from wild-type S49 lymphoma cells and in membranes from the UNC mutant, in which the β-adrenergic receptors are not coupled to G_s and AC, ethanol increased guanine nucleotide and fluoride ion–stimulated AC activity. However, little effect was seen in membranes prepared from the cyc⁻ mutant of the S49 cells that have no α_s (11,12). In contrast to G_s, there is little evidence for a role of G_i in ethanol's acute actions on AC. For example, in mouse striatal tissue, ethanol did not alter the inhibition of forskolin-stimulated AC activity by opiates and cholinergic agonists acting through G_i-coupled receptors (13,14). Furthermore, although high concentrations of ethanol did inhibit forskolin-stimulated AC activity in S49 lymphoma cells, this action of ethanol was not affected by treatment of the cells with pertussis toxin under conditions expected to inactivate G_i/G_o protein function (11). Pertussis toxin adenosine diphosphate (ADP)-ribosylates the α subunit of G_i/G_o proteins, blocks the ability of receptors to promote the exchange of guanosine triphosphate (GTP) for guanosine diphosphate (GDP) on the G_i/G_o α subunit, and locks the G_i/G_o proteins in their inactive heteromeric states (15). There is a report that ethanol *reduced* the magnitude of inhibition of cerebral cortical membrane AC activity produced by addition of adenosine to

AC assay mixtures containing rat cerebral cortical membranes (16). Both A_1 and A_2 adenosine receptors are present in cerebral cortical membranes, and although activation of A_1 adenosine receptors is inhibitory to AC activity, the presence of the AC stimulatory receptors (A_2 adenosine receptors), the simultaneous activation of these receptors by the added adenosine, and the coupling of these receptors to AC through G_s make interpretation of the result of Bauché and co-workers (16) difficult. The bulk of current evidence therefore suggests a selective involvement of G_s rather than G_i/G_o in the acute actions of ethanol on AC activity.

Further evidence for the involvement of G_s in the actions of ethanol comes from studies on the binding of isoproterenol to the β-adrenergic receptor. Studies that measured the displacement of the antagonist iodocyanopindolol by the β-adrenergic receptor agonist isoproterenol demonstrated that the displacement curves can be best described by mathematical models that postulate the presence of two states of the β-adrenergic receptor with different affinities for agonist. The high-affinity form of the receptor represents a complex of the receptor and the G_s heterotrimer, whereas the low-affinity form of the β-adrenergic receptor is considered to be the nascent (uncomplexed) receptor (17,18). The binding of guanine nucleotides to α_s leads to a dissociation of the G_s protein heterotrimer and the destabilization of receptor/G_s complexes. The effect of guanine nucleotides on isoproterenol binding parameters is to produce a reduction in the number of high-affinity sites for isoproterenol and to increase the number of low-affinity sites for this agonist. Ethanol at concentrations of less than 100 mM mimicked the actions of guanine nucleotides in assays that measured the high-affinity and low-affinity binding of isoproterenol to the brain β-adrenergic receptor (19).

Adenylyl cyclase activation caused by nonhydrolyzable guanine nucleotide analogs, such as Gpp(NH)p, exhibits a "lag period" prior to attainment of a steady-state phase of enzyme activity (20). This lag period is a reflection of a guanine nucleotide–induced conformational event wherein the α subunit of the G protein adopts an active conformation (21). The lag phase can be significantly modulated by addition of agonist for receptors that couple to either G_s or G_i (20). Kinetic analysis of the effects of isoproterenol on the activity of cerebral cortical AC after addition of Gpp(NH)p demonstrated that isoproterenol significantly reduced the $t_{1/2}$ for activation of AC by the added Gpp(NH)p (10). The addition of moderate concentrations of ethanol (50 mM) to assay mixtures containing isoproterenol accentuated isoproterenol's ability to reduce the $t_{1/2}$ for activation of AC. One can postulate from such results that a possible mechanism of ethanol's stimulatory actions on AC is to promote the dissociation of the subunits of G_s and, in the presence of guanine nucleotides, to promote the GTP for GDP exchange (α_s activating) reaction.

The level and duration of activity of heteromeric G proteins is not simply determined by the association of GTP with the α subunit of these proteins and by the dissociation of the α and βγ subunits. It is also determined by the GTPase activity of the G protein α subunit and the dissociation rate of GDP from the α subunit. The hydrolytic activity varies among the various α subunits of the G proteins and is critical in controlling the duration of the active Gα·GTP conformation and the life cycle of both the α and βγ subunits of the G proteins (22). No direct evidence, however, currently exists regarding ethanol's actions on actual guanine nucleotide binding parameters to recombinant or purified G_s, or on the hydrolysis rates of GTP by α_s in the presence of ethanol.

Cholera toxin ADP-ribosylates the α_s protein and blocks its ability to hydrolyze GTP. Therefore in the presence of GTP, cholera toxin–treated α_s is not subject to the cycle of shuttling between the inactive complex with $\beta\gamma$ and the activated α_s monomer and is primarily present in its active, GTP-bound form. The pretreatment of cells (6) or brain membranes (4) with cholera toxin reduces the extent of ethanol-induced stimulation of membrane-bound AC activity in the presence of GTP but does not eliminate fully the acute stimulatory actions of ethanol on AC activity. Such data, and data showing further stimulation by ethanol of brain membrane AC activity after pretreatment of membranes with a maximally stimulating concentration of nonhydrolyzable guanine nucleotide analogs (4,10), indicate the presence of two components of ethanol's actions. The first component is the previously described enhancement of the activation of α_s. The second component seems to be related to sensitization of the catalytic unit of AC to the activated α_s. The presence of two distinct components to ethanol's actions on the AC signal transduction system is supported by the examination of chloroform's action on the AC system. Chloroform, added at a concentration that produces an identical membrane concentration of chloroform to the membrane concentration of *ethanol* produced by 200 mM of ethanol, *inhibited* GTP-stimulated striatal membrane AC activity, whereas ethanol further enhanced the effects of GTP (4). On the other hand, both chloroform and ethanol had similar *stimulatory* actions on AC activity when assays were performed with cholera toxin–treated membranes and GTP. More recently, distinctive actions of chloroform versus the effects of ethanol and another short-chain primary alcohol (butanol) were observed in HEK-293 cells transiently expressing the type VII isoform of AC (AC7) (6). At equivalent membrane concentrations of chloroform, ethanol, and butanol, both ethanol and butanol increased prostaglandin E_1 (PGE_1)-stimulated AC activity in these cells, whereas chloroform had no effect.

The demonstration of the cholera toxin–insensitive component of ethanol's actions, as well as the cholera toxin–sensitive component, on AC activity seems to be related to the isoform of AC present in an assay system (6). Currently, nine mammalian isoforms of AC have been cloned and completely sequenced (23,24). Significant structural and regulatory diversity is evident in this family of enzymes. The AC isoforms differ in their regulation by α_s, α_i, and $\beta\gamma$ subunits of the G proteins (23,25). The AC isoforms also differ in their responsiveness to calcium, calcium/calmodulin, and phorbol esters (protein kinase C [PKC]) (25–27). With regard to ethanol's actions, the divergent abilities of the various AC isoforms to integrate coincident signals (28), and the ability of one modifier of enzyme activity to influence the enzyme's response to another modulatory agent (29), are of significant interest. In examining the effects of ethanol on AC activity in HEK-293 cells transiently transfected with complementary DNA (cDNA) for the type I (AC1), II (AC2), III (AC3), V (AC5), VI (AC6), and VII (AC7) isoforms, it was noted that the stimulatory actions of ethanol on AC activity depended on concurrent addition of PGE_1 to the assay system, and depended on which isoform of AC was being expressed in the HEK-293 cells (6). Even in the presence of PGE$_1$, no effects of ethanol were noted with HEK-293 cells expressing the AC1 and AC3 isoforms (6). In the presence of PGE$_1$, cells expressing the AC7 isoform were most affected by ethanol (128% stimulation of AC activity by 200 mM of ethanol), whereas more modest ethanol-induced stimulation (50–60%) was noted in cells expressing the AC2, AC5, and AC6 isoforms (6). Interestingly, in cells expressing AC2, pretreatment with cholera toxin eliminated the stimulatory actions of ethanol (200 mM), whereas ethanol still produced about 50% stimulation of AC activity in cells transfected with AC7 and treated with cholera toxin. Ethanol was also able to further stimulate AC activity in HEK-293 cells co-transfected with AC7 and the constitutively active form of α_s (Q227L). These experiments (6) indicate that ethanol has actions on the AC7 isoform that are independent of its actions in promoting the activation of G$_s$. The results also indicate that the characteristics of each of the AC isoforms play a role in determining the extent of ethanol's actions on the activity of the AC signal transduction system.

Currently, several of the AC isoforms have been shown to be *sensitized* to the actions of α_s-GTP by simultaneous exposure to forskolin (29), or to $\beta\gamma$ subunit proteins (28,30), and by putative post-translational modification (via PKC) initiated by treatment of cells with phorbol esters (27). Given the differential response of the AC2 isoform versus the AC5 and AC6 isoforms to $\beta\gamma$ (i.e., AC2 isoforms are conditionally activated by $\beta\gamma$ [23], whereas AC5 and AC6 isoforms are insensitive to $\beta\gamma$); one is currently hard pressed to explain the *equivalent* responses of the AC2, AC5, and AC6 isoforms to ethanol by invoking $\beta\gamma$-related mechanisms. It is of interest, however, that in the studies of McHugh-Sutkowski and co-workers (29), a synergistic activation of AC2, AC5, and AC6 was witnessed in the presence of forskolin and activated α_s-GTPγS, whereas the effects of forskolin and α_s-GTPγS on the activity of the AC1 were simply additive. Although the focus of the studies by McHugh-Sutkowski and co-workers (29) was the effect of α_s on the binding and biologic activity of forskolin, one can surmise the reverse situation (i.e., that the presence of forskolin could alter the sensitivity of AC2, AC5, and AC6 to the actions of α_s-GTPγS). Whether ethanol, possibly like forskolin, can sensitize particular isoforms of AC to the actions of activated α_s needs to be further explored. The data militating against direct sensitization of AC to the actions of α_s by ethanol are derived from experiments with membrane preparations of the HEK-293 cells transfected with particular ACs, versus the data derived from whole-cell preparations. In assays of membrane preparations, the effects of ethanol on PGE$_1$-stimulated AC activity (percent further stimulation) were quite small, and membranes of cells transfected with either AC2, AC5, or AC7 showed no differential sensitivity to ethanol (M. Yoshimura and B. Tabakoff, unpublished observations).

The feature linking the ethanol-sensitive isoforms of AC is their sensitivity to phorbol esters and their possible phosphorylation by PKC. The type II subfamily enzymes, including AC7, show significant modulation of their activity and their responsiveness to α_s by treatment with phorbol esters (27,31). Recent evidence also demonstrates PKC-mediated modulation of the activity of AC5 (albeit to a significantly lesser extent than enzymes of the type II family) (32). One can speculate that ethanol's actions in whole-cell preparations are, in part, mediated through activation of PKC and subsequent modification (phosphorylation) of particular ACs, which would sensitize the enzymes to activation by α_s. Recent studies of cerebellar granule cell preparations suggest that certain actions of ethanol are mediated by activation of PKC (33), and current evidence on ethanol's actions on PGE_1-stimulated AC activity in HEL cells demonstrates that a significant inhibition of ethanol's effects can be produced by staurosporine and other, more specific inhibitors of PKC (33a). If one is to invoke PKC-related mechanisms to explain even a portion of ethanol's actions on the AC signal transduction system, one is, however, confronted with the data that ethanol has effects that significantly differ in magnitude on two closely related isoforms of AC, the type II and type VII enzymes (6). Both enzymes are sensitized to the actions of α_s by pretreatment of cells expressing these enzymes with phorbol esters (27,31). On the other hand, the kinetics of the sensitization process differ for AC2 and AC7 (31), and these differences in the responsiveness of the AC2 and AC7 isoforms to the actions of PKC (phorbol esters) may provide an explanation for the observed quantitative differences in ethanol sensitivity.

If the presence of various isoforms of AC can contribute to a cell's sensitivity or resistance to ethanol-induced perturbations in the initial stages of cAMP signaling, the differential distribution of the AC isoforms in various cells of the body (23), and the differential distribution of the ACs in the neurons of brain (e.g., ref. 34), may contribute to the characteristic spectrum of ethanol's actions in an intoxicated individual. For instance, neurons expressing the particularly ethanol-sensitive isoform, AC7, may be preferentially affected by ethanol. *In situ* hybridization studies identifying the location of AC7 messenger RNA (mRNA) have demonstrated a concentration of this message in the Purkinje/granule cell layer of the cerebellum (31).

Cyclic AMP–mediated signaling has been shown to be a controlling element in the modulation of the ethanol sensitivity of certain $GABA_A$ receptor–coupled ion channels and nucleoside transporters (35,36). Particularly interesting in this regard is the proposed "feed-forward" sensitization of $GABA_A$ receptors on cerebellar Purkinje cells by, and to, ethanol's actions. The Purkinje cells of the cerebellum modulate a number of functions related to body equilibrium. The $GABA_A$ receptors on the Purkinje cells can, when activated, produce hyperpolarization and decrease the firing rates of Purkinje neurons. In a number of experimental systems, ethanol has been shown to potentiate the activation of $GABA_A$ receptors by GABA (37). On the other hand, little effect of ethanol was evidenced on $GABA_A$ receptor function on the cerebellar Purkinje cells unless the β-adrenergic receptor agonist, isoproterenol, was simultaneously applied to the Purkinje cells (35). In the presence of applied isoproterenol, ethanol potentiated GABA actions and mediated a profound inhibition of Purkinje cell firing *in vivo* in rat brain. More current work (38) has demonstrated that activation of protein kinase A (PKA) in the Purkinje cells by 8-bromo-cAMP could also sensitize the $GABA_A$ receptor to ethanol's actions. It is of interest that *in situ* hybridization studies identifying the location of AC7 mRNA have demonstrated a concentration of this message in the Purkinje/granule cell layer of the cerebellum, and the cerebellar Purkinje cells contain significant quantities of AC7 protein (38a,b). As discussed, this form of AC is particularly sensitive to activation by ethanol if simultaneously activated via G_s-coupled receptors (e.g., β-adrenergic receptors). One can therefore propose that ethanol has a two-pronged action on the Purkinje cells. First, ethanol could potentiate the function of a particularly ethanol-sensitive form of AC in the Purkinje cell. The formed cAMP would activate PKA, and PKA activity would sensitize the Purkinje cell $GABA_A$ receptor to ethanol's actions. Ethanol-induced potentiation of GABA effects would lead to accentuated inhibition of Purkinje cell firing, and this "feed-forward" cascade of events may be responsible for the incoordination and postural anomalies seen in ethanol-intoxicated individuals.

CHRONIC ACTIONS OF ETHANOL

As discussed earlier, the acute effect of ethanol is usually to increase agonist-stimulated AC activity. However, in cells or tissues that have been exposed chronically to ethanol, there is often a reduced responsiveness of AC to agents that usually enhance this enzyme's activity. Early studies using brain tissue of mice and rats treated chronically with ethanol indicated decreased dopamine-stimulated AC activity in striatal tissue (39), and decreased norepinephrine-stimulated activity in cerebral cortical membranes (40). Although it was reported that chronic ethanol treatment also decreased cerebral cortical β-adrenergic receptor number (41), which could account for the decreased stimulation of AC activity by β-adrenergic agonists, later investigations revealed significantly reduced stimulation of AC by guanine nucleotides, fluoride, forskolin, Mn^{2+}, and vasoactive intestinal peptide, as well as isoproterenol, in cerebral cortical membrane preparations from ethanol-treated mice (42). These results suggested that the change in AC activity produced by chronic ethanol intake had the characteristics of heterologous desensitization (43). Heterologous desensitization refers to the situation in which, after prolonged exposure of cells to an agonist, AC becomes refractory to stimulation by multiple activators acting via various receptors. This phenomenon is distinguished from homologous desensitization, in which only the response to the particular agonist to which the cells were exposed is reduced (43).

Decreased agonist and guanine nucleotide–stimulated AC activity was also observed in hippocampal tissue of ethanol-treated C57BL/6 mice but not in cerebellar tissue of either C57BL/6 mice or Sprague-Dawley rats (44,45). On the other hand, more recently, Wand and co-workers (46) reported decreased stimulation of cerebellar AC activity by agonists (prostaglandin E_1, isoproterenol, and corticotropin-releasing hormone), guanine nucleotides, and forskolin when tissue was obtained from chronically ethanol-treated lines of mice that had been selectively bred for sensitivity (LS) or resistance (SS) to the hypnotic effect of ethanol. These investigators also found decreased pituitary AC activity in the chronically ethanol-treated LS mice (47). The effect on AC activity of chronic administration of ethanol in nonneuronal cells has also been investigated. Desensitization of the AC response was found in regenerating liver of chronically ethanol-treated rats (48), but it was reported that forskolin-stimulated pancreatic AC activity was not altered in rats treated chronically with ethanol (49). However, there was a decreased ability of somatostatin and guanine nucleotides to *inhibit* AC activity in pancreatic tissue of ethanol-treated rats (49).

The preponderance of evidence indicates that chronic ethanol treatment of animals results in decreased responsiveness of AC in brain, and possibly other tissues, to stimulation by agonists or other agents that act through the stimulatory G protein G_s. Differences in results among laboratories (e.g., with respect to cerebellar AC activity) may reflect differences in the duration of ethanol exposure and/or in the development of tolerance to or physical dependence on ethanol in the animals, which is not always measured. The chronic effect of ethanol on AC activity may also vary, depending on the tissue or cellular localization of the enzyme. This variation may reflect the form(s) of G proteins or ACs that are expressed in a particular tissue or cell type.

"Desensitization" of AC activity after chronic ethanol exposure has also been a frequent finding when activity is measured in cultured neuronal or nonneuronal cells, although there are some exceptions. For example, chronic *in vitro* exposure of N1E-115 or NG108-15 neuroblastoma cells, S49 lymphoma cells, or primary cultures of cerebellar granule neurons to concentrations of ethanol ranging from 25 mM to 200 mM for several days resulted in reduced responsiveness of AC activity to stimulation by prostaglandin E_1, isoproterenol, and adenosine analogs (50–53). Although initial studies in which AC activity was measured in membrane preparations of PC12 cells did not reveal a chronic effect of ethanol, later studies, in which cAMP production in intact cells was assessed, demonstrated a decreased response of AC to stimulation by an adenosine analog (54,55).

There have, however, also been reports that chronic ethanol treatment does *not* alter the sensitivity of AC to agonists in embryonic chick myocytes or in mouse neuroblastoma N18TG2 cells (56,57) and that chronic exposure of cultured hepatocytes to ethanol resulted in an *increase* in agonist-stimulated cAMP production (58). As discussed earlier, these differences might result if there is differential expression among the cell types investigated of elements of the AC signal transduction pathway (e.g., G proteins, isoforms of AC) and if the quantity or function of these elements is differentially modulated in different cell types (22,23). One can also postulate differential expression of modulators of AC activity, such as protein kinases or phosphatases (59), which could contribute to differing responses to chronic ethanol treatment in the cultured cells. The exceptions to the "rule" of desensitization of AC activity following chronic ethanol exposure (e.g., responses of the hepatocytes or the N18TG2 cells) may prove useful in assessing the role of particular proteins in generating the desensitization response to ethanol treatment that occurs in a majority of cells and tissues. For example, if there is evidence that expression of a specific isoform of AC is necessary to observe ethanol-induced "desensitization," one could determine whether that isoform is absent in a cell that does not demonstrate desensitization.

Because the change in AC activity produced by chronic ethanol exposure in many cases resembles that produced by agonist-induced heterologous desensitization, as described earlier, many of the ethanol studies to date have been influenced by the proposed mechanisms for such desensitization. Heterologous desensitization is characterized by a reduction in responsiveness of AC not only to agonists, but also to receptor-independent stimulation (i.e., to stimulation by guanine nucleotides, fluoride, and forskolin), and therefore quantitative or qualitative changes in G proteins have been suggested as an underlying mechanism for heterologous desensitization (60). In fact, there is substantial evidence that prolonged stimulation of AC by agonists can lead to changes in the quantities of G proteins (both G_s and G_i) (61). As a result, there have been a number of studies of the effects of chronic ethanol treatment on G protein quantity, both in cultured cells and in animal tissue. In one of the first published investigations of chronic ethanol effects on G protein levels, a 30% decrease in the mRNA for α_s, and an equivalent decrease in α_s protein, was found in NG108-15 cells that had been chronically exposed to ethanol (62). The decrease in α_s was accompanied by a decrease in agonist-stimulated AC activity, but there was no change in the response to forskolin. In later studies, ethanol-treated PC12 cells that showed reduced adenosine-stimulated AC activity also displayed a decrease in α_s (54), and there was a decrease in α_s in pituitary tissue of ethanol-treated LS mice, where, as described earlier, agonist-stimulated AC activity was reduced (47). An effort was made in the investigation of NG108-15 cells (as well as S49 lymphoma cells) to address the mechanism of the ethanol-induced changes in AC activity and G protein levels. It was suggested (52) that the effect of chronic ethanol exposure to decrease AC stimulation was in fact an example of *agonist-induced* heterologous desensitization. Evidence was presented that ethanol inhibited adenosine transport into NG108-15 cells, resulting in an extracellular accumulation of this transmitter (52,63). Prolonged exposure of the cells to the accumulated high concentration of adenosine was proposed to result in the observed decrease in α_s and AC activity. However, in a later study of NG108-15 cells, although chronic ethanol exposure was again found to reduce the amount of α_s (and α_i), AC activity was *increased* instead of decreased, and these changes were reported *not* to be due to accumulation of extracellular adenosine (were not reversed by adenosine deaminase added to the medium) (64). Similarly, decreased AC activity and α_s levels in PC12 cells were reported not to be a result of accumulation of extracellular adenosine (65). Recent reports have also indicated that chronic ethanol treatment of S49 cells can result in a decrease in α_s with no change in fluoride-, isoproterenol-, or forskolin-stimulated AC activity (66). These findings do not support a proposition that chronic ethanol treatment induces an agonist (adenosine)-mediated heterologous desensitization. Furthermore, the results suggest that changes in the total level of α_s do not necessarily predict changes in AC activity. This conclusion is supported by studies of the stoichiometry of the proteins involved in regulation of AC activity (in NG108-15 cells) (67). Evidence is accumulating that G proteins do not represent the rate-limiting elements in activation (or inhibition) of AC activity. Quantitative studies suggest that α_s is present in cells at a higher level than either G protein–coupled receptors or AC (67). To lower agonist-stimulated AC activity, large reductions (90% or greater) in α_s must occur (67), whereas chronic ethanol-induced decreases in α_s content, when they occur, are in the range of 30–40%.

Another postulated change that could result in decreased AC activity is an *in-*

crease in the inhibitory G protein, G_i. In ethanol-exposed N1E-115 cells, where prostaglandin E_1 stimulation of AC activity was decreased, there was a threefold increase in α_i and a smaller decrease in α_s after prolonged ethanol treatment (56). Wand and co-workers (46) found two- to fourfold increases in α_{i1} and α_{i2} in cerebellar tissue of ethanol-treated LS and SS mice, in concert with the decreased stimulation of AC activity by various agonists, and only a 30% decrease in α_s in pituitary tissue of LS mice (47). It was reported that treatment of the cerebellar tissue with pertussis toxin (to block α_i activity) normalized AC activity (46). This experiment supports the possibility that the large increases in α_i may contribute to the observed lowering of stimulated AC activity. However, increases in α_i levels have *not* been observed in other instances where stimulation of AC activity was reduced by chronic ethanol exposure. For example, in NG108-15 or PC12 cells, decreased responsiveness to adenosine was not accompanied by an increase in α_i (54,56), nor was there an increase in α_i in anterior pituitary tissue of LS mice (47), where agonist-stimulated AC activity was decreased (see earlier). It is also important to note that studies of agonist-induced changes in α_i content have questioned the relationship between levels of α_i and AC activity, since increases in α_i have been associated with both higher and lower enzyme activity (60). It is surprising that there has not been a determination of changes in *inhibition* of AC by inhibitory agonists when α_i is increased after chronic ethanol exposure. Such a change has been demonstrated following stimulatory receptor agonist-induced increases in α_i. For example, prolonged activation of AC by isoproterenol or forskolin induced a large increase in α_{i2} in S49 cells, through a PKA-dependent mechanism, and this change was accompanied by an increased inhibitory effect of somatostatin (68).

Overall, it is clear that chronic ethanol treatment can produce, in some cells and tissues, an increase in α_i concurrent with a decrease in stimulation of AC activity. However, the causal relationship of the ethanol-induced changes in α_i and AC activity has not been established. It is possible that, at least in some cases, the increase in α_i produced by chronic ethanol treatment is not the underlying mechanism for the chronic ethanol-induced desensitization of AC activity. However, proteins in the G_i family also regulate K^+ channels (69), and α_{i2} in particular is a transducer of mitogenic signals in a number of cell types, via activation of the MAP kinase pathway (70). In mature neurons, the mitogenic activity of α_i may not be relevant, but the effect of chronic ethanol exposure on activity of the ion channels through the influence of α_i (and possibly through the influence of β/γ [71]) has not been well investigated and deserves attention.

There have also been instances where chronic ethanol treatment had no effect on the levels of G proteins, although the treatment resulted in decreased stimulation of AC activity. When C57BL/6 mice were fed ethanol chronically in a regimen that produces functional tolerance to and physical dependence on ethanol, there was no significant change in the content of α_s, α_{i2}, α_{i3}, or β subunits of the G proteins in several brain areas, even though stimulation of AC by various stimulatory receptor agonists, guanine nucleotides, and forskolin was shown to be reduced (42). Similarly, chronic ethanol ingestion using a paradigm that generates tolerance and physical dependence did not alter the content of α_s, α_i, or α_o in cortex, striatum, or nucleus accumbens of rats (72).

The overall conclusion that can be drawn from the described studies is that changes in the *total quantity* of G proteins cannot well account for the "desensitization" of AC activity that is produced by chronic ethanol exposure. First, this desensitization can occur in the absence of alterations in G protein content; second, the changes in G protein content that have been observed do not always correlate with the changes in AC activity. A role for G proteins in the ethanol-induced "heterologous desensitization" should not, however, be completely discounted. When estimates of the quantity of α_s-AC complexes were made by assaying high-affinity forskolin binding on cell membranes in the presence of maximally stimulating concentrations of nonhydrolyzable guanine nucleotide analogs, it was found that the level of the complex (forskolin bound) was substantially lower than the total level of α_s (67). This phenomenon may at least partially explain why small changes in the overall level of α_s do not necessarily correspond to altered stimulation of AC activity. It has, however, been demonstrated that G proteins can interact with tubulin (73) and that microtubule-disrupting reagents can increase agonist-stimulated cAMP production, suggesting enhanced accessibility of α_s to AC in the presence of the microtubule-disrupting agents (74). It may be relevant to note that chronic treatment of rats with various antidepressants also resulted in *increased* guanine nucleotide–stimulated AC activity in cerebral cortical tissue, without any change in the *total* levels of α_s, α_i, α_o, or β subunits of G proteins, but with an apparent increase in the amount of α_s-AC complex (75). An explanation provided by the authors was that the antidepressant treatment might alter the cellular cytoskeleton, resulting in changes in the interaction of G_s with tubulin and AC (75). The previously mentioned studies, and those demonstrating an influence of the cytoskeleton on ethanol's actions (76), suggest the possibility that cytoskeletal components may be important in mediating the chronic actions of ethanol.

Another possible basis for the "desensitization" of AC activity after chronic ethanol treatment, and one that has not been addressed in any detail, is a quantitative change in the catalytic unit of AC. In some instances, Mn^{2+}-stimulated AC activity, which is believed to be a direct measure of catalytic unit function, has been shown to be decreased in tissue of chronically ethanol-treated animals (42). Similarly, stimulation of AC by high concentrations of forskolin, which has been proposed to reflect a direct interaction of forskolin with the catalytic unit, is reduced in brain tissue of chronically ethanol-treated animals (42,46), as are various parameters of forskolin binding to brain membranes (77). In this regard, it may be relevant to note that chronic *morphine* treatment of mice, which, in contrast to ethanol treatment, results in *increased* AC activity in brain areas such as the nucleus accumbens and locus ceruleus (78), was reported to result in increased mRNA for AC8 in particular brain areas (locus ceruleus and amygdala) (79). Similar studies have not yet been carried out to determine whether AC expression is altered in brains of chronically ethanol-treated animals, but the results with morphine indicate the potential for localized changes in expression of particular types of AC.

One piece of evidence that may indirectly support the possibility of changes in the level or characteristics of AC catalytic units is the finding of virtual elimination of high-affinity β-adrenergic agonist binding sites, with no change in the total number of antagonist binding sites, in brain tissue of chronically ethanol-treated mice, where agonist-stimulated AC activity was also lowered (19). High-affinity agonist binding has been suggested to reflect the formation of a receptor-α_s complex (17,22, and see earlier). Levitzki and co-workers, however, have postulated and provided evidence for a "collision-coupling" model of activation of AC in which α_s and the catalytic unit behave kinetically as a single entity (80). In this model, as opposed to the more commonly accepted "G-shuttle" model in which the G protein is not closely associated with either the receptor or the catalytic unit, the receptor transiently interacts with the α_s-catalytic unit complex (80). Based on the "collision-coupling" model, one can postulate that the loss of high-affinity agonist binding sites would reflect a decrease not in α_s per se but in the α_s-catalytic unit complex. Because recent data (67) suggest that the level of the catalytic unit may be the rate-limiting factor in the stimulation of AC activity, a decrease in the level of the enzyme would also lower the α_s-catalytic unit complex and could account for the loss of high-affinity agonist binding produced by chronic ethanol treatment. Given this model, of course, any change in the level of the α_s-catalytic unit complex, whether produced by an altered level of AC or an altered availability of α_s (see earlier), would produce a similar effect on agonist binding.

A qualitative change in the AC molecule could also produce the observed decrease in stimulated AC activity. Recently, recombinant AC5 was shown to be phosphorylated by protein kinase A (81), and this phosphorylation resulted in reduced activation of the enzyme by forskolin and α_s proteins. This desensitization would represent a feedback control mechanism that could be activated by stimulation of AC activity by agonists and that would be enhanced by acute ethanol exposure. In brain, AC5 is primarily expressed in striatum, but mRNA for this form of AC is also present in other brain areas where chronic ethanol-induced desensitization of AC has been observed, including cerebral cortex (82). A statement has also appeared that the activity of AC6 can be inhibited by PKA-mediated phosphorylation (see ref. 81), but it is unknown if other forms of AC can be regulated in this manner. Given that chronic ethanol treatment in general produces about a 30% decrease in stimulation of AC activity in brain, one could speculate that ethanol treatment results in modification of certain forms of AC by PKA, leaving the activity of other forms intact. Depending on the isoform(s) of AC expressed in various cells, such an explanation could also account for the fact that in some cells, chronic ethanol treatment does not produce desensitization of AC activity (see earlier). It is also of interest that PKA activation has been implicated in the "desensitization" of AC activity produced by ethanol in S49 cells and in PC12 cells (36,83) and that S49 cells have been reported to express AC6 (as well as AC7) (84,85).

However, some caveats are associated with consideration of post-translational modification of proteins as an explanation for chronic ethanol-induced desensitization of AC activity. First, the time course of development and loss of this desensitization in brain appears to be inconsistent with a mechanism involving a rapid process such as phosphorylation/dephosphorylation. The ethanol-induced decrease in high-affinity agonist binding in cerebral cortical tissue of mice became apparent after 2 days of ethanol ingestion (19), and the decrease in guanine nucleotide–stimulated AC activity in cortical tissue took between 3 and 5 days after ethanol withdrawal to normalize (40). The time course of these changes is more consistent with a mechanism involving altered transcription, translation, or protein degradation than with a post-translational modification such as phosphorylation. Another issue is that changes in enzyme activity caused by post-translational modifications such as phosphorylation would be expected to be detectable when AC activity is measured in intact cells but may not be evident in membrane preparations, unless significant care is taken to preserve the phosphorylation state of the proteins. Studies using brain tissue and some cultured cell preparations, where AC desensitization has been noted, have been carried out using membrane preparations, and the methods used for these preparations would not be expected to maintain protein phosphorylation. The desensitization of AC activity measured in these instances is more likely to reflect a change in protein quantity or sequestration than a change in phosphorylation or other unstable post-translational modifications. However, it is noteworthy, as mentioned earlier, that AC desensitization was observed only in intact PC12 cells, and not in membrane preparations from these cells (54,55). Similarly, desensitization was observed in intact S49 cells (36) but was minimal when AC activity was measured in a membrane preparation (66). These results could be consistent with a postulate that phosphorylation of AC5 or AC6 by PKA in these cells may lead to the desensitization phenomenon.

It is, of course, possible that chronic ethanol treatment can induce AC desensitization by more than one mechanism. Under some conditions, and in some cell types, changes in phosphorylation may predominate, leading to changes in activity that can only be detected in intact cells. In other cell types, and with other conditions (e.g., longer ethanol exposure), more stable changes in protein levels produced by altered transcription or translation, or changes in protein sequestration, may occur, resulting in desensitization that can be observed in membrane preparations and for longer periods of time.

Regardless of the molecular mechanism underlying chronic ethanol-induced desensitization of AC activity, this change is likely to be of physiological/pharmacological significance in the generation of tolerance to certain effects of ethanol. A role for the AC signal transduction system in the development of functional tolerance to the hypnotic effect of ethanol has been demonstrated in mice (86). In these animals, depletion of brain norepinephrine by 6-hydroxydopamine (6-OHDA) prevented the development of ethanol tolerance, while leaving the development of physical dependence unaffected. Thus, the presence of ethanol in the lesioned mice was necessary but not sufficient to produce ethanol tolerance as it did in intact mice. However, when the lesioned mice were treated with forskolin during ethanol exposure, tolerance developed normally (86). These findings can be interpreted to indicate that ethanol potentiation of noradrenergically induced increases in cAMP production may be required for ethanol tolerance to develop. Chronic activation of AC by ethanol in the presence of an agonist (e.g., norepinephrine), or in the presence of forskolin, presumably results in the desensitization of the AC system that is associated with ethanol tolerance.

A relationship of desensitization of agonist-stimulated cAMP generation to ethanol tolerance can again be postulated to involve the $GABA_A$ receptor system in cerebellar Purkinje cells that is described earlier. Following chronic ethanol exposure, desensitization of the AC system would lead to decreased β-adrenergic stimulation of cAMP production in the cerebellar Purkinje cells, and a resulting decrease in ethanol potentiation of the response to GABA. Because ethanol-induced potentiation of GABA responses in these cerebellar cells is believed to play a key role in mediating certain behavioral effects of ethanol, including ataxia and incoordination, the reduced ability of ethanol to potentiate the GABA response in an animal treated chronically with ethanol would generate tolerance to the incoordinating and ataxic effects of ethanol.

ADENYLYL CYCLASE AS A STATE OR TRAIT MARKER FOR ALCOHOL DEPENDENCE

The suggestion that chronic ethanol-induced desensitization of AC activity plays a role in the development or expression of functional ethanol tolerance implies that AC desensitization would be demonstrable in humans who display ethanol tolerance. Valverius and co-workers (87) reported a decrease in high-affinity agonist binding to β-adrenergic receptors in brain tissue of diagnosed alcoholics who had measurable blood or urine alcohol at the time of death. This change was similar to that found in brains of chronically ethanol-treated rodents (see earlier). There has also been a report of a decrease in α_s (30%) in temporal cortex of alcoholics, compared to controls (88), when measured post mortem.

Blood cells, and platelets in particular, have often been used as an accessible tissue to study neuronal events in humans. AC activity in platelets and lymphocytes of alcoholics has therefore been investigated in several studies. Platelet AC activity was originally measured in a population of male alcoholics and age-matched control subjects (89), and it was found that stimulation of activity by prostaglandin E_1, fluoride, and GppNHp was lower in platelet membranes of alcoholic subjects than in controls. Low agonist (adenosine)–stimulated AC activity has also been reported in lymphocytes of alcoholics (90). These findings appeared to be similar to the changes in AC activity observed in brains of chronically ethanol-treated animals, and, as in the animal studies described earlier, investigation of the mechanism for decreased AC activity in platelets and lymphocytes of alcoholics has centered primarily on measurement of G protein levels. Waltman and co-workers (91) reported threefold-higher average levels of α_{i2} mRNA and protein in lymphocytes of alcoholics where AC activity was decreased. In *platelet* membranes of alcoholics, mean levels of α_{i2} and β subunits were also reported to be higher than in those of controls (92). However, in neither this latter study (92) nor others (93) was there a significant correlation *in individuals* of the levels of α_i or α_s with platelet AC activity (93). Thus, there may be overall differences in G protein levels in blood cells of alcoholics, compared with controls, but these differences do not appear to be associated with the decreased AC activity in a given individual.

The original studies of platelet and lymphocyte AC activity in alcoholics were undertaken, based on the animal studies, with the implicit assumption that chronic alcohol ingestion could produce a desensitized state of AC (i.e., that this enzyme activity might serve as a physiologic "state marker" of chronic alcohol consumption). However, it was noted that platelet AC activity remained low in alcoholics who had reportedly abstained from alcohol for up to 4 years (89), and Waltman and co-workers (91) reported that AC activity (basal and prostaglandin E_1– and forskolin-stimulated) was decreased in lymphocytes of alcoholics who had abstained from alcohol for about 3 weeks, but not in "actively drinking" alcoholics. In contrast, stimulation of AC in brains of chronically ethanol-treated animals returned to normal within several days after alcohol withdrawal (40). Furthermore, when lymphocytes obtained from control subjects and alcoholics were cultured for four to six generations, the AC activity of cells that originally came from the alcoholics was more sensitive to the chronic effect of ethanol than the enzyme in cells from control subjects (94). All these data suggested that low responsiveness of platelet and/or lymphocyte AC activity to stimulation by agonists could represent an inherent characteristic of the cells of the alcoholic individuals (i.e., could be a "trait marker" for a genetic predisposition to alcoholism) rather than a response of the cells to chronic alcohol ingestion. Further evidence for a genetically influenced difference in AC activity between alcoholics and nonalcoholics derives from the finding that platelet AC activity stimulated by Gpp(NH)p was lower in alcoholics with a positive family history of alcoholism than in control subjects (95). Even more pertinent is the report of Lex and co-workers (96) showing that fluoride-, prostaglandin E_1–, and guanine nucleotide–stimulated AC activities were lower in *control* women with a positive family history of alcoholism than in control women without such a history.

The results to date do not entirely resolve the issue of whether lowered AC activity in blood cells of alcoholic (alcohol-dependent) individuals represents a response to chronic alcohol intake or an inherent characteristic. It is possible that chronic alcohol intake does alter the AC pathway in humans, as suggested by the studies of Valverius and co-workers (87), but that the AC of individuals who are genetically predisposed to alcoholism is inherently less sensitive to agonist stimulation than that of nonpredisposed individuals prior to any alcohol exposure.

The suggestion that platelet AC activity could represent a genetically influenced "trait marker" for alcoholism has led to investigations of the inheritance patterns of AC activity. Devor and co-workers (97) found that fluoride-stimulated platelet AC activity is transmitted in a Mendelian fashion as a co-dominant locus in families. However, genetic linkage between low platelet AC activity and alcoholism was not established in these families (97). There are several possible explanations for such results. As has been pointed out (98), the occurrence of an association between a genetic marker and a disease or disorder, in the absence of a close linkage, can indicate that the marker is neither necessary nor sufficient for the occurrence of the disorder, but only increases the risk for that disorder. Another possibly confounding factor is the presence of alcoholic subtypes (e.g., the type I and type II alcoholic subtypes proposed by Cloninger [99]). One could postulate that the linkage might be stronger if AC activity were assessed according to these subtypes of alcoholism. In a recent study, however, although platelet fluoride–stimulated AC activity was confirmed to be low in male alcoholics, there was no difference in enzyme activity between type I and type II alcoholics (93). Therefore, it appears that low platelet AC activity may be related to characteristics not distinguished by the type I and type II subtyping criteria, and studies of AC in alcoholics subtyped according to other criteria are needed.

The molecular basis for the difference in agonist-stimulated AC activity between controls and alcoholics has yet to be determined. However, as discussed earlier, there is little evidence to support a key role for G protein levels in determining these differences. In one study, it was noted that Mn^{2+}-stimulated AC activity was low in platelets of alcoholics, compared with controls (91). Since, as mentioned, Mn^{2+} is believed to stimulate the catalytic unit of the enzyme directly directly, these data could implicate quantitative or qualitative differences in the AC catalytic unit between the groups. Adenylyl cyclase 7 was originally identified in the human erythroleukemia (HEL) cell line, a widely used model for studies of the characteristics of human platelets (100). Since AC7 appeared to be the major form of AC expressed in the HEL cells, it may also be a major form of the enzyme in platelets. If low-platelet AC activity does represent a trait marker for a genetic predisposition to alcoholism, AC7 could be considered a "candidate gene" for association with a predisposition to alcoholism. Therefore, polymorphisms in the gene for AC7 were examined in control and alcoholic subjects. One tetranucleotide repeat sequence located in the 3′-untranslated region of the gene was found to be polymorphic, but there was no obvious association between the polymorphic sequences and a diagnosis of alcohol dependence (101). However, only a small number of individuals was investigated in this study, and further investigations, with larger groups, will be necessary to assess possible associations definitively. Another approach to determining genes that are involved in a disorder such as alcoholism, which is most probably polygenic in its determinants, is the identification of quantitative trait loci (QTL). This approach assumes that the genetic contribution to a disorder such as alcoholism is the result of quantitatively variable contributions of multiple genes (102). Genes that contribute to genetic variance in quantitative traits are called QTLs and are detected by allelic association (i.e., a correlation in the population between a phenotype and a particular allele [102]). The Collaborative Study on the Genetics of Alcoholism (COGA) has identified several QTLs associated with the diagnosis of alcoholism by DSM-IV and other criteria. One of the most statistically significant QTLs was located on chromosome 16 (16p13.3-13.2) by use of the microsatellite marker UT5069 (103). The gene for AC7 has been localized to human chromosome 16 but is distant from this QTL (i.e., 16q12-13). However, recently the ninth form of AC was identified (AC type IX) (24). Interestingly, the chromosomal localization of this isoform of the enzyme (24) is in the region of the QTL on chromosome 16 identified by the COGA. Although the presence of AC9 in platelets has not been determined, it was reported to have a broad distribution and to be present in all other tissues tested, including brain (24). It will be of interest to determine the expression of this form of the enzyme in platelets and lymphocytes, and to identify polymorphisms in the gene that may be studied for their possible association with a diagnosis of alcohol dependence.

ACKNOWLEDGMENTS

This work was supported by NIAAA grant #AA09014 and The Banbury Foundation.

REFERENCES

1. Tabakoff B, Hoffman PL, McLaughlin A: Is ethanol a discriminating substance? *Semin Liver Dis* 1988;8:26–35.
2. Chin JH, Goldstein DB: Effects of low concentrations of ethanol on the fluidity of spin-labeled erythrocyte and brain membranes. *Mol Pharmacol* 1977;13:435–441.
3. Franks NP, Lieb WR: Molecular and cellular mechanisms of anesthesia. *Nature* 1994;367:607–614.
4. Luthin GR, Tabakoff B: Activation of adenylate cyclase by alcohol requires the nucleotide-binding protein. *J Pharmacol Exp Ther* 1984;228:579–587.
5. Rabin RA, Bode DC, Molinoff PB: Relationship between ethanol-induced alterations in fluorescence anisotropy and adenylate cyclase activity. *Biochem Pharmacol* 1986;35:2331–2335.
6. Yoshimura M, Tabakoff B: Selective effects of ethanol on the generation of cAMP by particular members of the adenylyl cyclase family. *Alcohol Clin Exp Res* 1995;19:1435–1440.
7. Rall TW: Hypnotics and sedatives; ethanol. In: Gilman AG, Rall TW, Nies AS, Taylor P, eds. *The pharmacologic basis of therapeutics*. 8th ed. New York: Pergamon, 1990;345–382.
8. Rabin RA: Chronic ethanol exposure of PC12 cells alters adenylate cyclase activity and intracellular cyclic AMP content. *J Pharmacol Exp Ther* 1990;252:1021–1027.
9. Gorman RE, Bitensky MW: Selective activation by short chain alcohols of glucagon responsive adenylyl cyclase in liver. *Endocrinology* 1970;87:1975–1081.
10. Saito T, Lee JM, Tabakoff B: Ethanol's effects on cortical adenylate cyclase activity. *J Neurochem* 1985;44:1037–1044.
11. Bode DC, Molinoff PB: Effects of ethanol *in vitro* on the beta adrenergic receptor-coupled adenylate cyclase system. *J Pharmacol Exp Ther* 1988;246:1040–1047.
12. Rabin RA, Molinoff PB: Multiple sites of action of ethanol on adenylate cyclase. *J Pharmacol Exp Ther* 1983;227:551–556.
13. Hoffman PL, Tabakoff B: Ethanol does not modify opiate receptor inhibition of striatal adenylate cyclase. *J Neurochem* 1986;46:812–816.
14. Rabin RA: Effect of ethanol on inhibition of striatal adenylate cyclase activity. *Biochem Pharmacol* 1985;34:4329–4331.
15. Clapham DE, Neer EJ: New roles for G-protein βγ-dimers in transmembrane signalling. *Nature* 1993;365:403–406.
16. Bauché F, Bourdeaux-Jaubert AM, Giudicelli Y, Nordmann R: Ethanol alters the adenosine receptor-N_i-mediated adenylate cyclase inhibitory response in rat brain cortex *in vitro*. *FEBS Lett* 1987;219:296–300.
17. Spiegel A, Carter A, Brann M et al: Signal transduction by guanine nucleotide-binding proteins. *Recent Prog Horm Res* 1987;44:337–375.
18. Contreras ML, Wolfe BB, Molinoff PB: Thermodynamic properties of agonist interactions with the beta adrenergic receptor-coupled adenylate cyclase system. I. High- and low-affinity states of agonist binding to membrane-bound beta adrenergic receptors. *J Pharmacol Exp Ther* 1986;237:154–164.
19. Valverius P, Hoffman PL, Tabakoff B: Effect of ethanol on mouse cerebral cortical β-adrenergic receptors. *Mol Pharmacol* 1987;32:217–222.
20. Spence S, Houslay MD: The role of G_i and the membrane-fluidizing agent benzyl alcohol in modulating the hysteretic activation of human platelet adenylate cyclase by guanylyl 5'-imidodiphosphate. *Biochem J* 1993;291:945–949.
21. Iyengar R, Birnbaumer L: Hysteretic activation of adenylyl cycloases. I. Effect of Mg ion on the rate of activation by guanine nucleotides and fluoride. *J Biol Chem* 1981;256:11036–11041.

22. Tang W-J, Iñiguez-Lluhi JA, Mumby S, Gilman AG: Regulation of mammalian adenylyl cycloases by G-protein α and βγ subunits. *Cold Spring Harbor Symp Quant Biol* 1992;57:135–144.
23. Sunahara RK, Dessauer CW, Gilman AG: Complexity and diversity of mammalian adenylyl cyclases. *Annu Rev Pharmacol Toxicol* 1996;36:461–480.
24. Premont RT, Matsuoka I, Mattei M-G et al: Identification and characterization of a widely expressed form of adenylyl cyclase. *J Biol Chem* 1996;271:13900–13907.
25. Iyengar R: Molecular and functional diversity of mammalian G_s-stimulated adenylyl cyclases. *FASEB J* 1993;7:768–775.
26. Yoshimura M, Cooper DMF: Cloning and expression of a Ca^{2+}-inhibitable adenylyl cyclase from NCB-20 cells. *Proc Natl Acad Sci U S A* 1992;89:6716–6720.
27. Yoshimura M, Cooper DMF: Type-specific stimulation of adenylylcyclase by protein kinase C. *J Biol Chem* 1993;268:4604–4607.
28. Bourne HR, Nicoll R: Molecular machines integrate coincident synaptic signals. *Cell* 1993;72:65–75.
29. McHugh Sutkowski E, Tang W-J, Broome CW et al: Regulation of forskolin interactions with type I, II, V, and VI adenylyl cyclases by $G_{s\alpha}$. *Biochemistry* 1994;33:12852–12859.
30. Yoshimura M, Ikeda I, Tabakoff B: μ-opioid receptors inhibit dopamine-stimulated activity of type V adenylyl cyclase but enhance dopamine-stimulated activity of type VII adenylyl cyclase. *Mol Pharmacol* 1996;50:43–51.
31. Hellevuo K, Yoshimura M, Mons N et al: The characterization of a novel human adenylyl cyclase which is present in brain and other tissues. *J Biol Chem* 1995;270:11581–11589.
32. Kawabe J-i, Ebina T, Toya Y et al: Regulation of type V adenylyl cyclase by PMA-sensitive and -insensitive protein kinase C isoenzymes in intact cells. *FEBS Lett* 1996;384:273–276.
33. Snell LD, Tabakoff B, Hoffman PL: Involvement of protein kinase C in ethanol-induced inhibition of NMDA receptor function in cerebellar granule cells. *Alcohol Clin Exp Res* 1994;18:81–85.
33a. Rabbani M, Tabakoff B: The involvement of protein kinase C in ethanol's actions on adenylyl cyclase. *Alcohol Clin Exp Res* 1997;21:40A.
34. Mons N, Yoshimura M, Cooper DMF: Discrete expression of Ca^{2+}/calmodulin-sensitive and Ca^{2+}-insensitive adenylyl cyclases in the rat brain. *Synapse* 1993;14:51–59.
35. Lin AM-Y, Freund RK, Palmer MR: Sensitization of γ-aminobutyric acid-induced depressions of cerebellar Purkinje neurons to the potentiative effects of ethanol by beta adrenergic mechanisms in rat brain. *J Pharmacol Exp Ther* 1993;265:426–432.
36. Nagy LE, Diamond I, Gordon AS: cAMP-dependent protein kinase regulates inhibition of adenosine transport by ethanol. *Mol Pharmacol* 1991;40:812–817.
37. Tabakoff B, Hoffman PL: Alcohol addiction: an enigma among us. *Neuron* 1996;16:909–912.
38. Freund RK, Palmer MR: 8-Bromo-cAMP mimics β-adrenergic sensitization of GABA responses to ethanol in cerebellar Purkinje neurons *in vivo*. *Alcohol Clin Exp Res* 1996;20:408–412.
38a. Mons N, Yoshimura M, Ikeda H, Tabakoff B: Immunological detection of the ethanol sensitive type 7 adenylyl cyclase. *Alcohol Clin Exp Res* 1997;21:39A.
38b. Mons N, Yoshimura M, Ikeda H, Hoffman PL, Tabakoff B: Immunological assessment of the distribution of type VII adenylyl cyclase in brain. Submitted.
39. Tabakoff B, Hoffman PL: Development of functional dependence on ethanol in dopaminergic systems. *J Pharmacol Exp Ther* 1979;208:216–222.
40. Saito T, Lee JM, Hoffman PL, Tabakoff B: Effects of chronic ethanol treatment on the β-adrenergic receptor-coupled adenylate cyclase system of mouse cerebral cortex. *J Neurochem* 1987;48:1817–1822.
41. Banerjee SP, Sharma VK, Khanna JM: Alterations in beta-adrenergic receptor binding during ethanol withdrawal. *Nature* 1978;276:407–409.
42. Tabakoff B, Whelan JP, Ovchinnikova L et al: Quantitative changes in G proteins do not mediate ethanol-induced downregulation of adenylyl cyclase in mouse cerebral cortex. *Alcohol Clin Exp Res* 1995;19:187–194.
43. Clark RB: Desensitization of hormonal stimuli coupled to regulation of cyclic AMP levels. *Adv Cyclic Nucleotide Protein Phosphor Res* 1986;20:155–209.
44. Valverius P, Hoffman PL, Tabakoff B: Hippocampal and cerebellar β-adrenergic receptors and adenylate cyclase are differentially altered by chronic ethanol ingestion. *J Neurochem* 1989;52:492–497.
45. Rabin RA, Baker RC, Deitrich RA: Effects of chronic ethanol exposure on adenylate cyclase activities in the rat. *Pharmacol Biochem Behav* 1987;26:693–697.

46. Wand GS, Diehl AM, Levine MA et al: Chronic ethanol treatment increases expression of inhibitory G-proteins and reduces adenylylcyclase activity in the central nervous system of two lines of ethanol-sensitive mice. *J Biol Chem* 1993;268:2595–2601.

47. Wand GS, Levine MA: Hormonal tolerance to ethanol is associated with decreased expression of the GTP-binding protein, $G_{s\alpha}$ and adenylyl cyclase activity in ethanol-treated mice. *Alcohol Clin Exp Res* 1991;15:705–710.

48. Diehl AM, Yang SQ, Cote P, Wand GS: Chronic ethanol consumption disturbs G-protein expression and inhibits cyclic AMP-dependent signalling in regenerating rat liver. *Hepatology* 1992;16:1212–1219.

49. Alvaro-Alfonso I, del Carmen Boyano-Adánez M, Arilla E: Ethanol-induced modification of somatostatin-responsive adenylyl cyclase in rat exocrine pancreas. *Biochim Biophys Acta* 1995;1268:115–121.

50. Gordon AS, Collier K, Diamond I: Ethanol regulation of adenosine receptor-stimulated cAMP levels in a clonal neural cell line: an *in vitro* model of cellular tolerance to ethanol. *Proc Natl Acad Sci U S A* 1986;83:2105–2108.

51. Richelson E, Stenstrom S, Forrary C et al: Effects of chronic exposure to ethanol on the prostaglandin E_1 receptor–mediated response and binding in a murine neuroblastoma clone (NIE-115). *J Pharmacol Exp Ther* 1986;239:687–692.

52. Nagy IE, Diamond I, Collier K et al: Adenosine is required for ethanol-induced heterologous desensitization. *Mol Pharmacol* 1989;36:744–748.

53. Bode DC, Molinoff PB: Effects of chronic exposure to ethanol on the physical and functional properties of the plasma membrane of S49 lymphoma cells. *Biochemistry* 1988;27:5700–5705.

54. Rabin RA: Ethanol-induced desensitization of adenylate cyclase: role of the adenosine receptor and GTP-binding proteins. *J Pharmacol Exp Ther* 1993;264:977–983.

55. Rabin RA: Differential response of adenylate cyclase and ATPase activities after chronic ethanol exposure of PC12 cells. *J Pharmacol Exp Ther* 1988;51:1148–1155.

56. Charness ME, Querimit LA, Henteleff M: Ethanol differentially regulates G proteins in neural cells. *Biochem Biophys Res Commun* 1988;155:138–143.

57. Blumenthal RS, Flinn IW, Proske O et al: Effects of chronic ethanol exposure on cardiac receptor-adenylate cyclase coupling: studies in cultured embryonic chick myocytes and ethanol fed rats. *Alcohol Clin Exp Res* 1991;15:1077–1083.

58. Nagy LE, DeSilva SEF: Ethanol increases receptor-dependent cyclic AMP production in cultured hepatocytes by decreasing G_i-mediated inhibition. *Biochem J* 1992;286:681–686.

59. Grabarek J, Raychowdhury M, Ravid K et al: Identification and functional characterization of protein kinase C isozymes in platelets and HEL cells. *J Biol Chem* 1992;267:10011–10017.

60. Reithmann C, Geirschik P, Werdan K, Jakobs KH: Role of inhibitory G protein α-subunits in adenylyl cyclase desensitization. *Mol Cell Endocrinol* 1991;82:C215–C221.

61. Milligan G, Green A: Agonist control of G-protein levels. *TIPS* 1991;12:207–209.

62. Mochly-Rosen D, Chang F-H, Cheever L et al: Chronic ethanol causes heterologous desensitization of receptors by reducing α_s messenger RNA. *Nature* 1988;333:848–850.

63. Nagy LE, Diamond I, Casso DJ et al: Ethanol increases extracellular adenosine by inhibiting adenosine uptake via the nucleoside transporter. *J Biol Chem* 1990;265:1946–1951.

64. Williams RJ, Veale MA, Horne P, Kelly E: Ethanol differentially regulates guanine nucleotide-binding protein α subunit expression in NG108–15 cells independently of extracellular adenosine. *Mol Pharmacol* 1993;43:158–166.

65. Rabin RA, Fiorella D, Van Wylen DGL: Role of extracellular adenosine in ethanol-induced desensitization of cyclic AMP production. *J Neurochem* 1993;60:1012–1017.

66. Nakamura J: Role of protein synthesis on ethanol regulation of adenylyl cyclase activity in wild-type S49 murine lymphoma cells. *Alcohol Clin Exp Res* 1996;20:302–306.

67. Milligan G: The stoichiometry of expression of protein components of the stimulatory adenylyl cyclase cascade and the regulation of information transfer. *Cell Signal* 1996;8:87–96.

68. Hadcock JR, Ros M, Watkins DC, Malbon CC: Cross-regulation between G-protein-mediated pathways. Stimulation of adenylyl cyclase increases expression of the inhibitory G-protein, G_i alpha 2. *J Biol Chem* 1990;265:14784–14790.

69. Schreibmayer W, Dessauer CW, Vorobiov D et al: Inhibition of an inwardly rectifying K^+ channel by G-protein α-subunits. *Nature* 1996;380:624–627.

70. Johnson GL, Gardner AM, Lange-Carter C et al: How does the G protein, G_{i2}, transduce mitogenic signals? *J Cell Biochem* 1994;54:415–422.

71. Herlitze S, Garcia DE, Mackle K et al: Modulation of Ca^{2+} channels by G-protein $\beta\gamma$ subunits. *Nature* 1996;380:258–262.
72. Pellegrino SM, Woods JM, Druse MJ: Effects of chronic ethanol consumption on G proteins in brain areas associated with the nigrostriatal and mesolimbic dopamine systems. *Alcohol Clin Exp Res* 1993;17:1247–1253.
73. Yan K, Greene E, Belga F, Rasenick MM: Synaptic membrane G proteins are complexed with tubulin *in situ. J Neurochem* 1996;66:1489–1495.
74. Popova JS, Johnson GL, Rasenick MM: Chimeric $G_{\alpha s}/G_{\alpha i2}$ proteins define domains on $G_{\alpha s}$ which interact with tubulin for the β adrenergic activation of adenylyl cyclase. *J Biol Chem* 1994;269: 21748–21754.
75. Chen J, Rasenick MM: Chronic antidepressant treatment facilitates G protein activation of adenylyl cyclase without altering G protein content. *J Pharmacol Exp Ther* 1995;275:509–517.
76. Whatley VJ, Brozowski SJ, Hadingham KL et al: Microtubule depolymerization inhibits ethanol-induced enhancement of $GABA_A$ responses in stably transfected cells. *J Neurochem* 1996;66: 1318–1321.
77. Valverius P, Hoffman PL, Tabakoff B: Brain forskolin binding in mice dependent on and tolerant to ethanol. *Brain Res* 1989;503:38–43.
78. Nestler EJ: Molecular mechanisms of drug addiction. *J Neurosci* 1992;12:2439.
79. Matsuoka I, Maldonado R, Defer N et al: Chronic morphine administration causes region-specific increase of brain type VIII adenylyl cyclase mRNA. *Eur J Pharmacol* 1994;268:215–221.
80. Levitzki A: Regulation of hormone-sensitive adenylate cyclase. *TIPS* 1987;8:299–303.
81. Iwami G, Kawabe J-i, Ebina T et al: Regulation of adenylyl cyclase by protein kinase A. *J Biol Chem* 1995;270:12481–12484.
82. Hellevuo K, Hoffman PL, Tabakoff B: Adenylyl cyclases: mRNA and characteristics of enzyme activity in three areas of brain. *J Neurochem* 1996;67:177–185.
83. Rabin RA, Edelman AM, Wagner JA: Activation of protein kinase A is necessary but not sufficient for ethanol-induced desensitization of cyclic AMP production. *J Pharmacol Exp Ther* 1992;262: 257–262.
84. Premont RT, Jacobowitz O, Iyengar R: Lowered responsiveness of the catalyst of adenylyl cyclase to stimulation by GS in heterologous desensitization: a role for adenosine 3',5'-monophosphate-dependent phosphorylation. *Endocrinology* 1992;131:2774–2784.
85. Watson PA, Krupinski J, Kempinski AM, Frankenfield CD: Molecular cloning and characterization of the type VII isoform of mammalian adenylyl cyclase expressed widely in mouse tissues and in S49 mouse lymphoma cells. *J Biol Chem* 1994;269:28893–28898.
86. Szabó G, Hoffman PL, Tabakoff B: Forskolin promotes the development of ethanol tolerance in 6-hydroxydopamine-treated mice. *Life Sci* 1988;42:615–621.
87. Valverius P, Hoffman PL, Tabakoff B: β-adrenergic receptor binding in brain of alcoholics. *Exp Neurol* 1988;105:280–286.
88. Ozawa H, Katamura Y, Hatta S et al: Alterations of guanine nucleotide-binding proteins in post-mortem human brain in alcoholics. *Brain Res* 1993;620:174–179.
89. Tabakoff B, Hoffman PL, Lee JM et al: Differences in platelet enzyme activity between alcoholics and nonalcoholics. *N Engl J Med* 1988;318:134–139.
90. Diamond I, Wrubel B, Estrin W, Gordon A: Basal and adenosine receptor-stimulated levels of cAMP are reduced in lymphocytes in alcoholic patients. *Proc Natl Acad Sci U S A* 1987;84: 1413–1416.
91. Waltman C, Levine MA, McCaul ME et al: Enhanced expression of the inhibitory protein $G_{i2\alpha}$ and decreased activity of adenylyl cyclase in lymphocytes of abstinent alcoholics. *Alcohol Clin Exp Res* 1993;17:315–320.
92. Lichtenberg-Kraag B, May T, Schmidt LG, Rommelspacher H: Changes of G-protein levels in platelet membranes from alcoholics during short-term and long-term abstinence. *Alcohol Alcoholism* 1995;30:455–464.
93. Parsian A, Todd RD, Cloninger CR et al: Platelet adenylyl cyclase activity in alcoholics and subtypes of alcoholics. *Alcohol Clin Exp Res* 1996;20:745–751.
94. Nagy LE, Diamond I, Gordon A: Cultured lymphocytes from alcoholic subjects have altered cAMP signal transduction. *Proc Natl Acad Sci U S A* 1988;85:6973–6976.
95. Saito T, Katamura Y, Ozawa H et al: Platelet GTP-binding protein in long-term abstinent alcoholics with an alcoholic first-degree relative. *Biol Psychiatry* 1994;36:495–497.
96. Lex BW, Ellingboe J, LaRosa K et al: Comparison of platelet monoamine oxidase and adenylate cy-

clase activities in female alcoholics and in non-alcoholic control women with and without a family history of alcoholism. *Harvard Rev Psych* 1993;1:229–237.

97. Devor EJ, Cloninger CR, Hoffman PL, Tabakoff B: A genetic study of platelet adenylate cyclase activity: evidence for a single major locus effect in fluoride-stimulated activity. *Am J Hum Genet* 1991;49:372–377.

98. Hodge SE: What association analysis can and cannot tell us about the genetics of complex disease. *Am J Med Genet* 1994;54:318–323.

99. Cloninger CR: Neurogenetic adaptive mechanisms in alcoholism. *Science* 1987;236:410–416.

100. Hellevuo K, Yoshimura M, Kao M et al: A novel adenylyl cyclase sequence cloned from the human erythroleukemia cell line. *Biochem Biophys Res Commun* 1993;192:311–318.

101. Hellevuo K, Welborn R, Menninger JA, Tabakoff B: Human adenylyl cyclase type 7 (ADCY7) contains polymorphic repeats in the 3'untranslated region. *Am J Med Gen* (*Neuropsych Gen*) 1997;47:95–98.

102. Plomin R, Owen MJ, McGuffin P: The genetic basis of complex human behaviors. *Science* 1994;264:1733–1739.

103. Reich T, Begleiter H, et al: A genomic survey of alcohol dependence and related phenotypes:results from the collaborative study on the genetics of alcoholism (COGA). *Alcohol Clin Exp Res* 1996;20(8 Suppl):133A–137A.

*Advances in Second Messenger and
Phosphoprotein Research,* Vol. 32,
edited by Dermot M. F. Cooper
Lippincott–Raven Publishers, Philadelphia © 1998

10

Simultaneous Fluorescence Ratio Imaging of Cyclic AMP and Calcium Kinetics in Single Living Cells

Maria A. DeBernardi and Gary Brooker

*Department of Cell Biology, Georgetown University School of Medicine,
Washington, DC 20007; and Atto Instruments, Inc., Rockville, Maryland 20850*

CALCIUM AND cAMP SECOND MESSENGER CROSS-TALK

A large body of experimental evidence is now available that indicates a crucial role for signal transduction cross-talk in integrating multiple extracellular signals and translating them into a final biologic response. In particular, two major intracellular second messengers, calcium (Ca^{2+}) and cyclic adenosine monophosphate (cAMP), have been shown to modulate each other, either synergistically or antagonistically, in a variety of cell systems. Interactions between these two signal transduction pathways can occur at different levels: (i) gene transcription: the promoter transcriptional activity of Ca^{2+}- and cAMP-inducible genes can be regulated by a common *trans*-acting factor, CREB (1); (ii) Ca^{2+} and cAMP, as second messengers, can modulate each other's intracellular levels (examples are provided in but not limited to refs. 2–17); and (iii) Ca^{2+} and cAMP-evoked cell responses can either reinforce or counteract each other to produce a final response that best fits the biologic dynamics of the system (e.g., glycogenolysis, cardiac muscle contraction).

CALCIUM INHIBITION OF cAMP ACCUMULATION IN C6–2B CELLS

We have a long-standing interest in the interactions between Ca^{2+} and cAMP as second messengers, and we have been studying the regulation of cAMP accumulation by Ca^{2+} in an *in vitro* model; namely, rat glioma C6–2B cells. In this cell line, activation of receptors that promote the hydrolysis of phosphoinositol 4,5-biphosphate (PI), with consequent formation of diacylglycerol (DAG) and 1,4,5-inositol trisphosphate (IP_3), (i) increases the cytosolic free Ca^{2+} concentrations ($[Ca^{2+}]_i$) by evoking Ca^{2+} release from intracellular stores, and (ii) potently inhibits cAMP accumulation elicited by the β-adrenergic receptor agonist, isoproterenol, or postreceptor adenylyl cyclase (AC) agonists, such as cholera toxin and forskolin (4,8–10). This negative effect of Ca^{2+} on cAMP formation is mediated neither by an increased inhibitory input on the AC through a pertussis toxin–sensitive $G_{i/o}$ protein or an enhanced cAMP degradation, nor by protein kinase C (PKC), the other final effector activated by DAG upon PI hydrolysis (4,8,9). However, sequestration of intracellular Ca^{2+} or blockade of Ca^{2+} mobilization from internal stores totally abolishes the inhibition of agonist-stimulated cAMP accumulation in C6–2B cells (8). Similar to PI-coupled receptor agonists, the microsomal Ca^{2+}–adenosine triphosphatase (ATPase) inhibitor, non–phorbol ester tumor promoter, thapsigargin (TG), which increases $[Ca^{2+}]_i$ without activating PI hydrolysis or PKC (8 and refs. within), potently inhibits C6–2B cell cAMP accumulation (8,10,16). Ca^{2+} ionophores also severely impair the cAMP response in these cells (4).

The preceding findings strongly suggest that the increased $[Ca^{2+}]_i$ is the primary mediator of the inhibition of cAMP accumulation in C6–2B cells. The inhibitory effect of submicromolar concentrations of Ca^{2+} on the AC activity of digitonin-permeabilized C6–2B cells (4) provides compelling evidence in support of a "direct" negative effect of Ca^{2+} upon the enzyme synthesizing cAMP in these cells. Interestingly, C6–2B cells express almost exclusively the Ca^{2+}-inhibitable type VI AC (AC6), along with trace amounts of type III (AC3) (10). Because AC6, whose activity is inhibited by submicromolar Ca^{2+} concentrations (18), is very abundant in heart and other cell types that display a Ca^{2+}-inhibited cAMP accumulation (18,19), it is very likely that this isoform is the target site for the Ca^{2+} regulation of the cAMP metabolic pathway in these systems (20). C6–2B cells then are an ideal model for studying the Ca^{2+} regulation of agonist-stimulated cAMP accumulation because virtually all cAMP produced, being synthesized by a Ca^{2+}-inhibitable AC, can be controlled by Ca^{2+}-mobilizing signals.

SINGLE-CELL IMAGING MEASUREMENTS OF CALCIUM AND cAMP

Until recently, all the studies involving measurement of cAMP levels in cells or tissues have relied upon the use of extracts or homogenates where the cAMP content was determined by a variety of methods such as radioimmunoassay (21), incorporation of radioactive adenine to label the ATP pool (22), and competitive binding to cAMP-dependent protein kinase A (PKA) (23). Any of these methods requires a large number of cells or amount of tissue to be destroyed for each time point studied, whereas no information on the intracellular compartmentation cAMP is provided. An analogous situation existed for the measurement of cytosolic Ca^{2+} before fluorescent Ca^{2+} indicators were developed. Immunocytochemistry has been used to address the issue of intracellular localization of cAMP in microwave-fixed tissues and cells (24). However, because (i) the mechanism of cross-linking of cAMP by microwaves is not known and (ii) several steps are required for visualization of the nucleotide, the results on cAMP distribution on such fixed specimens are the subject of questions about "redistribution" of cAMP during sample fixation and processing.

Because fluorescence is usually highly specific, extremely sensitive, and amenable to microscopic detection, fluorescent probes for intracellular second messengers are now becoming the tools of choice to study cell signal transduction *in vivo*. Cell-permeable optical probes for Ca^{2+} are now commonly used and, together with newly developed real-time imaging hardware and software, have allowed researchers to investigate the spatial (gradients of cytosolic $[Ca^{2+}]_i$; e.g., ref. 25) and temporal (oscillation in $[Ca^{2+}]_i$; e.g., ref. 26) dynamics of the Ca^{2+} signaling pathway in response to steady stimuli in single living cells from different sources. Among the fluorescent Ca^{2+} indicators, the most widely used is Fura-2 (27), a dual-excitation, single-emission dye whose excitation spectrum changes as a function of the cytosolic free Ca^{2+} concentration.

Upon the recent development of FlCRhR (cAMP Fluorosensor), the first cAMP probe available for the nondestructive measurement of cAMP in living cells (28) by digital video imaging, confocal microscopy and photometry, cAMP levels and intracellular distribution have been studied in a limited number of systems. It has thus been possible to follow in "real time" and continuously the dynamics of this important second messenger *in vivo* (28–36). The sensor consists of recombinant murine cAMP-dependent PKA in which the catalytic (C_α) and regulatory (R_{II}) subunits are labeled with a different fluorescent dye, fluorescein and rhodamine, respectively (28). In the holoenzyme configuration (C_2R_2), which represents inactive PKA, the proximity of the two fluorophores allows fluorescence resonance energy transfer (FRET) to occur, such that, when excited with blue light (488 nm), fluorescein transfers its energy (520 nm) to rhodamine, which emits orange light (>570 nm). As it occurs in the native PKA, when cAMP binds to the R subunit of FlCRhR, the subunits reversibly dissociate, the fluorophores become separated by an effectively infinite distance, and FRET therefore no longer occurs. Consequently, C-bound fluorescein now fully emits green light (thus the 520 nm emission signal is enhanced), whereas R-bound rhodamine, no longer excited by fluorescein, emits less energy (thus the >570 nm emission signal is reduced). Therefore, increases in intracellular free cAMP concentration ($[cAMP]_i$) in living cells microinjected with FlCRhR can be quantitatively measured as a change in the 520 nm and >570 nm emission wavelengths of the probe. It is important to note that because of the nature of the probe, upon elevation of $[cAMP]_i$ and dissociation of the subunits, the intracellular localization and possible migration of the individual FlCRhR subunits (fluorescein-labeled C_α and rhodamine-labeled R_{II}) can be investigated in living cells (28,30–34,37–41). In experiments where FlCRhR-labeled cells are treated with cAMP-elevating agents and imaged after 20–30 minutes, under 488 nm excitation light, the 520 nm emission signal generated by C-bound fluorescein localizes virtually only in the nucleus, reflecting the translocation of the C subunit to the nucleus (an example is shown in Fig. 3, lower panel; please see legend). Because FlCRhR is a *dual-emission single-excitation* dye, the analysis of $[cAMP]_i$ in single living cells is performed by *emission ratio* imaging. Likewise, because Fura-2 is a *dual-excitation single-emission* dye, analysis of $[Ca^{2+}]_i$ in living cells is performed by *excitation ratio* imaging. The theoretical advantage of ratio imaging is that the quotient of the two images gives a measure of cAMP (or Ca^{2+} or any other molecule being imaged) that is independent of the concentration of the dye, optical path length, thickness of the sample, absolute brightness of the excitation source, or absolute sensitivity of the detection.

SIMULTANEOUS FLUORESCENCE RATIO IMAGING OF Ca^{2+} AND cAMP IN SINGLE LIVING CELLS

Until very recently, the analyses of resting and stimulated $[Ca^{2+}]_i$ and $[cAMP]_i$ have been performed on separate biologic samples, both in cell/tissues extracts and in single living cells. Thus the simultaneous evaluation of the changes in the dynamics of either second messenger in response to external stimuli, or even in response to each other, has never been possible. Because it has become clear over the last several years that cross-talk between second messengers is responsible for translating multiple external signals into a final biologic response, the simultaneous measurement of two ubiquitously important second messengers such as Ca^{2+} and cAMP is highly desirable and likely to provide precious information on the intracellular biochemistry of these two regulatory molecules. As mentioned earlier, we have been interested in the negative regulation by Ca^{2+} of the cAMP metabolic pathway in C6–2B cells. Radioimmunoassay on extracts from populations of cells and Fura-2 imaging in single cells were used previously by us to separately measure cAMP and Ca^{2+}, respectively, in a number of cell types (4,8–10,42,43). By combining digital fluorescence ratio imaging of Ca^{2+} and cAMP with Fura-2 and FlCRhR, we have now investigated the spatial and temporal dynamics of both Ca^{2+} and cAMP at virtually the same time and in the same living cell (44). The methodology of simultaneous fluorescence ratio imaging of Ca^{2+} and cAMP as well as the results obtained in C6–2B cells and REF-52 rat embryonic fibroblasts are presented.

Cell Labeling and Fluorescence Microscopy-Imaging Configuration

C6–2B cells and REF-52 fibroblasts were grown on 25-mm round glass coverslips. Because FlCRhR is a protein complex with an aggregate molecular mass of 172 kDa, it needs to be pressure microinjected into individual cells. FlCRhR (cAMP Fluorosensor) is a product of Atto Instruments, Inc. (Rockville, MD), which can now be obtained from Molecular Probes (Eugene, OR). FlCRhR (average stock concentration: 20 μM) is injected into the cell cytoplasm where it reaches a final concentrations of 0.2–2 μM considering an injection volume of 1–10% of the cell volume (30). Microinjection was performed while bathing the cells in growth medium and using an Eppendorf Transjector 5246-Micromanipulator 5171 system (Eppendorf, Madison, WI) with a starting injection pressure of about 100–150 hPa and a capillary holding pressure of about 50–75 hPa. Both parameters need to be adjusted, depending on the pipet tip diameter, concentration of FlCRhR to be injected, cell resistance, and so on. We have found that when a minimum but continuous flow of the dye from the tip of the microinjection needle is allowed, FlCRhR microinjection is easier and pipet clogging is greatly minimized. One-millimeter thin-walled borosilicate glass with microfilament for back-filling (such as TW 100F capillaries from World Precision Instruments, Inc., Sarasota, FL) is suitable for preparing microinjection needles. A Flaming/Brown horizontal-type pipet puller, such as Model P-97 from Sutter Instrument Co. (Novato, CA), gives reproducible micropipets and offers the possibility of a variety of pulling programs to better fit the individual cell type injection needs. For general-purpose microinjection of cells >20 μm in diameter, a pipet tip opening of about 0.5 μm is usually appropriate. Although thinner tips would be preferable, concentrated FlCRhR tends to clog them more easily (30). However, depending upon the size and the sensitivity to the injection procedure of the cell type being studied, the tip opening needs to be adjusted accordingly. Premade micropipets, such as Femtotips (Eppendorf; opening diameter = 0.5 μm $\pm$ 0.2 μm) might be an alternative, although expensive, to home-made microinjection pipets. Microloaders (Eppendorf) are needed to load Femtotips whereas standard microtips are suitable for back-loading custom pulled glass pipets. With both types of pipets, a volume of <0.5 μl FlCRhR is generally used to load the tip, and with optimal microinjection parameters, an average of 100 cells can be injected with such amounts of dye. To minimize possible photobleaching of FlCRhR during the microinjection procedure, we use a >600-nm long pass red filter positioned above the microscope condenser to block wavelengths <600 nm from reaching the sample (FlCRhR excitation = 488 nm). If desired, FlCRhR-labeled cells, either in resting conditions or upon drug treatments, can be fixed and stored for later microscopic analysis. In this case, the coverslips are rinsed thoroughly with phosphate buffer solution, fixed with 100% ethyl alcohol for 10 minutes at room temperature, rehydrated with phosphate buffer solution, and mounted onto slides in 20% 0.5 M Tris pH 8.2, 80% glycerol and 2% DABCO (1,4-diazabicyclo-[2.2.2]octane, an antifading agent). Slides are kept at 4°C and protected from light and moisture. For imaging analysis of FlCRhR and Fura-2 in living cells, after microinjection of FlCRhR, the cells were loaded with 5 μM Fura-2AM (Molecular Probes) at room temperature for 30 minutes in serum-free Ham's F-10 medium supplemented with 20 mM Na-HEPES, pH 7.4, washed, and then imaged in the same medium at 22°C, as previously described (42,43). Because Fura-2 is a dual-excitation single-emission dye and FlCRhR is a single-excitation dual-emission dye, simultaneous imaging of Ca^{2+} and cAMP was accomplished by a combination of dual-excitation and dual-emission fluorescence ratio imaging using the Carl Zeiss Attofluor RatioVision workstation (Carl Zeiss Inc., Thornwood, NY). This system is equipped with a Zeiss Axiovert 135 microscope (Fluar 40, 1.3 na, oil immersion objective) and dual-emission Attofluor ICCD cameras. Illumination is provided by a 100-W mercury arc lamp. A Zeiss FT510 dichroic reflector is used to separate the excitation light, which is below 510 nm, from the emission light at >516 nm (Fig. 1). Excitation filters (10 nm bandpass filters of 334, 380, 488 nm) located in the filter changer are automatically and alternately placed into the excitation light path while proper emission wavelengths are monitored with emission filters (510–530 nm bandpass and >570 nm long-pass) inserted in the respective Attofluor ICCD cameras. Dual-emission ICCD cameras are aligned, within one pixel of one another, before each day's experiments using a mixture of fluorescein and rhodamine to mimic the two emission wavelengths generated by FlCRhR. Ca^{2+} and cAMP were simultaneously imaged by (i) exciting Fura-2 at 334 and 380 nm with its

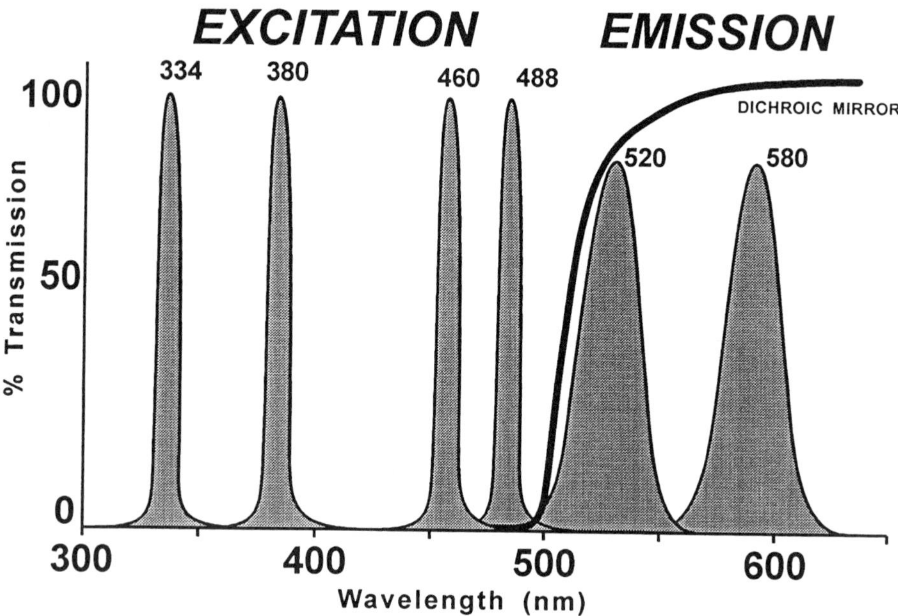

FIG. 1. Spectral characteristic of the excitation and emission filters and dichroic beamsplitter used in the Carl Zeiss Attofluor RatioVision workstation. When imaging Ca^{2+} and cAMP simultaneously, Fura-2 is excited at 334 and 380 nm (with its single emission wavelength monitored at 520 nm) and FlCRhR is excited at 488 nm while the dual emission signals are measured at 520 and 580 nm. In the microscope (Zeiss Axiovert 135), the epi excitation is isolated from the fluorescent emission by a Zeiss FT510 dichroic mirror such that more than 90% of the emission light is directed to the detectors. The 460 nm excitation filter is not used for Ca^{2+} and cAMP imaging described in this paper, but, together with the 488 nm excitation filter, it is used to image BCECF, a fluorescent pH indicator.

emission monitored at 510–530 nm, and (ii) exciting FlCRhR at 488 nm with the emission signals monitored at 510–530 nm and >570 nm, respectively. The 334/380 nm excitation ratio for Fura-2 increases as a function of the $[Ca^{2+}]_i$ whereas the FlCRhR 510–530/>570 nm emission ratio (referred to as 520/580 nm emission ratio) increases upon $[cAMP]_i$ elevation (Fig. 2). The data relative to $[Ca^{2+}]_i$ and $[cAMP]_i$ changes in the same cell were collected simultaneously, although the frequency of sample illumination differed between the two dyes, with FlCRhR excited generally once every 30 seconds while Fura-2 was excited every 5 seconds. The reason for this difference in the frequency of illumination is that fluorescein, the dye used to label the catalytic subunit of FlCRhR, photobleaches quite rapidly under 488 nm excitation light. The development of a second-generation cAMP Fluorosensor, where fluorescein is replaced by a fluorescent dye with improved photostability is thus highly desirable, so that cAMP kinetics could be followed at higher speeds and for longer time periods. In this regard, it needs to be pointed out that, by virtue of the good image resolution and "quietness" (high sensitivity and low noise) of the Attofluor ICCD cameras of the system used in this study, a single-frame image of the cells

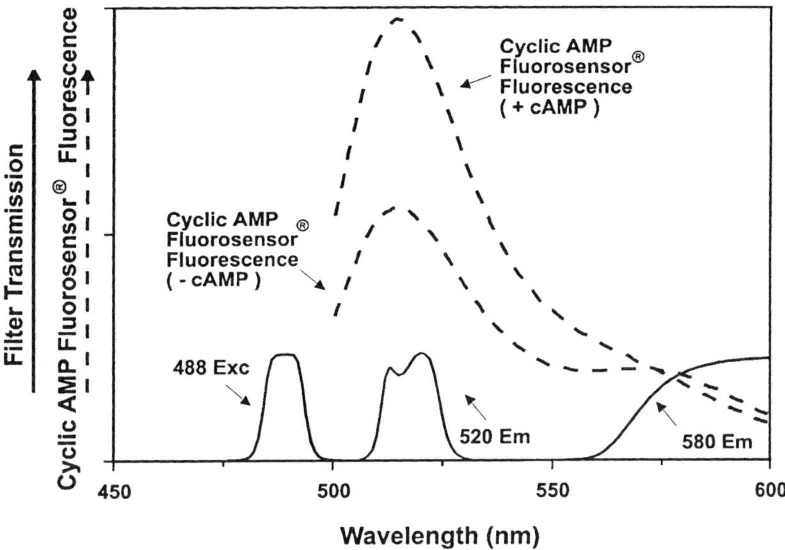

FIG. 2. Fluorescence emission spectra of FlCRhR (cAMP Fluorosensor) in the absence or presence of cAMP and transmission properties of the filters used to image FlCRhR. FlCRhR consists of fluorescently labeled PKA with C-bound fluorescein (emitting at approximately 520 nm) and R-bound rhodamine (absorbing 520 nm light from nearby fluorescein and emitting >570 nm). As it occurs in the native PKA, FlCRhR undergoes dissociation of its labeled subunits upon cAMP binding. This results in a change in the emission spectrum of the dye with increased 520 nm emission and reduced 580 nm emission signals (for details please refer to the text). Since FlCRhR is a single-excitation (488 nm), dual-emission (510–530 nm and >570 nm) dye, cAMP levels are monitored by FlCRhR emission ratio imaging. Here the transmission properties of the filters used to image FlCRhR are shown. The same filter configuration is maintained when performing simultaneous imaging of Ca^{2+} and cAMP with Fura-2 and FlCRhR described in this study.

(1/30 second excitation and image capture) provided sufficient resolution for microscopic analysis. This eliminates the need to average images, thus reducing light exposure and photobleaching of FlCRhR. Calibration for Fura-2 was performed as described (45).

SINGLE-CELL Ca^{2+}/cAMP CROSS-TALK IN C6–2B CELLS

Pivotal studies, performed to assess the feasibility of the Ca^{2+}/cAMP simultaneous fluorescence imaging, demonstrated that the detection of the fluorescence generated by FlCRhR or Fura-2, either in resting or in stimulated state (Figs. 3 and 4), is not appreciably affected by the simultaneous presence of both dyes in the same cell (44). No interference between the fluorescent signals of the two dyes was observed when cells, microinjected with FlCRhR and loaded with Fura-2, were imaged (see Fig. 3). Moreover, the extent of the increase in the FlCRhR 520/580 nm emission ratio evoked by the cAMP-generating agent, isoproterenol (ISO), was virtually identical in C6–2B cells labeled with FlCRhR only (Fig. 4A) and in cells co-labeled with

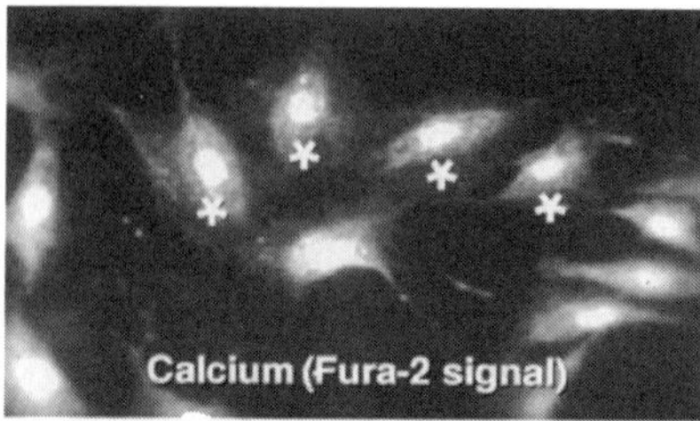

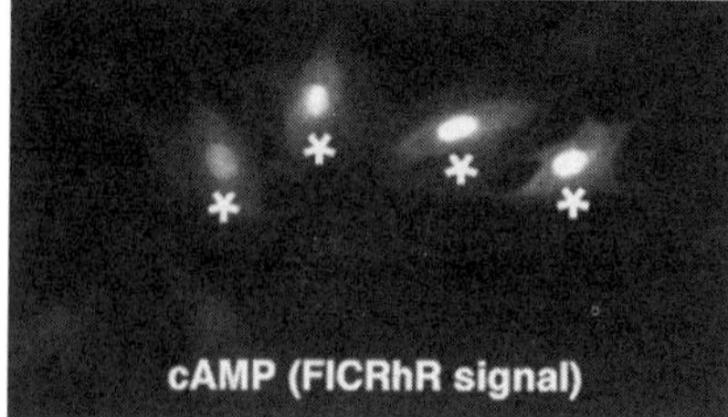

FIG. 3. Simultaneous fluorescence imaging of Ca^{2+} and cAMP in single living REF-52 fibroblasts. Cells were grown on a glass coverslip, loaded with Fura-2, and four of them were microinjected with FICRhR (marked with the asterisk). Cells were analyzed by fluorescence digital imaging microscopy as described in the text. Upper panel: Ca^{2+} is monitored with Fura-2 (334 and 380 nm excitation, 520 nm emission); lower panel: cAMP is monitored with FICRhR (488 nm excitation, 520 and 580 nm emission). Shown are 520 nm emission images (with Fura-2 excited at 334 nm and FICRhR excited at 488 nm) taken 20 minutes after addition of the cAMP-generating, adenylyl cyclase stimulator, forskolin (50 μM), which induced changes in the fluorescence spectrum of FICRhR only. Similarly, the Ca^{2+} ionophore, ionomycin (5 μM) added 5 minutes after forskolin, affected Fura-2 fluorescent signal only. Notice that each dye can be viewed independently of the other, even though both are present in the marked cells. The FICRhR image also shows the nuclear translocation of the 520 nm-emitting fluorescein-labeled catalytic subunit of the dye, a result of cAMP action *in vivo*.

FICRhR and Fura-2 (Fig. 4B). Likewise, the change in the Fura-2 334/380 nm excitation ratio mediated by the ionomycin-induced $[Ca^{2+}]_i$ increase was comparable both in kinetics and magnitude in cells labeled with Fura-2 only (Fig. 4C) and in cells co-labeled with Fura-2 and FICRhR (Fig. 4D). Single-cell cAMP imaging revealed that individual cells respond to cAMP-elevating agents in a heterogeneous fashion, similar to what we previously reported for Ca^{2+} responses. It is important to emphasize that because (i) FICRhR is a dual-emission dye and (ii) the change in its emission ratio upon cAMP formation is relatively small (average 40% increase in 520/580 nm ratio) when compared with the Ca^{2+}-mediated increase in Fura-2 excitation ratio (average 200% increase in 334/380 nm ratio) (compare *y* axis values in Fig. 4), the two emission cameras used for the detection of FICRhR fluorescence changes have to be perfectly registered and very sensitive with a high signal-to-noise ratio.

Previous findings (4,8–10,16) have shown that, in C6–2B cells, agents promoting increased $[Ca^{2+}]_i$ inhibit agonist-induced cAMP accumulation measured in cell extracts. In these cells, for example, a short exposure to TG ($IC_{50} = 19.3 \pm 0.2$ nM) would reduce by about 60% the cAMP formation stimulated by a simultaneous or subsequent challenge with isoproterenol (ISO) or forskolin (FO) (8–10). A similar experimental protocol was used for Ca^{2+}/cAMP simultaneous fluorescence studies in living C6–2B cells (44). Interestingly, single-cell Fura-2 imaging had revealed a heterogeneous Ca^{2+} response to TG such that, among the cells being imaged in a given field, some would show a massive increase in $[Ca^{2+}]_i$ whereas others would exhibit only a modest $[Ca^{2+}]_i$ rise. We exploited this feature to investigate the regu-

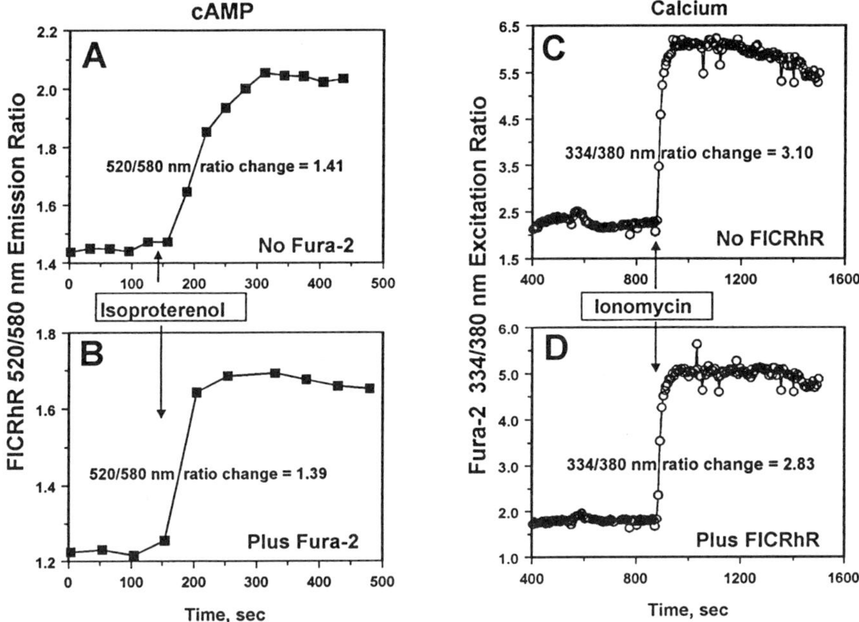

FIG. 4. Simultaneous cAMP/Ca^{2+} imaging feasibility studies. (**A**) Single C6–2B cells were microinjected with FlCRhR and imaged as described in the text. After the baseline FlCRhR fluorescence emission ratio (indicating basal $[cAMP]_i$) was recorded, the β-adrenergic receptor agonist isoproterenol (ISO, 10 μM) was added. The elevation in $[cAMP]_i$ elicited by ISO resulted in a 1.41-fold increase in FlCRhR 520/580 nm emission ratio. When a parallel set of cells microinjected with the same batch of FlCRhR was also labeled with Fura-2 (**B**), ISO-induced cAMP accumulation resulted in a virtually identical, 1.39-fold, increase in FlCRhR 520/580 nm emission ratio. (**C**) C6–2B cells labeled with Fura-2 were challenged with ionomycin (IONO, 1 μM), which evoked a 3.10-fold increase in the Fura-2 334/380 nm excitation ratio. In cells co-labeled with Fura-2 and FlCRhR (**D**), a comparable (2.83-fold) increase in the Fura-2 excitation ratio was observed upon IONO treatment. (Adapted with permission from DeBernardi MA, Brooker G: Single cell Ca^{2+}/cAMP cross-talk monitored by simultaneous Ca^{2+}/cAMP fluorescence ratio imaging. *Proc Natl Acad Sci U S A* 1996;93:4577–4582.) Although the extent of the increase in Fura-2 excitation ratio in cells co-labeled with FlCRhR and Fura-2 was found to be overall slightly smaller than that in cells labeled with Fura-2 only, the kinetics of the Ca^{2+} responses were virtually identical in both cell populations in all the experiments. In this and following figures, the experimental results showing single-cell $[cAMP]_i$ and $[Ca^{2+}]_i$ are presented as population means (2–10 cells being imaged per field) of the FlCRhR 520/580 nm emission ratio and Fura-2 334/380 nm excitation ratio, respectively, plotted against time.

lation by Ca^{2+} of C6–2B cell cAMP production by comparing the single-cell cAMP responses in a population of cells being exposed to FO and TG and imaged at the same time. As shown in Fig. 5, an inverse relationship between TG-elevated $[Ca^{2+}]_i$ and FO-stimulated cAMP accumulation was observed. Among the cells followed throughout the experiment, those that displayed a lower $[Ca^{2+}]_i$ increase in response to TG exhibited a greater FO-elicited cAMP accumulation measured as an increase in FlCRhR 520/580 nm emission ratio (Fig. 5A). On the other hand, in cells most re-

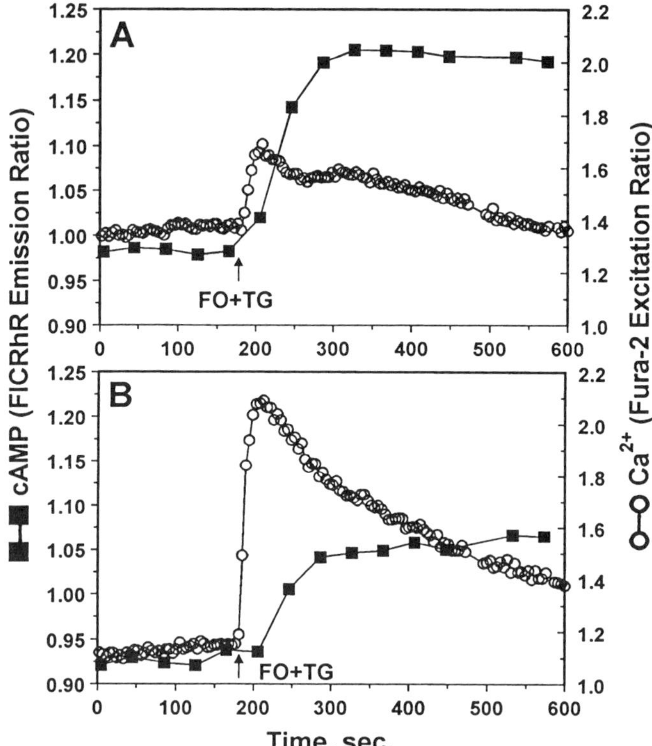

FIG. 5. Increased Ca^{2+}_i levels proportionally down-regulate cAMP response in single C6–2B cells. Cells were labeled with Fura-2, microinjected with FlCRhR, and imaged as described in the text. FO (50 μM) and TG (1 μM) were added together to the cells to evoke changes in $[cAMP]_i$ and $[Ca^{2+}]_i$, respectively. Seven single C6–2B cells, co-labeled with both dyes, were simultaneously imaged and the data relative to each dye's response in individual cells were separately and automatically collected. An analysis of the data revealed that, upon simultaneous addition of the drugs, cells that displayed a smaller increase in TG-induced $[Ca^{2+}]_i$ (reflected in a smaller Fura-2 334/380 nm excitation ratio change) exhibited a greater FlCRhR 520/580 nm emission ratio, indicating that a greater cAMP accumulation occurred after FO stimulation (**A**). Conversely, FO-evoked FlCRhR response was smaller in cells where TG induced a greater increase in $[Ca^{2+}]_i$ (**B**). Here the average FlCRhR and Fura-2 responses from four (**A**) and three (**B**) single cells from the same microscopic field are shown. (Modified with permission from DeBernardi MA, Brooker G: Single cell Ca^{2+}/cAMP cross-talk monitored by simultaneous Ca^{2+}/cAMP fluorescence ratio imaging. *Proc Natl Acad Sci U S A* 1996;93:4577–4582.)

sponsive to TG, and showing a greater increase in Ca^{2+}-mediated Fura-2 334/380 nm excitation ratio, the FO-induced cAMP formation was reduced as reflected by 40–50% decrease in FlCRhR emission ratio (Fig. 5B). These results, showing that Ca^{2+} inhibits cAMP formation in single living C6–2B cells, are in line with our previous *in vitro* finding (4,8–10) that originally proposed Ca^{2+} as an intracellular regulator of the cAMP metabolic pathway in these cells.

The cell specificity of the Ca^{2+}/cAMP cross-talk observed in C6–2B cells is supported by the lack of effect of elevated $[Ca^{2+}]_i$ on the cAMP response of REF-52 cells, a rat embryonic fibroblast cell line, where agonist-induced cAMP synthesis is not under the negative regulation of Ca^{2+} (44). REF-52 fibroblasts also display a heterogeneous Ca^{2+} response (Color Plate 1, upper left panel) to TG, although not as pronounced as that exhibited by C6–2B cells. As shown in Color Plate 1, REF-52 cell cAMP/FlCRhR response to FO is not affected by a simultaneous TG-evoked increase in $[Ca^{2+}]_i$ (see details in Color Plate 1 legend). In control experiments, where FO was added to the cells in the absence of TG, a comparable FlCRhR emission ratio increase was detected (data not shown). It is important to note that, contrary to what we found in C6–2B cells, the magnitude of the Ca^{2+} increase evoked by TG in different fibroblasts being imaged in the same microscopic field did not inversely correlate with the magnitude of the cAMP response elicited by FO added together with TG. Indeed, in the experiment shown in Fig. 6, cells with a smaller Ca^{2+} response to TG (measured as an approximately 1.3-fold increase in Fura-2 excitation ratio [Fig. 6A] versus about a 1.6-fold ratio change in cells shown in Fig. 6B) displayed a slightly smaller FO-stimulated cAMP accumulation (measured as an approximately 1.45-fold increase in FlCRhR emission ratio [Fig. 6A] versus about a 1.54-fold ratio change in cells shown in Fig. 6B).

In C6–2B cells, the elevation of $[Ca^{2+}]_i$ is inhibitory not only to a simultaneous cAMP response but also to a subsequent activation of cAMP synthesis. Indeed, TG or IONO reduced and sometimes completely prevented the cAMP accumulation stimulated by a subsequent challenge of the cells with FO or ISO (44). However, consistent with the results described in Color Plate 1 and Fig. 6, agonist-stimulated cAMP production in REF-52 fibroblasts was not significantly affected by a prior increase in $[Ca^{2+}]_i$ (44). Thus, the data presented so far provide novel *in vivo* evidence confirming the working hypothesis we had formulated early on based upon previous *in vitro* results (where Ca^{2+} and cAMP were measured in separate cell populations, with cAMP being determined in pooled samples and Ca^{2+} in single cells) which postulates a negative cross-talk between Ca^{2+} and cAMP in C6–2B cells (4,8–10).

Our earlier studies, however, do not provide information on the potential effect of Ca^{2+} upon an ongoing cAMP response evoked by an agonist applied to the cells prior to elevating $[Ca^{2+}]_i$. Therefore the acute consequences of increased $[Ca^{2+}]_i$ on the ongoing, agonist-induced cAMP production in C6–2B cells were explored during the course of this study and using this novel simultaneous Ca^{2+} and cAMP fluorescence ratio imaging (44). The experiment was designed to first stimulate cAMP accumulation with ISO and subsequently elevate $[Ca^{2+}]_i$ with TG in single C6–2B cells co-labeled with FlCRhR and Fura-2 (Fig. 7A). ISO evoked an immediate 1.4-fold increase in the FlCRhR 520/580 nm emission ratio that remained at peak level for at least 2 minutes (*solid squares*). However, when $[Ca^{2+}]_i$ was increased with TG (*open circles*; basal $[Ca^{2+}]_i$: ~70 nM; peak after TG: ~300 nM), the FlCRhR signal progressively decreased to about 50% of its peak value by about 9 minutes from TG addition (or 13 minutes after ISO addition).

In a parallel experiment, in the absence of Ca^{2+}-generating stimuli, the ISO-induced 1.4-fold increase in FlCRhR 520/580 nm emission ratio was longer-lasting such that, after about 15 minutes from ISO addition, the FlCRhR signal was still 85% of its peak value (*solid diamonds*). On the other hand, in REF-52 fibroblasts

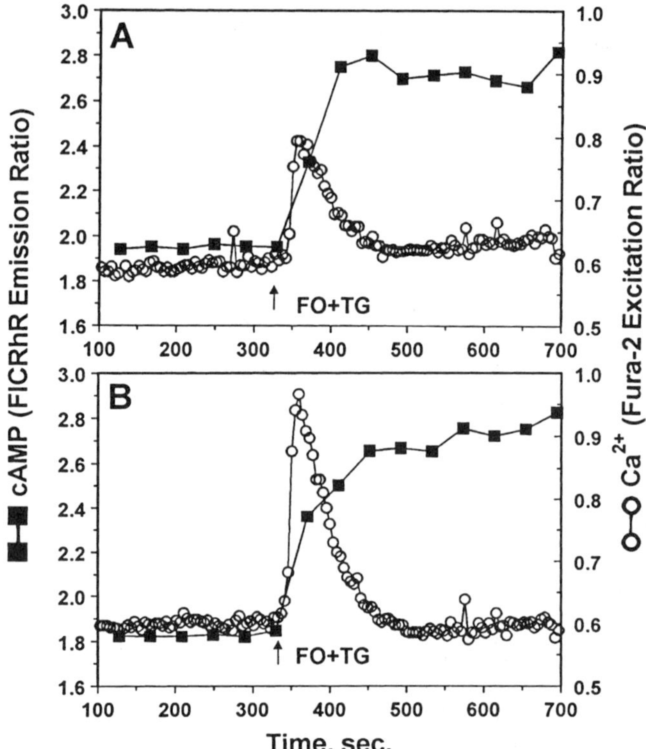

FIG. 6. Increased Ca^{2+}_i levels do not affect cAMP-mediated FlCRhR response in REF-52 fibroblasts. REF-52 cell labeling and imaging configuration as well as experimental protocol and drug concentrations are identical to those used C6–2B cells and described in the legend to Fig. 5. Briefly, the data relative to the responses of 12 FlCRhR- and Fura-2-labeled fibroblasts to the simultaneous addition of TG and FO were collected. Subsequently, the FO-evoked cAMP responses of two fibroblasts that, among the 12 cells imaged in the field, increased $[Ca^{2+}]_i$ the least upon TG (**A**) were analyzed and found to be comparable with the cAMP/FlCRhR responses of two other fibroblasts that increased $[Ca^{2+}]_i$ the most upon TG (**B**). Results are shown as the average emission and excitation ratios of FlCRhR and Fura-2, respectively, of these two sets of fibroblasts. The lack of inhibitory effect of increased $[Ca^{2+}]_i$ on the cAMP/FlCRhR response in REF-52 fibroblasts supports the cell specificity of the negative regulatory role played by Ca^{2+} on the C6–2B cell cAMP metabolic pathway.

neither the kinetics nor the magnitude of the FlCRhR response elicited by FO was affected by a subsequent $[Ca^{2+}]_i$ increase evoked by TG or IONO (44). An example is shown in Fig. 7B, where fibroblasts were challenged with FO and, while cAMP/FlCRhR response was still at plateau, TG was added. Despite the marked rise in $[Ca^{2+}]_i$, the FlCRhR emission ratio enhanced by FO was not affected and remained at maximal level for the whole duration of the TG-evoked Ca^{2+} response (about 5 minutes from the initial Ca^{2+} rise to the return to baseline value). Thus, whereas the ongoing, agonist-promoted cAMP synthesis in C6–2B cells is inhibited by Ca^{2+}-generating signals, in REF-52 fibroblasts the activation as well as the main-

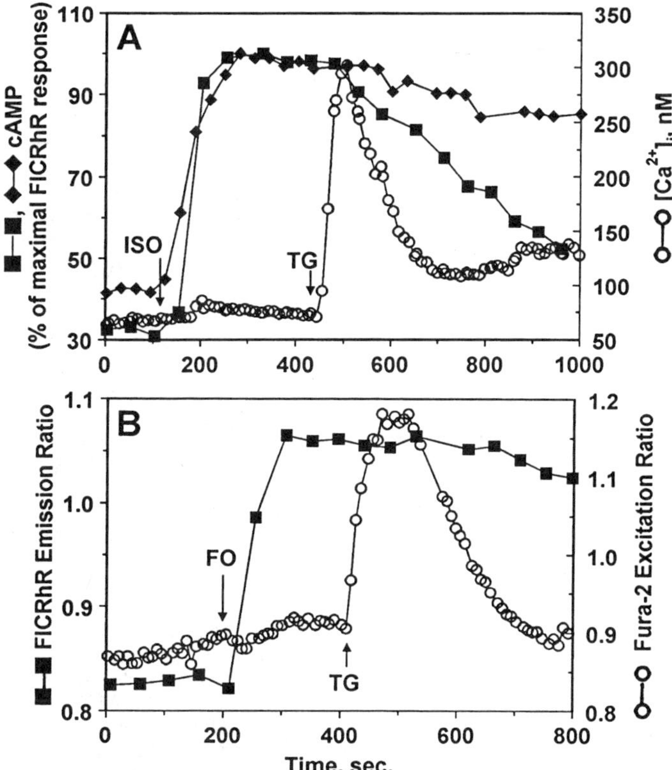

FIG. 7. Differential effect of Ca^{2+} on ongoing, agonist-stimulated cAMP synthesis in C6–2B and REF-52 cells. (**A**) C6–2B cells microinjected with FlCRhR and loaded with Fura-2 were imaged and challenged with ISO (10 µM). Upon elevation of [cAMP]$_i$, FlCRhR emission ratio (*solid squares*) rapidly increases, remains elevated to peak level for about 2 minutes, and, only when [Ca^{2+}]$_i$ is elevated with TG (1 µM, *open circles*), progressively decreases. *Solid diamonds* represent the time course of the FlCRhR response induced by ISO in the absence of Ca^{2+}-mobilizing agents from a parallel experiment, performed the same day on the same batch of C6–2B cells microinjected with the same lot of FlCRhR. The steady FlCRhR signal can be appreciated and compared with the short-lasting response detected instead in the presence of TG (*solid squares*). (**B**) REF-52 fibroblasts, co-labeled with FlCRhR and Fura-2, were challenged with FO (50 µM), which elicited a rapid increase in FlCRhR emission ratio. The subsequent addition of TG (1 µM) evoked a robust [Ca^{2+}]$_i$ rise that did not affect the FO-stimulated FlCRhR/cAMP response. (Modified with permission from DeBernardi MA, Brooker G: Single cell Ca^{2+}/cAMP cross-talk monitored by simultaneous Ca^{2+}/cAMP fluorescence ratio imaging. *Proc Natl Acad Sci U S A* 1996;93:4577–4582.)

tenance of the cAMP synthesis appears to be independent of changes in the intracellular Ca^{2+} homeostasis.

Reduction of the cAMP-initiated FlCRhR response in C6–2B cells was also observed in experiments where the cells were challenged first with ISO (Color Plate 2) or FO (Fig. 8) and, subsequently, with IONO, which elevates [Ca^{2+}]$_i$ to putatively maximal levels (resting [Ca^{2+}]$_i$ was 50–100 nM; [Ca^{2+}]$_i$ peak after IONO was

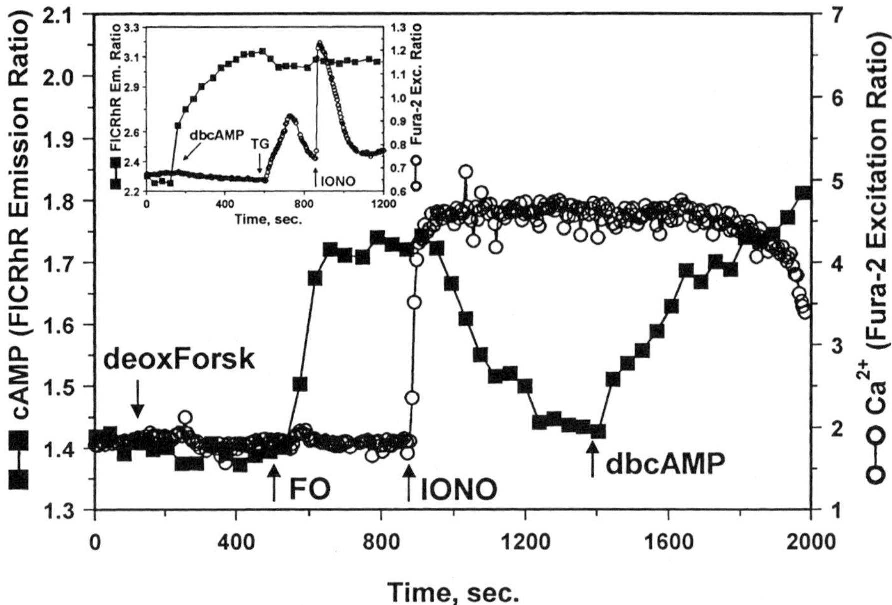

FIG. 8. Lack of effect of Ca^{2+} on the increase in FlCRhR emission ratio elicited by N^6, $O^{2'}$-dibutyryl cAMP (dbcAMP) in single C6–2B cells. Cells co-labeled with FlCRhR and Fura-2 were treated first with 1-deoxyforskolin (50 μM), an inactive FO analog that, as expected, failed to affect the FlCRhR signal, and then with FO (50 μM), which increased the FlCRhR emission ratio reflecting increased cAMP content. FlCRhR peak response lasted for about 4 minutes and thereafter, at the same time as $[Ca^{2+}]_i$ was greatly elevated by IONO (1 μM), underwent an abrupt reduction to virtually baseline value. However, when $[cAMP]_i$ was elevated by exogenous addition of dbcAMP (250 μM), the increase in FlCRhR emission ratio was completely restored, proving that, under conditions where intracellular Ca^{2+} level was above homeostatic values, the dye was still functional. Inset: C6–2B cells, co-labeled with FlCRhR and Fura-2, were treated with dbcAMP (250 μM) to elevate $[cAMP]_i$ without activating new cAMP synthesis. FlCRhR emission ratio in single C6–2B cells slowly increased and remained at peak level even when $[Ca^{2+}]_i$ was elevated by TG (1 μM) and, later, IONO (1 μM), showing that dbcAMP-evoked FlCRhR response in not affected by Ca^{2+}. (Modified with permission from DeBernardi MA, Brooker G: Single cell Ca^{2+}/cAMP cross-talk monitored by simultaneous Ca^{2+}/cAMP fluorescence ratio imaging. *Proc Natl Acad Sci U S A* 1996;93:4577–4582.)

500–1000 nM). These results indicate that in C6–2B cells, Ca^{2+} can not only inhibit agonist-stimulated cAMP accumulation on its start but also modulate a previously activated cAMP transduction signal by instantaneously inhibiting further cAMP synthesis.

Because the possibility existed that the reduced FlCRhR/cAMP response observed in C6–2B cells upon elevation of $[Ca^{2+}]_i$ was due to an impaired performance of the dye itself (rather than to a real inhibition of agonist-promoted cAMP accumulation), we performed the following experiments. C6–2B cells co-labeled with FlCRhR and Fura-2AM were treated with the cell-permeable cAMP analog N^6, $O^{2'}$-dibutyryl cAMP (dbcAMP), which exogenously increases $[cAMP]_i$ and therefore induces a change in FlCRhR emission ratio without newly stimulated cAMP synthesis

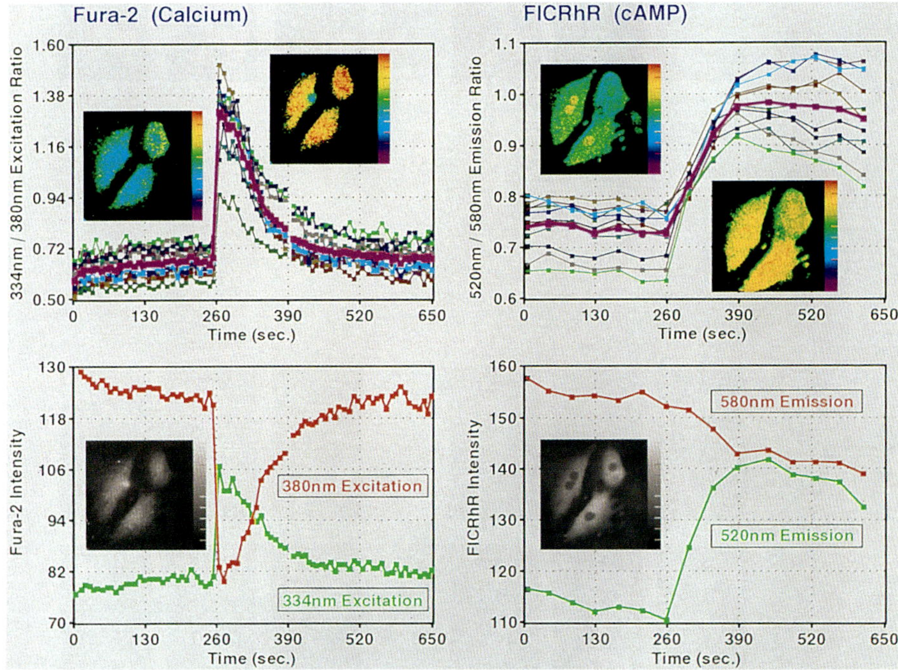

Color Plate 1. Ratiometric analysis of Ca^{2+} and cAMP kinetics in single REF-52 fibroblasts. Three fibroblasts, microinjected with FICRhR and loaded with Fura-2, were imaged as described in the text. Multiple regions of interest were set on each cell and followed throughout the experiment for simultaneous Ca^{2+} and cAMP measurements. The data collected from each region of interest are represented by the individual traces drawn in the upper graphs, with the average shown in purple. The heterogeneity in single-cell Ca^{2+} and cAMP basal and stimulated levels can be appreciated. The lower panels show the average Fura-2 excitation (lower left graph) and FICRhR emission (lower right graph) wavelength intensities upon which the respective ratios, shown in the upper-panel graphs, were calculated. After recording baseline fluorescence signals for Ca^{2+} and cAMP, FO (50 μM) and TG (1 μM) were added together to the cells at 260 seconds. A rapid increase in $[Ca^{2+}]_i$ was measured as indicated by the immediate crossover of the Fura-2 334 and 380 nm excitation wavelengths and the resulting increase in the 334/380 nm ratio. Likewise, upon FO-elicited $[cAMP]_i$ elevation, FICRhR 520 nm emission signal gradually increased whereas the 580 nm emission intensity decreased in parallel, leading to an enhanced 520/580 nm emission ratio. Images in upper panels are pseudocolor ratio images (334/380 nm excitation ratio for Fura-2 with emission monitored at 520 nm; 520/580 nm emission ratio with excitation at 488 nm for FICRhR) of the same fibroblasts taken before and after addition of the drugs. Fura-2 and FICRhR ratios, reflecting $[cAMP]_i$ and $[Ca^{2+}]_i$, are coded in pseudocolor hues ranging from violet (low) to red (high), as shown in the color scales on the right side of each image. The black-and-white images of REF-52 cells shown in the lower panels are 520 nm emission images (with Fura-2 excited at 334 nm and FICRhR excited at 488 nm) taken prior to the addition of the drugs. The cytoplasmic localization of the microinjected FICRhR is evident in the image shown in the lower right panel. Nuclei are virtually signal-free because in resting conditions (low intracellular cAMP concentration) FICRhR cannot cross the nuclear membrane due to the holoenzyme (C_2R_2) high molecular weight. Moreover, by comparing this image with the FICRhR 520 nm emission image of REF-52 fibroblasts taken 20 minutes after FO addition (see Fig. 3, lower panel), the nuclear translocation of the 520 nm emission signal deriving from C-bound fluorescein can be appreciated.

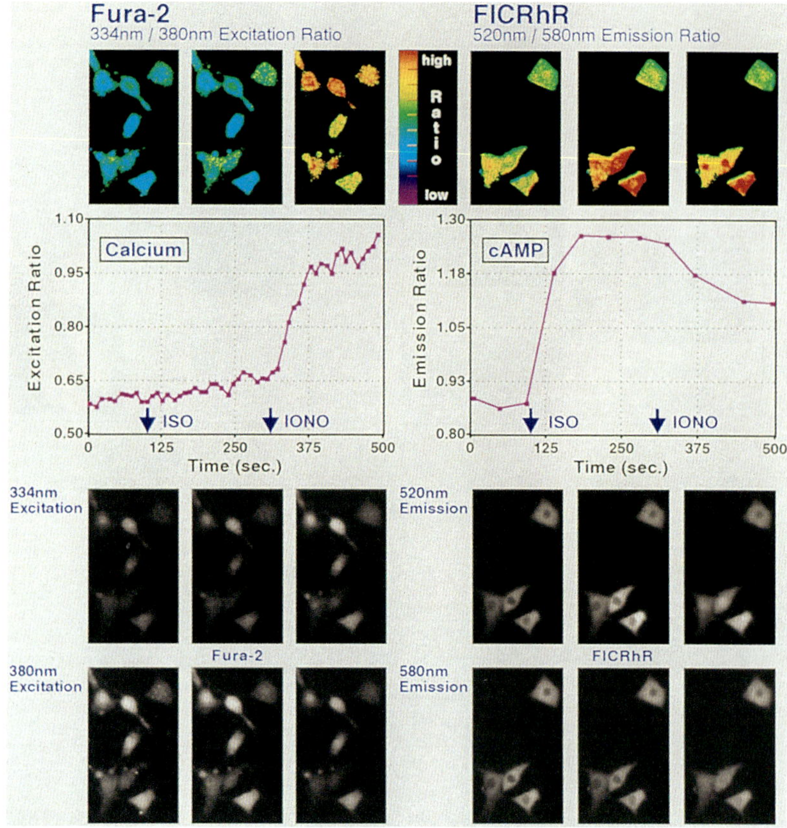

Color Plate 2. Ratiometric analysis of Ca^{2+} and cAMP kinetics in single C6–2B cells. In the microscopic field followed during the experiment, eight Fura-2-loaded cells (shown in the 334/380-nm excitation ratio color images on top of the calcium-related graph) were imaged, four of which had also been microinjected with FICRhR (in the 520/580 nm emission ratio color images shown on top of the cAMP-related graph only these four cells are detected). In these color images, increasing Fura-2 and FICRhR ratios (reflecting increasing $[cAMP]_i$ and $[Ca^{2+}]_i$) are coded in pseudocolor hues ranging from violet (low) to red (high), as indicated in the color scale shown between the two image sets. The graphs show the average Fura-2/Ca^{2+} and FICRhR/cAMP profiles from the four cells labeled with both dyes. Baseline Fura-2 and FICRhR fluorescent signals were recorded for about 1 minute (see first image in each color ratio image set) and then ISO (10 μM) was added. Whereas no significant changes in $[Ca^{2+}]_i$ were detected, a rapid rise in $[cAMP]_i$ was observed, as indicated by an increased FICRhR 520/580 nm emission ratio (see second image in each color ratio image set). Upon addition of IONO (1 μM), $[Ca^{2+}]_i$ markedly increased and, in parallel, FO-enhanced FICRhR/cAMP response, which had been at plateau for about 90 seconds, gradually decreased (see third image in each color ratio image set). The two sets of black and white Fura-2 images of C6–2B cells shown below the calcium-related graph are 520 nm emission images with Fura-2 excited at 334 nm (top row) and 380 nm (bottom row). The first image in each row shows the basal Fura-2 334 and 380 nm fluorescence. As expected, an increase in the 334 nm signal and a decrease in the 380 nm signal are observed upon IONO addition (third image of each row) whereas no significant changes in either wavelength are detected following ISO-induced cAMP increase (second image of each row). The two sets of black and white FICRhR images of C6–2B cells shown below the cAMP-related graph are 520 nm (top row) and 580 nm (bottom row) emission images with FICRhR excited at 488 nm. The first image of each row shows the basal FICRhR 520 nm and 580 nm fluorescence. An increase in the 520 nm emission signal is observed after ISO addition accompanied by a minor decrease in the 580 nm emission signal (second image of each row). A decreased fluorescence upon IONO addition and an initial cAMP-mediated nuclear localization of the 520 nm-emitting C-bound fluorescein signal can be appreciated in the third top image. From these images (334 and 380 nm excitation for Fura-2; 520 and 580 nm emission for FICRhR) the respective ratio images shown in pseudocolor on the top of the figure were derived.

being involved. In the presence of highly elevated $[Ca^{2+}]_i$, which had either prevented (44) or antagonized ISO- (Color Plate 2) and FO-induced (Fig. 8) FlCRhR emission ratio change, dbcAMP was able to completely restore the FlCRhR response (Fig. 8). Moreover, the dbcAMP-elicited increase in $[cAMP]_i$ and FlCRhR emission ratio was not affected by a subsequent $[Ca^{2+}]_i$ rise induced by TG and IONO (Fig. 8, inset). These results thus show that (i) changes in the Fura-2 excitation spectrum per se do not impair FlCRhR response in C6–2B cells, and (ii) FlCRhR responsiveness to increased $[cAMP]_i$ caused by providing cells with preformed cAMP is indeed insensitive to elevated $[Ca^{2+}]_i$.

The alternative possibility for a Ca^{2+}-mediated decrease in cAMP/FlChRh response could be that the degradation of cAMP is enhanced rather than its synthesis being impaired. In fact, a Ca^{2+}-dependent stimulation of phosphodiesterase activity appears to be the primary mechanism for the Ca^{2+}-mediated inhibition of cAMP accumulation in certain cell systems (46). However, extensive studies using a variety of phosphodiesterase inhibitors in C6–2B cells as well as other cell lines and tissues (4–6,8–12,16) mostly expressing the type VI, Ca^{2+}-inhibitable AC, have precluded a Ca^{2+}-induced enhancement in phosphodiesterase activity as the main mechanism responsible for Ca^{2+} inhibition of agonist-stimulated cAMP formation. In these cases, a direct negative effect of Ca^{2+} on AC has been demonstrated to be the likely mechanism. Also, during the course of this study, phosphodiesterase inhibitors, such as 3-isobutyl-1-methylxanthine and Ro20-1724, were not found to prevent the negative effect of Ca^{2+} on cAMP accumulation (data not shown). Thus, in light of these considerations, it is appropriate to interpret the inhibitory effect of elevated $[Ca^{2+}]_i$ on the FlCRhR response elicited in single C6–2B cells by cAMP-generating signals as the result of a true negative regulation by Ca^{2+} of the cAMP metabolic pathway, whose dynamics could now be followed *in vivo*.

CONCLUSION

The development of optical probes for biologic activity has progressed at a remarkable pace during the last decade. Currently, a wide variety of fluorescent, bioluminescent, and chemiluminescent light-emitting molecules are available that can be accurately targeted to cell membranes, intracellular compartments, and second messengers (reviewed in ref. 47). Besides often displacing radioisotopes as standard research tools, these probes have provided great improvements both in speed resolution and in ease of detection. Similarly, conventional light microscopy has been brought to a new level of sophistication and power by technical developments, such as real-time fluorescence ratio imaging and confocal and video intensification microscopy. Such technical advances have resulted in the fascinating ability to investigate various components of life and their dynamic changes inside single cells.

In this study, a combination of dual-excitation and dual-emission fluorescence ratio imaging of two fluorescent probes, FlCRhR and Fura-2, has been used to simultaneously image two intracellular second messengers, cAMP and Ca^{2+}, in the same living cell. This novel dual-dye ratio imaging technique (i) allowed for an immediate and continuous analysis of the kinetics of these two ubiquitously important second messengers, (ii) provided *in vivo* evidence for a dynamic regulation by Ca^{2+} of agonist-induced cAMP production in specific cell types, (iii) demonstrated the utility of this optical approach for the study of the spatial and temporal interactions between Ca^{2+} and cAMP, and potentially other intracellular regulatory messengers, at the single-cell and subcellular level.

Among the future applications of the simultaneous imaging of Ca^{2+} and cAMP is the study of the intracellular localization and Ca^{2+} sensitivity of different AC isoforms associated with the plasma membrane. Recent immunohistochemical studies have revealed a highly selective concentration of AC in neurons, particularly in dendritic spines (48,49). Ca^{2+} channels and pumps as well as PKA anchoring proteins are also present at high concentration in the dendritic spines (50,51). Because brain AC is Ca^{2+}-stimulable (2), the coexistence of Ca^{2+} entry sites with AC and putative targets for PKA phosphorylation in these regions suggests the presence of a "microdomain" of independent neuronal activity in the dendritic spines. A functional co-localization of Ca^{2+}-stimulable ACs with capacitative Ca^{2+} entry sites in discrete areas of the plasma membrane of nonexcitable cells has also been suggested by Fagan and co-workers (52). A similar compartmentation might be envisioned for Ca^{2+}-inhibitable AC (16). In this regard, particularly interesting for their physiological relevance are the recent findings that (i) Ca^{2+} inhibits catecholamine-stimulated cAMP production and AC activity in cardiac myocytes (12) and (ii) Ca^{2+}-inhibitable AC6 is the predominant isoform expressed in cardiac tissue (18,19). These observations, together with the long-standing evidence that cAMP levels oscillate during the myocardial contraction cycle (53,54), has prompted the suggestion that, in heart, Ca^{2+}/cAMP reciprocal effect (cAMP modulating Ca^{2+} homeostasis and Ca^{2+} inhibiting cAMP synthesis) could be the key mechanism for regulating cardiac rhythmicity and contractility (20). Therefore, subcellular pools of AC6 and Ca^{2+} entry sites would make sense for the efficacy and economy of the cAMP and Ca^{2+} signaling meet-point in the myocytes, on which the mechanical properties of the heart largely depend. Simultaneous imaging of the two major cardiac second messengers might reveal crucial information on their intracellular location and coordinate interplay as well as shed light on potential abnormal accumulation of either of them and/or impaired cross-regulation in pathologic samples.

To conclude, it certainly can be foreseen that the applications of this powerful optical approach to study simultaneously multiple aspects of the cell biochemistry *in situ* and in real time are endless and likely to provide exciting new insights and promising breakthroughs in virtually all biologic disciplines.

ACKNOWLEDGMENTS

This work was supported by grant 1R43GM50642 from the National Institute of General Medical Sciences to Atto Instruments, Inc., and grant RO1 HL28940 from the National Institute of Health to G.B. The authors are grateful to J. Scott McDonald for his precious help with the software and graphics.

REFERENCES

1. Sheng M, Thompson MA, Greenberg ME: CREB: a Ca-regulated transcription factor phosphorylated by calmodulin-dependent kinases. *Science* 1991;252:1427–1430.
2. Brostrom CO, Huang C, Breckenridge BM, Wolff DJ: Identification of a calcium-binding protein as a calcium dependent regulator of brain adenylate cyclase. *Proc Natl Acad Sci U S A* 1975;72:64–68.
3. Pepperell JR, Behrman HR: The calcium-mobilizing agent, thapsigargin, inhibits progesterone production in rat luteal cells by a calcium-independent mechanism. *Endocrinology* 1990;127:1818–1824.
3. Giannatasio G, Bianchi R, Spada A, Vallar L: Effect of calcium on adenylate cyclase of rat pituitary gland. *Endocrinology* 1987;120:2611–2619.
4. DeBernardi MA, Brooker G: Inhibition of cAMP accumulation by intracellular calcium mobilization in C6–2B cells stably transfected with substance K receptor DNA. *Proc Natl Acad Sci U S A* 1991;88:9257–9261.
5. Boyajian CL, Garritsen A, Cooper DMF: Bradykinin stimulates Ca^{2+} mobilization in NCB-20 cells leading to direct inhibition of adenylyl cyclase. *J Biol Chem* 1991;266:4995–5003.
6. Colvin RA, Oibo JA, Allen RA: Calcium innhibition of cardiac adenylyl cyclase: evidence for two distinct sites of inhibition. *Cell Calcium* 1991;12:19–27.
7. Caldwell KK, Boyajian CL, Cooper DMF: The effects of Ca^{2+} and calmodulin on adenylyl cyclase activity in plasma membranes derived from neuronal and non-neuronal cells. *Cell Calcium* 1992;13:107–121.
8. DeBernardi MA, Munshi R, Brooker G: Ca^{2+} inhibition of β-adrenergic receptor- and forskolin-stimulated cAMP accumulation in C6–2B rat glioma cells is independent of protein kinase C. *Mol Pharmacol* 1993;43:451–458.
9. Munshi R, DeBernardi MA, Brooker G: P_{2U}-purinergic receptors on C6–2B rat glioma cells: modulation of cytosolic Ca^{2+} and cAMP levels by protein kinase C. *Mol Pharmacol* 1993;44:1185–1191.
10. DeBernardi MA, Munshi R, Yoshimura M et al: Predominant expression of type-VI adenylate cyclase in C6–2B rat glioma cells may account for inhibition of cAMP accumulation by calcium. *Biochem J* 1993;293:325–328.
11. Altiok N, Fredholm BB: Bradykinin inhibits cAMP accumulation in D384-human astrocytoma cells via a calcium-dependent inhibition of adenylyl cyclase. *Cell Signal* 1993;3:279–288.
12. Yu HJ, Ma H, Green RD: Calcium entry via L-type calcium channels acts as a negative regulator of adenylyl cyclase activity and cAMP levels in cardiac myocytes. *Mol Pharmacol* 1993;44:689–693.
13. Ebersole BJ, Diglio CA, Kaufman DW, Berg KA: 5-Hydroxytryptamine 1-like receptors linked to increases in intracellular calcium concentration and inhibition of cAMP accumulation in cultured vascular smooth muscle cells derived from bovine basilar artery. *J Pharmacol Exp Ther* 1993;266:692–699.
14. Baukal AJ, Hunyady L, Catt KJ, Balla T: Evidence for participation of calcineurin in potentiation of agonist-stimulated cAMP formation by the calcium-mobilizing hormone, angiotensin II. *J Biol Chem* 1994;269:24546–2449.
15. Jaworsky DE, Matsuzaki O, Borisy FF, Ronnet GV: Calcium modulates the rapid kinetics of the odorant-induced cAMP signal in rat olfactory cilia. *Neuroscience* 1995;15:310–318.
16. Chiono M, Mahey R, Tate G, Cooper DMF: Capacitative Ca^{2+} entry exclusively inhibits cAMP synthesis in C6–2B glioma cells. *J Biol Chem* 1995;270:1149–1155.
17. Vajanaphanica M, Schultz C, Tsien RY et al: Cross-talk between calcium and cAMP-dependent intracellular signaling pathways. Implication for synergistic secretion in T84 colonic epithelial cells and rat pancreatic acinary cells. *J Clin Invest* 1995;96:386–393.
18. Yoshimura M, Cooper DMF: Cloning and expression of a Ca^{2+}-inhibitable adenylyl cyclase isoform from NCB-20 cells. *Proc Natl Acad Sci U S A* 1992;89:6716–6720.
19. Katsushika S, Chen L, Kawabe J et al: Cloning and characterization of a sixth adenylyl cyclase isoform: type V and VI constitute a subgroup within the mammalian adenylyl cyclase family. *Proc Natl Acad Sci U S A* 1992;89:8774–8778.
20. Cooper DMF, Brooker G: Ca^{2+}-inhibited adenylyl cyclase in cardiac tissue. *Trends Pharmacol Sci* 1993;14:34–35.
21. Brooker G, Harper JF, Terasaki WL, Moylan RD: Radioimmunoassay of cyclic AMP and cyclic GMP. In: Brooker G, Greengard P, Robinson GA, eds. *Advances in cyclic nucleotide research.* Vol 10. New York: Raven Press, 1979:1–33.
22. Evans T, Smith MM, Tanner LI, Harden TK: Muscarinic cholinergic receptors of two cell lines that regulate cAMP metabolism by different molecular mechanisms. *Mol Pharmacol* 1984;26:395–404.

23. Gilman AG: A protein binding assay for adenosine $3':5'$- cyclic monophosphate. *Proc Natl Acad Sci U S A* 1970;67:305–312.

24. Barsony J, Marx SJ: Immunocytology on microwave-fixed cells reveals rapid and agonist-specific changes in subcellular accumulation patterns for cAMP and cGMP. *Proc Natl Acad Sci U S A* 1990; 87:1188–1192.

25. Brooker G, Seki T, Croll D, Wahlstedt C: Calcium wave evoked by activation of endogenous or exogenously expressed receptors in *Xenopus* oocytes. *Proc Natl Acad Sci U S A* 1990;87:2813–2817.

26. Harootunian AT, Kao JP, Paraniape S, Tsien RY: Generation of calcium oscillation in fibroblasts by positive feedback between calcium and IP3. *Science* 1991;251:75–78.

27. Grynkiewicz G, Poenie M, Tsien RY: A new generation of Ca^{2+} indicators with greatly improved fluorescence properties. *J Biol Chem* 1985;260:3440–3450.

28. Adams SR, Harootunian AT, Buechler YJ et al: Fluorescence ratio imaging of cyclic AMP in single cells. *Nature* 1991;349:694–697.

29. Sammak PJ, Adams SR, Harootunian AT et al: Intracellular cyclic AMP, not calcium, determines the direction of vesicle movement in melanophores: direct measurement by fluorescence ratio imaging. *J Cell Biol* 1992;117:57–72.

30. Adams SR, Bacskai BJ, Taylor SS, Tsien RY: Optical probes for cyclic AMP. In: Mason WT ed. *Fluorescent and luminescent probes for biological activity.* London: Academic Press, 1993:133–149.

31. Bacskai BJ, Hochner B, Mahaut-Smith M et al: Spatially resolved dynamics of cAMP and protein kinase A in *Aplysia* sensory neurons. *Science* 1993;260:222–226.

32. Schultz C, Vajanaphanica M, Harootunian AT et al: Acetoxymethyl esters of phosphates, enhancement of the permeability and potency of cAMP. *J Biol Chem* 1993;268:6316–6322.

33. Hagiwara M, Brindle P, Harootunian A et al: Coupling of hormonal stimulation and transcription via the cyclic cAMP responsive factor CREB is rate limited by nuclear entry of protein kinase A. *Mol Cell Biol* 1993;13:4852–4859.

34. Harootunian AT, Adams SR, Wen W et al: Movement of the free catalytic subunit of cAMP-dependent protein kinase into and out of the nucleus can be explained by diffusion. *Mol Biol Cell* 1993; 4:993–1002.

35. Gurantz D, Harootunian AT, Tsien RY et al: VIP modulates neuronal nicotinic acetylcholine receptor function by a cyclic AMP-dependent mechanism. *J Neurosci* 1994;14(6):3540–3547.

36. Civitelli R, Bacskai BJ, Mahaut-Smith MP et al: Single cell analysis of cyclic AMP response to parathyroid hormone in osteoblastic cells. *J Bone Miner Res* 1994;9:1407–1417.

37. Meinkoth JL, Ji Y, Taylor SS, Tsien RY: Dynamics of the distribution of cyclic AMP-dependent protein kinase in living cells. *Proc Natl Acad Sci U S A* 1991;88:9595–9599.

38. Meinkoth JL, Alberts AS, Wen W et al: Signal transduction through the cAMP-dependent protein kinase. *Mol Cell Biochem* 1993;127–128:179–186.

39. Solberg R, Tasken K, Wen W et al: Human regulatory subunit RI beta of cAMP-dependent protein kinase: expression, holoenzyme formation and microinjection into living cells *Exp Cell Res* 1994; 214:595–605.

40. Fantozzi DA, Harootunian AT, Wen W et al: Thermostable inhibitor of cAMP-dependent protein kinase enhances the rate of export of the kinase catalytic subunit from the nucleus. *J Biol Chem* 1994;269:2676–2686.

41. Adams S, Bacskai B, Harootunian AT et al: Imaging of cAMP signals and A-kinase translocation in single living cells. In: Brown BL, Dobson PRM, eds. *Advances in second messengers and phosphoprotein research.* Vol 28. New York: Raven Press, 1993:167–170.

42. Mattie M, Brooker G, Spiegel S: Sphingosine-1-phosphate, a putative second messenger, mobilizes calcium from internal stores via an inositol trisphosphate-independent pathway. *J Biol Chem* 1994; 269:3181–3188.

43. DeBernardi MA, Rabin SJ, Colangelo AM et al: trkA mediates the nerve growth factor-induced intracellular calcium accumulation. *J Biol Chem* 1996;271:6092–6098.

44. DeBernardi MA, Brooker G: Single cell Ca^{2+}/cAMP cross-talk monitored by simultaneous Ca^{2+}/cAMP fluorescence ratio imaging. *Proc Natl Acad Sci U S A* 1996;93:4577–4582.

45. de Erausquin G, Brooker G, Costa E, Wojcik WJ: Stimulation of high affinity gamma-aminobutyric acid receptors potentiates the depolarization-induced increase of intraneuronal ionized calcium content in cerebellar granule neurons. *Mol Pharmacol* 1992;42:407.

46. Harden TK, Evans T, Hepler JR et al: Regulation of cyclic AMP metabolism by muscarinic cholinergic receptors. In: Cooper DMF, Seamon KB, eds. *Advances in cyclic nucleotide and protein phosphorylation research.* Vol 19. New York: Raven Press, 1985:207–220.

47. Tsien RY: Fluorescence imaging creates a window on the cell. *Chem Eng News*, July 18, 1994: 34–44.

48. Mons N, Harry A, Dubourg P et al: Immunohistochemical localization of adenylyl cyclase in rat brain indicates a highly selective concentration of synapses. *Proc Natl Acad Sci U S A* 1995;92: 8473–8477.

49. Mons N, Cooper DMF: Adenylate cyclases: critical foci in neuronal signaling. *Trends Neurosci* 1995;18:536–542.

50. Glantz SB, Amat JA, Rubin CS: cAMP signaling in neurons: patterns of neuronal expression and intracellular localization for a novel protein, AKAP 150, that anchors the regulatory subunit of cAMP-dependent protein kinase II beta. *Mol Biol Cell* 1992;3:1215–1228.

51. Carr DW, Stofko-Hahn RE, Fraser IDC et al: Localization of the cAMP-dependent protein kinase to the postsynaptic densities by A-kinase anchoring proteins. Characterization of AK AP79. *Proc Natl Acad Sci U S A* 1992;89:16816–16823.

52. Fagan KA, Mahey R, Cooper DMF: Functional co-localization of transfected Ca^{2+}-stimulable adenylyl cyclases with capacitative Ca^{2+}-entry sites. *J Biol Chem* 1996;271:12438–12444.

53. Brooker G: Oscillation of cyclic adenosine monophosphate concentration during the myocardial contraction cycle. *Science* 1973;182:933–934.

54. Brooker G: Implications of cyclic nucleotide oscillations during the myocardial contraction cycle. In: Drummond GI, Greengard P, Robinson GA, eds. *Advances in cyclic nucleotide research*. Vol 5. New York: Raven Press, 1975:435–452.

Subject Index

A

ABC-binding cassette superfamily, 24

ACA, 123

AC1 (adenylyl cyclase 1)

  inhibition by Ca^{2+}, 68–69

  pattern of expression of mRNA in, 43

  patterns of regulation of, 91, 92*f*

  stimulation by $G_{s\alpha}$ and forskolin, 3

AC2 (adenylyl cyclase 2)

  activation of by insect PKC, 111

  cAMP response of, 154, 156

  and $G_{\beta\gamma}$, 7–8

  and G protein subunits, 88–89

  mRNA in hippocampus, 44–45, 166*f*

  patterns of regulation of, 93*f*

  and PKC, 108–110

  structure of, 3–4, 4*f*

AC3 (adenylyl cyclase 3)

  inhibition of, 33

  long-term inhibition of, 40

  patterns of regulation of, 93–94, 94*f*

  tissue distributions of, 43

AC4 (adenylyl cyclase 4)

  patterns of regulation of, 93*f*

AC5 (adenylyl cyclase 5), 44, 112

  patterns of regulation of, 92*f*

  phosphorylation of, 184

  and PKC, 70–71

  regulation of, 71–72

  stimulation of by PKC, 70–71

AC6 (adenylyl cyclase 6)

  patterns of regulation of, 92*f*

AC7 (adenylyl cyclase 7)

  cAMP increases in, 70

  in human erythroleukemia (HEL) cell line,
    188

  mRNA in, 45, 178–179

  patterns of regulation of, 93*f*

AC8 (adenylyl cyclase 8)

  and CaM, 87

  patterns of regulation of, 91, 92*f*

  tissue distribution of, 43–44

  variant proteins of, 63–64

AC9 (adenylyl cyclase 9), 64–65, 114

  Ca^{2+}/calcineurin inhibition of, 165*f*

  cAMP response of, 154

  functional properties of, 157–164

    Ca^{2+} interaction in cell free system, 162–164

    pharmacology of in intact cells, 160–162

    and primary amino acid structure, 157–160

  inhibition by Ca^{2+}, 114, 157–158

  and mRNA in brain, 45, 165–166, 166*f*

  patterns of regulation of, 93–94, 94*f*

  phosphorylation of, 164

  tissue distributions of, 165–167

  uniqueness of, 168

ACG, 123–24

adenohypophysial corticotrophs, 167

adenosine, 11, 174–175, 181

adenylyl cyclases, class III

  catalytic domains of, 141–145

  molecular designs of, 138–139

  properties of, 140*t*

  and regulation, 138–39, 141, 145–147

adenylyl cyclases (ACs), 107–114

  amino acid sequence alignment in, 54,
    56–59*f,* 60*f,* 83*f,* 83–84, 124–125,
    143*f,* 157*f,* 157–160

  Ca^{2+}-inhibitable. *See* calcium (Ca^{2+})

  classes of, 137

  dimerization of, 26–27

  diversity of, 53, 107–108, 138*f,* 168–169

  effects of ethanol on. *See* ethanol

  G protein regulation of. *See names of*
    individual G proteins

  homology of, 55, 158*f*

"

AGC (*cont.*)

   localization of, 64–65, 82*t*. See also tissue distributions, of adenylyl cyclases

   mutant. *See* mutations

   phylogeny of cyclase domains of, 142*f*

   physiologic roles for, 72–74

      heart, 72–73

      long-term potentiation, 73–74

   previous studies of, 106–107

   regulation of. S*ee* regulation, of adenylyl cyclases

   regulatory properties of, 65–72. S*ee also* regulation, of adenylyl cyclases

   signaling pathways for, 1–2, 6*f*, 32

   structure of, 3–4, 5, 24*t*, 24–33, 28, 55, 59–62, 61*f*, 81–84, 82*t*

   subfamilies of, 23, 53, 84, 85*f*

   tissue distributions of, 12, 41–45, 42*t*, 64–65, 122, 165–167. S*ee* tissue distributions, of adenylyl cyclases

β-adrenergic receptor kinase (βARK)

   and β-adrenoceptor, 101–103

   differences from protein kinase A, 103

   and ethanol, 175

   role of *in vivo*, 102–103

β-adrenoceptor, 100–105

   and β-adrenergic receptor kinase, 101–103

   in alcoholics, 186

   desensitization of, 103

   mutations of, 103

   and protein kinase A, 101

   signaling pathway of, 100

α-adrenoceptor signaling, 103

aequorin, 32

AKAP79, 158. *See also* protein kinase A

alcoholics, AC changes in, 186–189

alternative splicing, of adenylyl cyclases, 2, 62–64, 82

AMPA/NMDA-receptors, 44

anterior pituitary gland, 167

arginine, 145

β-arrestin mutants, 103

asparagine, 145

AtT20 cells, 162–163, 163*f*

B

bacteria, 87, 125

BAPTA, 30–31

BAPTA-AM, 162

basal activities, 4–5, 12, 123

binding sites

   adenylyl cyclases, 5

   forskolin, 86

   nucleotide, 39, 141–142, 144

brain, 41, 42*t*, 43–45, 64. *See also* hippocampus

   and ethanol, 174–176, 178

   mRNA in, 44–45, 165–166, 166*f*

   variant proteins in, 63–64

Brevibacterium liquefaciens (Bl AC), 139, 141

C

CAAX motif, 102

Ca^{2+}/CaM (calcium/calmodulin), 2. *See also* calmodulin

   and adenylyl cyclases in long-term potentiation, 73–74

   and regulation, 8–9, 28, 68

   stimulation of adenylyl cyclases activity by, 90

$CaCl_2$, 163*f*

cactus, 130

calbindin, 35

calcineurin, 23, 69, 114, 157–160, 164–165

calcium-binding domain (C_1), putative, 108. *See also* cytoplasmic domains

calcium (Ca^{2+}), 2

   adenylyl cyclases *in vitro* sensitivity to, 27–29

   and cAMP cross-talk, 195, 201–210

   direct effects of, 29–33

   effect on cAMP synthesis in C6-2B and REF-52 cells, 195–196, 207*f*

   entry channels of, 29–30

   indirect effects of, 33

   inhibition by, 29–33, 44–47, 68–69, 72–73, 89–90, 205, 206*f*

   measurement of in living cells, 39–40

and oscillations of cAMP, 33–41, 72–73

and regulation, 8–9, 23–47

calmodulin (CaM), 1–2, 2, 8–9, 87. *See also* Ca^{2+}/calmodulin

and adenylyl cyclases in long-term potentiation, 73–74

and PKC, 108–109, 112

regulation by, 8–9, 28, 68, 90

-stimulated adenylyl cyclases in long-term potentiation, 73–74

CaM kinase, regulation by, 10–11, 113–114

CaM kinase II, 41, 68

cAMP (cyclic adenosine monophosphate), 121

and β-adrenoceptor, 100

and basal activities, 4–5, 12

in cardiac myocytes, 65

effect of exposure to TG on formation of, 202–204

effects of signals on, 2–3

and female sterility in *dunce,* 126

as frequency encoder, 41

generation of, 34*f,* 155–156*f,* 167

inhibition of, 6–7, 13, 69, 163*f,* 195

and long-term potentiation, 14–15, 73–74

models for signaling, 154–157

oscillations in, 33–41

assumptions about, 35–36

importance of, 41

mechanisms for, 33–35

requirements for, 36

sustained, 35–36, 37–38*f*

and PKC, 106

role in *Dictyostelium discoideum,* 122–123

role in ethanol tolerance, 185–186

roles of, 122

synthesis of, 69–70, 108–109, 162–164

cardiac myocytes, cAMP in, 65

catalysis

model for, 144*f*

sites for, 5, 11, 60–61, 84

catecholamines, 100, 103

caveolae, 32

central nervous system (CNS), 41

channel activity, 55

cholera toxin, 106, 176

cholesterol, 32

chloroform, action on ACs, 176

CMT cells, 162

coelenterazine, 32

Collaborative Study on the Genetics of Alcoholism (COGA), 188

"collision-coupling" model, 184

complementary deoxyribonucleic acids (cDNA), 26, 54

cloning of type V, 62–63, 111

corticotrophin releasing factor (CRF41), 41-residue, 160–161, 162–163, 167

COS cells, 110, 111

CRE (cAMP-responsive elements), 16*f*

CREB (cAMP-response element binding), 15, 16*f,* 73. *See also* cAMP (cyclic adenosine monophosphate)

cross-talk, 53, 74, 91, 195, 201–210

C6-2B cells, 195–196, 201–210, 207*f*

cyclic adenosine monophosphate. *See* cAMP

cystic fibrosis transmembrane conductance regulator (CFTR), 25*f,* 83

cytoplasmic domains (C$_1$ and C$_2$). *See also* adenylyl cyclases (ACs): structure of

and Ca^{2+} regulation, 160

forskolin binding sites of, 86

functions of, 144–145

inactivation of, 145

and PKC isozymes, 107–108

and sequence conservation, 60–61, 61*f,* 65

cytoplasmic regulator of adenylyl cyclase (CRAC), 123

D

DAG, 195–196

dendritic spines, 45, 64

diacylglycerol, 103

Dictyostelium discoideum (Dd ACG), 84, 122–124, 138, 141

dideoxy adenosine analogs, as inhibitors, 11

digitonin, 31

dimerization, 26–27

diterpene. *See* forskolin

dog, 71

dorsal-ventral patterning, 130

Drosophila melanogaster, 3, 40, 73, 74, 84, 124–128

   amino acid sequence alignment in, 124–125

   and homologs of mammalian AC, 127

   memory and learning in, 125

   phylogenetic analysis of cloned, 125, 125*f,* 127

   roles of protein kinase A in

      development of, 128–133

      oogenesis, 128–129

dunce, 40, 125–127, 128. *See also Drosophila melanogaster*

E

E. coli, 144

EGTA, 30–31, 126

ethanol, 173–189

   ACs as receptive element for, 173–174

   ACs as trait marker in dependence on, 186–189

   acute actions of, 174–179

   chronic actions of, 179–186

   explanation for reduction of AC activity of, 184–186

   reduction of AC activity of by, 180–184

F

FBKP12, 69

FICRhR (cAMP Fluorosensor), 197–201, 201*f,* 209–210

FK506, 158

FKBP12, 158–159

fluorescence resonance energy transfer (FRET), 197–198

FO, 206

forskolin, 1, 27, 66–67

   and AC9, 160, 162

   and AC desensitization by ethanol, 183

and amino acid sequence determination, 54

   regulation by, 5–6, 65, 85*t,* 86–87

Fura-2, 32–33, 209

fura 2-AM 162

G

$G_{\beta\gamma}$, 7–8, 67–68, 88–89, 123, 139, 141

$GABA_A$, 178–179, 186

gene expression, in LTP, 15–17

gene therapy, and proliferative disorders, 13

$G_{i\alpha}$, 6–7, 23, 66–67, 88, 105–106, 110, 174

glutathione-S-transferase (GST), 160, 164

glycoproteins, 62, 83

G proteins. *See names of individual proteins*

G_q, 89–91, 139

G_q-coupled pathways, 86*t*

growth factors, 12, 13, 113

$G_{s\alpha}$, 1–2, 88

   regulation by, 1–2, 3, 5–6, 86–87, 88

   role in desensitization, 105

   role in stimulation of AC by ethanol, 174

"G-shuttle" model, 184

guanine nucleotides, 175

guanosine diphosphate (GDP), 66

guanosine triphosphate (GTP), 66, 175–176

guanylyl cyclases, 138*f,* 139, 141, 142*f,* 143*f*

G_Z, 89

H

heart, 71

   Ca^{2+}-inhibitable adenylyl cyclases in, 72–73

   cAMP in, 65

Hedgehog (Hh) signal transduction pathway, 131, 132*f*

HEK-293 cells, 9, 64, 70–71, 108–109, 112, 176–177

heterologous desensitization, 101, 179–181, 183

hippocampus, 13, 43, 44, 73–74, 166

homologous desensitization, 101, 179–180

hormonal regulation, of adenylyl cyclases, 29, 53, 81, 87–89, 91

human erythroleukemia (HEL) cell line,
 188
human platelet membranes, 107
hydrophilic domains, 55, 60–61
hydrophobic domains, 55
hypothalamus, 43–44

I

I_{CRAC}, 30–31
imaging
 applications of, 210
 Ca^{2+} and cAMP, 198–201, 202*f*, 203*f*
 cell labeling and fluorescence microscopy-
  imaging configuration, 198–201
 measurements of cAMP and Ca^{2+},
  196–198
immunosuppression, 162
$InsP_3$, 31
insulin, 104
interleukin-1 (IL-1), 130
ionomycin, 29, 30
isoproterenol, 175

L

learning and memory, 40–41, 125–127
Leishmania donovani (Ld-RAC), 138, 141
leucine-zipper motif, 27
lipid microenvironment, 31–32, 86, 173
long-term potentiation, 13–15, 15*f*, 167
 Ca^{2+}/CaM-stimulated adenylyl cyclases in,
  73–74
 early phase of, 14–15, 73
 late phase of, 15–17, 16*f*, 73–74
low-stringency hybridization, 82

M

magnesium (Mg^{2+}), 4–5, 109
manganese (Mn^{2+}), 4–5, 183
MAP kinase pathway
 and $G_{\beta\gamma}$, 102
 suppression of, 12

M domains, 61–62
membrane-binding domain (C_1), putative, 108.
  See also cytoplasmic domains (C_1 and
  C_2)
membrane-spanning domains, 24–26
messenger RNA (mRNA)
 and AC9 in brain, 165–166, 166*f*
 and AC7 in cerebellum, 45
 and AC2 in hippocampus, 44–45
 in *Drosophila* oogenesis, 129
 pattern of expression in AC1, 43
mice, knockout, 40
models
 AC isotypes, suggested for study of, 168–169
 for cAMP signaling, 154–157
 catalysis, 144*f*
 "collision-coupling," 184
 "G-shuttle," 184
 for regulation of class III adenylyl cyclases,
  145–147
 for study of Ca^{2+} regulation of cAMP
  accumulation, 196–198
morphine, 183
mossy fiber synapse, 74
multimolecular arrays, 26–27
muscle, airway smooth, 104
mutations
 of adenylyl cyclases, 124
 of β-adrenoceptor, 103
 of β-arrestin, 103
 effects of protein kinase A, 131–132
 point, 5–6, 8, 144
 truncation, 26

N

N domain, 55, 59–60
negative feedback loops, 34
neuroblastoma-glioma cells, 106
N-glycosidase F, 62, 64
N-methyl D-aspartate (NMDA), 13–14, 15*f*, 16*f*,
 43, 73
nuclear factor of activated T-lymphocytes
 (NFAT1), 158

O

okadaic acid, 106
oscillations. *See* cAMP: oscillations in

P

PACAP$_1$, 154, 167
Paramecia, 25–26, 55, 83
patched (Ptc), 131
PCR, reverse-transcription, 28, 82
pelle kinase, 130
pertussis toxin-sensitive pathways, 87–89, 174
phorbol ester treatment, 69, 90, 107
  and AC sensitivity to ethanol, 177–178
  limitations of, 108
phosphodiesterase (PDE) activity, 33–34, 36,
    38, 40, 154, 156, 162, 209
phosphorylation, 71–72. *See also names of*
    specific kinases
  of AC5, 184
  of AC9, 164
  and β-adrenoceptor, 100, 104
  and βARK, 102
  of G$_{i\alpha}$, 105–106
  and oscillations, 39
  by PKC, 90, 107–108
  and protein kinase A, 101, 113
  sites for, 10–11
  and tyrosine kinases, 104, 105
PKA-anchoring proteins (AKAPs). *See* protein
    kinase A: anchoring proteins of
platelet-derived growth factor (PDGF), 13
pleckstrin homology (PH) domain, 102
point mutations. *See* mutations: point
PP2B, 69
PPi, 145
Premont, R. T., 162
proliferation, 12–13, 14*f*
protein kinase A (PKA)
  and β-adrenoceptor, 101
  and β-adrenoceptor signaling pathway, 100
  anchoring proteins of, 41, 64–65, 158
  differences from βARK, 103
  and *Drosophila* development, 128–133

  embryonic patterning, 129–130
    larval patterning, 130–133
    oogenesis, 128–129
  effects of mutations of, 131–132
  and ethanol, 178
  and phosphorylation, 101, 113
  regulation by, 10, 71–72, 113
protein kinase C (PKC), 1, 103–104
  and AC2, 108–109
  and AC5, 70–71
  conditional regulation by, 110–111
  diversity in, 107–108
  effect on cAMP signaling, 105–106
  and ethanol action on ACs, 178
  pathways of, 45
  regulation by, 9–10, 69–71, 90–91, 106–107
  regulation by atypical, 112
protein kinase C-α (PKC-α), 111
protein kinases. *See names of individual protein*
    kinases
P-site ligands, 11, 145–146
purification studies, 62
purine nucleotides. *See* P-site ligands
Purkinje cells, 178–179

Q

quantitative trait loci (QTL), 188
QXXER motif, 7

R

rat, 71
reagents, isotype-specific, 169
recoverin, 28
REF-52 cells, 204–205
regulation, of adenylyl cyclases, 2*t,* 23, 54,
    65–72, 139, 154*t*
  basal activities, 4–5, 123
  by Ca^{2+}/CaM, 8–9, 28, 68
  by class III, 138–139, 141
  of cloned, by Ca^{2+}, 29–33
    direct effects, 29–33
    indirect effects, 33

conditional, 110–112

effect of phorbol esters on, 69–70

by forskolin, 3, 86–87

functional consequences of differential, 12–17

by G protein subunits, 5–8, 66–68, 84–87, 85*t*, 90–91, 105–106, 109*t*, 182–183. *See also under names of individual G proteins*

patterns of, 91–94

   AC2, AC4, and AC7, 93*f*

   AC1 and AC8, 91, 92*f*

   AC3 and AC9, 93–94, 94*f*

   AC5 and AC6, 92*f*

by pertussis toxin-sensitive pathways, 87–89

by phosphorylation, 107–114

and proliferation, 12–13

by protein kinases. *See* protein kinases: regulation

by P-site ligands, 11, 145–146

of synaptic plasticity, 13–17, 15*f*, 46*f*

Rhizobium meliloti (Rm AC), 139

rhodamine, 197

rhodopsin kinase (GRK-1), 101, 102

Runge-Kutta method, 36

rutabaga (also rutabaga and Rutabaga), 3, 40, 74, 124, 125–127, 128. *See also Drosophila melanogaster*

S

Saccharomyces cerevisiae, 122, 139

Saccharomyces klyuyveri, 139

Schizosaccharomyces pombe, 122, 138–139

sequence diversity. *See* adenylyl cyclases (ACs): diversity of

serine residues, 104

signaling pathways. *See* adenylyl cyclases (ACs): signaling pathways for

skeletal muscle, 64

smoothened (Smo), 131

sterility, female *dunce,* 126–127, 128

striatin, 26

strontium (Sr^{2+}), 31

structural features. *See* adenylyl cyclases: structure of

supraoptic nucleus, 167

T

thapsigargin, 29, 30, 196

time-dependent variables, 34–35, 184–185

tissue distributions, of adenylyl cyclases, 2, 12, 41–45, 42*t*, 64–65, 122, 165–167. *See also names of specific organs and body sites*

toll protein, 130

transfection, 85, 90, 112

transport activity, 55

truncation analysis, 8–9

truncation mutants, 26

Trypanosoma brucei (TB ESAG), 138, 141

tyrosine kinases, 13, 104, 105, 112, 113

U

uridine triphosphate (UTP), 29

V

V-α "half molecule," 62–63

W

"Walker" motif, 83